WATER STRESS

Edited by **Ismail Md. Mofizur Rahman**
and **Hiroshi Hasegawa**

Water Stress
Edited by Ismail Md. Mofizur Rahman and Hiroshi Hasegawa

Published by InTech
Janeza Trdine 9, 51000 Rijeka, Croatia

Publishing Process Manager Martina Durovic

Technical Editor Teodora Smiljanic

Cover Designer InTech Design Team

Additional hard copies can be obtained from orders@intechopen.com

Water Stress, Edited by Ismail Md. Mofizur Rahman and Hiroshi Hasegawa
p. cm.
ISBN 978-953-307-963-9

For EU product safety concerns:
IN TECH d.o.o., Prolaz Marije Krucifikse Kozulić 3, 51000 Rijeka, Croatia,
info@intechopen.com or visit our website at intechopen.com.

Meet the editors

Dr. Ismail Md. Mofizur Rahman received his PhD degree in 2011 from Kanazawa University, Japan. He has published more than 28 original research articles, six book chapters, and one book. He is currently working as Associate Professor at the Department of Chemistry, Faculty of Science, University of Chittagong, Bangladesh.

Dr. Hiroshi Hasegawa received his DSc degree in 1997 from Kyoto University, Japan. He has published more than 79 original research articles, several book chapters and a number of books. He is currently working as a Professor at the Faculty of Chemistry, Institute of Science and Engineering, Kanazawa University, Japan.

Contents

Preface

Water, which makes up 80 to 90 per cent of the biomass of non-woody plants, is the central molecule in all physiological processes of plants. Water availability and water-use efficiency are among the important abiotic factors that have had and continue to have a decisive influence on plant evolution. The growth and distribution of natural and artificial vegetation, and the performance of cultivated plants is dependent on the permanent or temporary water deficit stress more than any other environmental factor. Water stress, in its broadest sense, encompasses both drought and flooding stress. Factors controlling stress conditions alter the normal equilibrium, and lead to a series of morphological, physiological, biochemical, and molecular changes in organisms, which adversely affect their growth and productivity. The average yields from the major crop plants may be reduced by more than 50 per cent as a result of water stress. However, plants have developed innate adaptations to stress conditions, with an array of biochemical and physiological interventions that involve the function of many stress-associated genes. On the other hand, productive and sustainable agriculture necessitates growing plants in arid and semiarid regions with less input of precious resources, such as fresh water.

This multi-authored book has been designed to provide comprehensive information on key contemporary issues associated with water stress to the plants. The book consists of both review-like studies, and the results of some new researches and case studies highlighting the research and practices aimed at improving water stress resistance and water-use efficiency and related issues. The reader will find viewpoints that are challenging and sometimes discerning, but motivating and thought-provoking in dealing with the challenges of water stress.

We hope that this publication will be a useful document for students, professionals, and researchers working on the water stress in plants. We extend our sincere appreciation to the authors, who are from different countries, for their contribution to the book. We thank Ms. Ana Nikolic from InTech Open Access Publisher for inviting us to be editors of this book. We would like to extend our special appreciation to the

Publishing Process Manger Ms. Martina Durovic for her encouragement and superb support throughout this whole process.

Dr. Ismail Md. Mofizur Rahman
University of Chittagong
Bangladesh

Dr. Hiroshi Hasegawa
Kanazawa University
Japan

Introductory Chapter

Water Stress in Plants: Causes, Effects and Responses

Seyed Y. S. Lisar[1], Rouhollah Motafakkerazad[1], Mosharraf M. Hossain[2] and Ismail M. M. Rahman[3,*]
[1]*Department of Plant Sciences, Faculty of Natural Sciences, University of Tabriz, Tabriz*
[2]*Institute of Forestry and Environmental Sciences, University of Chittagong, Chittagong*
[3]*Department of Chemistry, Faculty of Science, University of Chittagong, Chittagong*
[1]*Iran*
[2,3]*Bangladesh*

1. Introduction

Biosphere's continued exposure to abiotic stresses, for example, drought, salinity, extreme temperatures, chemical toxicity, oxidative stress, etc., cause imbalances in the natural status of the environment. Each year, stresses on arable plants in different parts of the world disrupt agriculture and food supply with the final consequence - famine. Factors controlling stress conditions alter the normal equilibrium, and lead to a series of morphological, physiological, biochemical and molecular changes in plants, which adversely affect their growth and productivity. The average yields from the major crop plants may reduce by more than 50% owing to stresses. However, plants also have developed innate adaptations to stress conditions with an array of biochemical and physiological interventions that involves the function of many stress-associated genes. In this chapter, we aim at the stresses related to water and the expression 'drought' which is derived from the agricultural context, is used as equal to water stress throughout the article.

Water, comprising 80-90% of the biomass of non-woody plants, is the central molecule in all physiological processes of plants by being the major medium for transporting metabolites and nutrients. Drought is a situation that lowers plant water potential and turgor to the extent that plants face difficulties in executing normal physiological functions. However, a few groups of animals and a wide variety of plants are known for their tolerance to desiccation during the adult stages of their life cycle. Though our knowledge on plant's drought tolerance is ancient, the modern scientific study of drought tolerance started in 1702 with Anthony von Leeuwenhoek's discovery of the survival of rotifers without water for months.

---

*Corresponding Author

2. Water stress – Why and how?

Plants experience water stress either when the water supply to their roots becomes limiting or when the transpiration rate becomes intense. Water stress is primarily caused by the water deficit, *i.e.* drought or high soil salinity. In case of high soil salinity and also in other conditions like flooding and low soil temperature, water exists in soil solution but plants cannot uptake it – a situation commonly known as 'physiological drought'. Drought occurs in many parts of the world every year, frequently experienced in the field grown plants under arid and semi-arid climates. Regions with adequate but non-uniform precipitation also experience water limiting environments.

Since the dawn of agriculture, mild to severe drought has been one of the major production-limiting factors. Consequently, the ability of plants to withstand such stress is of immense economic importance. The general effects of drought on plant growth are fairly well known. However, the primary effect of water deficit at the biochemical and molecular levels are not considerably understood yet and such understanding is crucial. All plants have tolerance to water stress, but the extent varies from species to species. Knowledge of the biochemical and molecular responses to drought is essential for a holistic perception of plant resistance mechanisms to water limited conditions in higher plants.

3. Effects of water stress on plants

Drought, as an abiotic stress, is multidimensional in nature, and it affects plants at various levels of their organization. In fact, under prolonged drought, many plants will dehydrate and die. Water stress in plants reduces the plant-cell's water potential and turgor, which elevate the solutes' concentrations in the cytosol and extracellular matrices. As a result, cell enlargement decreases leading to growth inhibition and reproductive failure. This is followed by accumulation of abscisic acid (ABA) and compatible osmolytes like proline, which cause wilting. At this stage, overproduction of reactive oxygen species (ROS) and formation of radical scavenging compounds such as ascorbate and glutathione further aggravate the adverse influence. Drought not only affects plant water relations through the reduction of water content, turgor and total water, it also affects stomatal closure, limits gaseous exchange, reduces transpiration and arrests carbon assimilation (photosynthesis) rates. Negative effects on mineral nutrition (uptake and transport of nutrients) and metabolism leads to a decrease in the leaf area and alteration in assimilate partitioning among the organs. Alteration in plant cell wall elasticity and disruption of homeostasis and ion distribution in the cell has also been reported. Synthesis of new protein and mRNAs associated with the drought response is another outcome of water stress on plants. Under the water stress cell expansion slows down or ceases, and plant growth is retarded. However, water stress influences cell enlargement more than cell division. Plant growth under drought is influenced by altered photosynthesis, respiration, translocation, ion uptake, carbohydrates, nutrient metabolism, and hormones.

3.1 Photosynthesis

Photosynthesis is particularly sensitive to the effects of water deficiency. Plants' resistance to water deficiency yields metabolic changes along with functional and structural rearrangements of photosynthesizing apparatus. Photosynthesis of higher plants decreases with the reduction in the relative water content (RWC) and leaf water potential. Lower photosynthesis rate is a usual effect of water stress in plants and has been attributed primarily to stomatal limitation and secondarily to metabolic impairment. However, metabolic impairment is the more complex phenomenon than the stomatal limitation though the relative importance of stomatal or metabolic inhibitions is unclear. Some studies blamed stomatal closure for the inhibition of C_4 photosynthesis under water stress while others concluded that non-stomatal factors play the major role.

The photosynthesis rate of leaves in both C_3 and C_4 plants decrease under the drought conditions. Evidence indicates that C_4 photosynthesis is more sensitive to water stress and C_4 plants, such as corn (*Zea mays* L.) are more susceptible to water deficiency than C_3 plants, such as wheat. It explains the predominance of C_4 plants in hot, arid regions - areas prone to frequent drought. C_3 and C_4 plants are alike in the basic process of photosynthesis like Calvin cycle and electron transport chain components, yet significant differences exist between them, which make their responses to water stress differ at a number of levels.

There are some co-factors, which decrease plants' photosynthesis under water stress. Of them, qualitative and quantitative changes in the pool of photosynthesizing pigments, low CO_2 uptake due to stomatal closure and resistance, poor assimilation rates in photosynthetic leaves are prominent. Assimilation rates in photosynthetic leaves decreases due to reduced photosynthetic metabolites and enzymes activity, low carboxylation efficiency and inhibition of chloroplast activity at low water potential. Among other co-factors of water stress, the damage of the photosynthetic apparatus through the production of ROS such as superoxide and hydroxyl radicals, worth special mention.

Decrease in chlorophyll content of leaves under water stress is well known. Water stress inhibits chlorophyll synthesis at four consecutive stages: (I) the formation of 5-aminolevuliniuc acid (ALA); (II) ALA condensation into porphobilinogen and primary tetrapyrrol, which is transformed into protochlorophyllide; (III) light-dependent conversion of protochlorophyllide into chlorophyllide; and (IV) synthesis of chlorophylls a and b along with their inclusion into developing pigment–protein complexes of the photosynthetic apparatus. In the majority of cases, carotenoids are less sensitive to water stress than chlorophyll, which has been demonstrated for several species of agricultural plants. However, unlike chlorophyll, increase in xanthophyll pigments such as zeaxanthin and antheraxanthin in plants under water stress have been reported. Xanthophyll pigments have a protective role on plants under stress, and some of these pigments are involved in the xanthophyll cycle which has inhibitory role on ROS production.

RuBisCO, the key enzyme for carbon metabolism in leaves, acts as a carboxylase in the Calvin cycle and as an oxygenase in the photorespiration which, however, frequently is viewed as an adverse process. RuBisCO is the most critical player influencing the physiology of plants under water-stressed conditions. Under the conditions of water stress, a rapid decrease in the amount of RubisCO takes place in most plants which in turn leads to lower activity of the enzyme. This effect is evident in all plants studied though the extent is species-dependent. Water deficiency reduces the supply of carbon dioxide from the environment due to the closure of stomata. Consequently, photorespiration increases which ensure partial substrate replenishment and maintain the carboxylating function of RuBisCO. The end result is the utilization of excess reducing equivalents in chloroplast that causes a reduction in the oxygen-free radicals' production leading to the oxidative damage in chloroplasts. The reduction in chloroplast volume can also be linked to the desiccation within the chloroplast that leads to the conformational changes in RuBisCO. Moreover, drought stress conditions acidify the chloroplast stroma causing inhibition to the RuBisCO activity. In addition, decline in RuBisCO activity is also caused by the lack of the substrate for carboxylation, reduction in the amount and/or activity of the coupling factor - ATPase, loss of RBP recognition sites in RuBisco, structural alterations of chloroplasts and RuBisCO, and release of RuBisCO from damaged plastids. In addition to RuBisCO, water stress can reduce activity of other photosynthetic enzymes to different extents such as NADP-dependent glyceraldehyde phosphate dehydrogenase, phosphoenolpyruvate carboxylase, NAD-dependent malate dehydrogenase, phosphoribulos kinase, fructose-1,6-bisphosphatase and sucrose phosphate synthase.

In addition to its negative effects on dark reactions of photosynthesis, water stress also disrupts the cyclic and non-cyclic types of electron transport during the light reaction of photosynthesis. The disruption is clearer in the oxygen-releasing complex and electron transfer from protochlorophyllide to P700. Lower electron transport rate negatively affects photophosphorylation process and decrease ATP synthesis as well as $NADP^+$ reduction. ATPase inhibition under water deficiency is also responsible for the reduction in ATP levels in chloroplasts. All these factors cumulatively affect the intensity of photo-assimilation and the stability of the photosynthetic apparatus under the conditions of water stress. Both of the PSs in chloroplasts are affected by water deficiency, however, PS I of some plants are more severely damaged compared to PS II, though there is an opposite conclusion as well.

3.2 Protein synthesis

Drought conditions bring about quantitative and qualitative changes in plant proteins. In general, proteins in the plant leave decrease during water deficiency due to the suppressed synthesis, more pronouncedly in C_3 than in C_4 plants. Water stress alters gene expression and consequently, the synthesis of new proteins and mRNAs. The main proteins those synthesized in response to water stress are LEA, desiccation stress protein, proteins those respond to ABA, dehydrins, cold regulation proteins, proteases, enzymes required for the biosynthesis of various osmoprotectants, the detoxification enzymes (SOD, CAT, APX, POD, GR). In addition, protein factors involved in the regulation of signal transduction and gene expression, such as protein kinases and transcription factors are also synthesized. The majority of these stress response proteins are dehydrin-like proteins, which accumulate during seed production and embryo maturation of many higher plants as well as in water stressed seedlings. These proteins have highly conserved domain that linked to hydrophobic interactions needed for macromolecular stabilization.

Heat-shock proteins (Hsps) and late embryogenesis abundant (LEA)-type proteins are two major types of stress-induced proteins during different stresses including water stress. Protection of macromolecules such as enzymes, lipids and mRNAs from dehydration are well known functions of these proteins. LEA proteins accumulate mainly in the embryo. The exact functions and physiological roles of these proteins are unknown. Hsps act as molecular chaperones and are responsible for protein synthesis, targeting, maturation and degradation in many cellular processes. They also have important roles in stabilization of proteins and membranes and in assisting protein refolding under stress conditions. Expression of LEA-type genes under osmotic stress is regulated by both ABA-dependent and independent signaling pathways. Genes encoding LEA-type proteins are diverse - RD (responsive to dehydration), ERD (early response to dehydration), KIN (cold inducible), COR (cold regulated), and RAB (responsive to ABA) genes.

3.3 Lipids

Water stress can lead to a disturbance of the association between membrane lipids and proteins as well as enzymes activity and transport capacity of membranes. Drought results in the variation of fatty acid composition, for example, an increase in fatty acids having less than 16 carbons in chloroplasts. Lipid peroxidation is the well-known effect of drought and many other environmental stresses via oxidative damage.

3.4 Morphological, anatomical and cytological changes

In the majority of the plant species, water stress is linked to changes in leaf anatomy and ultrastructure. Shrinkage in the size of leaves, decrease in the number of stomata; thickening of leaf cell walls, cutinization of leaf surface, and underdevelopment of the conductive system - increase in the number of large vessels, submersion of stomata in succulent plants and in xerophytes, formation of tube leaves in cereals and induction of early senescence are the other reported morphological changes.

The root-to-shoot ratio increases under water-stress conditions to facilitate water absorption and to maintain osmotic pressure, although the root dry weight and length decrease as reported in some plants like sugar beet and Populus. Higher root-to-shoot ratio under the drought conditions has been linked to the ABA content of roots and shoots. Water stress is linked to decrease in stem length in plants such as Albizzia, Erythrina, Eucalyptus and Populus with up to 25% decrease in plant height in citrus seedling. Decreased leaf growth, total leaf area and leaf-area plasticity were observed under the drought conditions in many plant species, such as peanut and *Oryza sativa.* Although water saving is the important outcome of lower leaf area, it causes reduced crop yield through reduction in photosynthesis. Decrease in plant biomass consequences from the water deficit in crop plants, mainly due to low photosynthesis and plant growth and leaf senescence during the stress conditions. However, in some plants, higher yield was reported under-water deficit condition.

3.5 ABA accumulation

The plant hormone ABA accumulates under-water deficit conditions and plays a major role in response and tolerance to dehydration. Closure of stomata and induction of the expression of multiple genes involved in defense against the water deficit are known functions of ABA. The amount of ABAs in xylem saps increases substantially under reduced water availability in the soil, and this results in an increased ABA concentration in different compartments of the leaf. Another well-known effect of drought in plants is the decrease in PM-ATPase activity. Low PM-ATPase increases the cell wall pH and lead to the formation of ABA^- form of abscisic acid. ABA^- cannot penetrate the plasma membrane and translocate toward the gourd cell by the water stream in the leaf apoplasm. High ABA concentration around guard cell results in stomata closure and help to conserve water.

3.6 Mineral nutrition

Water stress also affects plant mineral nutrition and disrupts ion homeostasis. Calcium plays an essential role in structural and functional integrity of plant membrane and other structures. Decrease in plant Ca^{2+} content was reported in many plants, for example, approximately 50% decrease in Ca^{2+} in drought stressed maize leaves, while in roots Ca^{2+} concentration was higher compared to control. Potassium is an important nutrient and plays an essential role in water relation, osmotic adjustment, stomatal movement and finally plant resistance to drought. Decrease in K^+ concentration was reported in many plant species under water deficient condition, mainly due to membrane damage and disruption in ion homeostasis. K^+ deficient plant has lower resistance to water stress. Nitrogen metabolism is the most important factor that influences plant growth and performance. Disruption in N-metabolism is a crucial in-plant injury under the water deficit conditions. Some studies showed the reduction of nitrate uptake and decrease in nitrate reductase activity under water stress.

4. Drought and oxidative stress in plants

Oxidative stress, which frequently accompanies many abiotic stresses like high temperature, salinity, or drought stress, causes a serious secondary effect on cells. Oxidative stress is accompanied by the formation of ROSs such as O_2^-, $^1O_2$, H_2O, and OH^-. ROSs damage membranes and macromolecules affect cellular metabolism and play a crucial role in causing cellular damage under drought stress.

Drought creates an imbalance between light capture and its utilization, which inhibits the photosynthesis in leaves. In this process imbalance between the generation and utilization of electrons is created. Dissipation of excess light energy in photosynthetic apparatus results in generation of reactive oxygen species (ROS). Denaturation of functional and structural macromolecules is the well-known results of ROS production in cells. DNA nicking, amino acids, protein and photosynthetic pigments oxidation, and lipid peroxidation are the reported effects of ROS. As a consequence, cells activate some responses such as an increase in the expression of genes for antioxidant functions and production of stress proteins, up-regulation of anti-oxidants systems, including anti-oxidant enzymes and accumulation of compatible solutes. All these responses increase scavenging capacity against ROSs.

5. Plant responses to water stress

Plants adapt themselves to drought conditions by various physiological, biochemical, anatomical, and morphological changes, including transitions in gene expression. The physiology of plants' response to drought at the whole plant level is highly complex and involves deleterious and/or adaptive changes. This complexity is due to some factors such as plant species and variety, the dynamics, duration and intensity of soil water depletion, changes in water demand from the atmosphere, environmental conditions, as well as plant growth and the phenological state in which water deficit is developed.

Plants' strategies to cope with drought normally involve a mixture of stress avoidance and tolerance strategies. Early responses of plants to drought stress usually help the plant to survive for some time. The acclimation of the plant to drought is indicated by the accumulation of certain new metabolites associated with the structural capabilities to improve plant functioning under drought stress. The main aspects of plant responses to water involve the maintenance of homeostasis (ionic balance and osmotic adjustment), counter action to resulted damages and their quick repair such as scavenging of ROS and decrease oxidative stress and the regulation and recovery of growth.

The complex plant response to water stress, alike other abiotic stresses, involves many genes and biochemical and molecular mechanisms. Sequentially, they are: signal sensing, perception and transduction by osmosensors like AtHK1, kinases and phospholipases as well as secondary messengers; transcriptional control by transcription factors such as DREB (dehydration-responsive transcription factors); and activation of stress responsive mechanisms such as detoxification of ROS by enzymes such as SOD and CAT; osmoprotection by compatible solutes and free radicals scavengers such as glutathione and proline; and water and ion homeostasis by aquaporins and ion transporters. The results of these responsive pathways are the re-establishment of cellular homeostasis and functional and structural protection and finally stress resistance or tolerance.

Stomata closure is the well-known first responsive event of plants to water deficiency. Stomatal closures are more closely related to soil moisture content than leaf water status, and it is mainly controlled by chemical signals such as ABA produced in dehydrating roots. A direct correlation between the xylem ABA content and stomatal conductance has been demonstrated. Changes in plant hydraulic conductance, plant nutritional status, xylem sap pH, farnesyl tranferase activity, leaf-to-air vapor pressure deficit and decrease in relative water content are other factors working in stomatal regulation plants. Although CO_2 assimilation and net photosynthesis decreases due to stomatal closure but attainment of low transpiration rate and prevention of water losses from leaves is a good tradeoff for survival in exchange of growth. Stomata can completely close in mild to severe stress depending on plant species, and tolerant species control stomata opening to allow some carbon fixation and improving water-use efficiency. The increased stomatal resistance under stress levels indicates the efficiency of a species to conserve water.

6. Plants resistance to water stress

Plants optimize the morphology, physiology and metabolism of their organs and cells in order to maximize productivity under the drought conditions. The reactions of the plants to water stress differs significantly at various organizational levels depending upon intensity and duration of stress as well as plant species and its stage of development. Stress resistance in plant is divided into two categories, including stress tolerance and stress avoidance. Drought avoidance is the ability of plant to maintain high tissue water potential under drought conditions, while drought tolerance is a plant's stability to maintain its normal functions even at low tissue water potentials. Drought avoidance is usually achieved through morphological changes in the plant, such as reduced stomatal conductance, decreased leaf area, development of extensive root systems and increased root/shoot ratios. On the other hand, drought tolerance is achieved by cell and tissue specific physiological, biochemical, and molecular mechanisms, which include specific gene expression and accumulation of specific proteins. The dehydration process of drought-tolerant plants is characterized by fundamental changes in water relation, biochemical and physiological process, membrane structure, and ultrastructure of sub cellular organelles. Some plants are able to cope with arid environments by mechanisms that mitigate drought stress, such as stomatal closure, partial senescence of tissues, reduction of leaf growth, development of water storage organs, and increased root length and density, in order to use water more efficiently. Water flux through the plant can be reduced or water uptake can be increased by several physiological adaptations. Many lichens, bryophytes, and a few ferns can survive in a dried state. Resistant dried structures like seeds and pollen grains are frequently present in most plant species. Among vascular plants, a small group of angiosperms known as poikilohydric or resurrection plants, such as *Craterostigma plantagineum* can tolerate extreme dehydration at the whole plant level. This suggests that the genetic properties required for drought tolerance are present in flowering plant.

6.1 Accumulation of compatible solutes

Osmolytes have some important role in plant responses to water stress and resistance. Many studies indicated that the accumulation of compatible solutes in plant's causes resistance to various stresses such as drought, high temperature and high salinity. Osmotic adjustment and turgor regulation are the well-illustrated functions of these compounds in plants and algae since their high solubility in water acts as a substitute for water molecules released from leaves. The primary function of compatible solutes is to prevent water loss to maintain cell turgor and to maintain the gradient for water uptake into the cell. These metabolite accumulations in cells leads to increase in the osmotic potential and finally resulted in higher water uptake capacity by roots and water saving in the cells. Natural osmolytes concentrations in plant cells can reach 200 mM or more, and such concentrations are osmotically significant. For example, under water stress the proline concentration can reach up to 80% of the total amino acid pool in some plants.

In addition to osmoregulation function, compatible solutes have some other functions in plants such as, protecting of enzymes and membrane structures and integrity, maintain protein conformation at low water potentials, scavenging free oxygen radicals and stabilizing cellular macromolecule's structures such as membrane components. A study also indicated the involvement of glycinebetaine in the protection of the transcriptional and translational machinery under stress conditions. Hydroxyl radicals are the most dangerous of all active oxygen species, but no enzyme has been shown to decompose hydroxyl radicals. Some compatible solutes function as scavengers of hydroxyl radicals such as proline, citrulline and mannitol.

Some aspects of these functions of compatible solutes are related to their extremely hydrophilic property, and hence might also replace water molecules around nucleic acids, proteins and membranes during water shortages. Cell water deficits cause an increase in the concentration of ions that destabilize macromolecules. Compatible solutes might prevent interaction between these ions and cellular components by replacing the water molecules around these components, thereby, protecting against destabilization during drought. For example, protection of RuBisCO by betaine and proline and stabilization of PSII super complex by betaine were shown in pervious works.

Osmolyte's accumulation in plants is caused, not only by the activation of biosynthesis, but also by the inactivation of degradation. Transformed plants with the higher ability in compatible solutes' production showed a significant increase in plant's tolerance to osmotic stress. All known compatible solutes was not accumulated in all plant species, for example, accumulation of GlyBet occurs in some, but not in all, higher plant species such as xerophytes and halophytes. Citrulline accumulates in leaves of wild watermelon plants under drought. Organisms other than plants also accumulate compatible solutes; for example, glycerol in yeast and phytoplanktons, ectoine in *Halomonas*, trimethylamine *N*-oxide and urea in shallow-sea animals, and di-*myo*-inositol-1,1'-phos-phate and related compounds in thermophilic and hyperthermophilic bacteria and Archea.

Compatible solutes are divided into three major groups - amino acids (e.g. proline), polyamins and quaternary amines (e.g. glycinebetaine, dimethylsulfoniopropionate), polyol (e.g. mannitol, trehalose) and sugars like sucrose and oligosacharids. Free proline is believed to play a key role in cytoplasmic tolerance in many species and, therefore, in the resistance of the whole plant to severe drought. Sugars can play a role in osmoregulation under a drought condition in many plants such as alfalfa and *Ziziphus mauritiana*. Many studies indicated that solute accumulation under water stress contributes to inhibition of shoot growth. It is clear because compatible solute synthesis and accumulation need high energy level.

6.2 Activation of antioxidant systems

Free oxygen radicals, produced as the usual secondary consequence of environmental stresses, are very dangerous for cell components and must be precisely regulated. All plants have developed several antioxidant systems, both enzymatic and non-enzymatic, to scavenge these toxic compounds. Among antioxidant enzymes are catalases (CAT), superoxide dismutase (SOD), peroxidases (POD), ascorbate peroxidases (APX), glutathione reductase (GR) and monodehydroascorbate reductase (MDAR) are prominent. Besides, there are antioxidant molecules such as ascorbic acid (AA), glutathione, tocopherols, flavanones, carotenoids and anthocyanins. Some other compounds like osmolytes (e.g. proline), proteins (e.g. peroxiredoxin) and amphiphilic molecules (e.g. tocopherol) also have ROS scavenging function and may act as the antioxidant. Non-enzymatic plant antioxidants are either AA-like scavengers or they are pigments. Some of these compounds are multifunctional; for example, AA can react with H_2O_2, O_2, OH^- and lipid hydroperoxidases, acts as the enzyme co-factor and as a donor/acceptor of electron. The degree of activities of antioxidant systems under drought stress is extremely variable. The defining factors include variation in plant species, in the cultivars of the same species, development and the metabolic state of the plant, and the duration and intensity of the stress, etc.

7. Biotechnology and water stress

Various approaches have been so far been tested to produce stress tolerant plants using classical genetic methods as well as improved plant breeding techniques. One approach to improve plant resistance and crop performance in water-limited environments is to select genotypes that have improved yield in dry environments. The approach is proven partially successful, but it is difficult to accomplish due to the variability of rainfall and the polygenic nature of drought tolerance. The strategy of gene transfer to crop plants from their more tolerant wild relatives using classical genetic methods has also been of limited success. A partial list of potentially important traits for plant breeding might include water-extraction efficiency, water-use efficiency, hydraulic conductance, osmotic and elastic adjustments, and modulation of leaf area.

Plant modification for enhanced tolerance is mostly based on the manipulation of genes that protect and maintain the function and structure of cellular components. The genetically complex responses to abiotic stress conditions are more difficult to control and engineer. Present engineering strategies rely on the transfer of one or several genes associated with stress responsive pathways. Although the current efforts to improve plant stress tolerance by gene transformation have resulted in important achievements; the nature of the genetically complex mechanisms of abiotic stress tolerance, and the potential detrimental side effects, make this task extremely difficult.

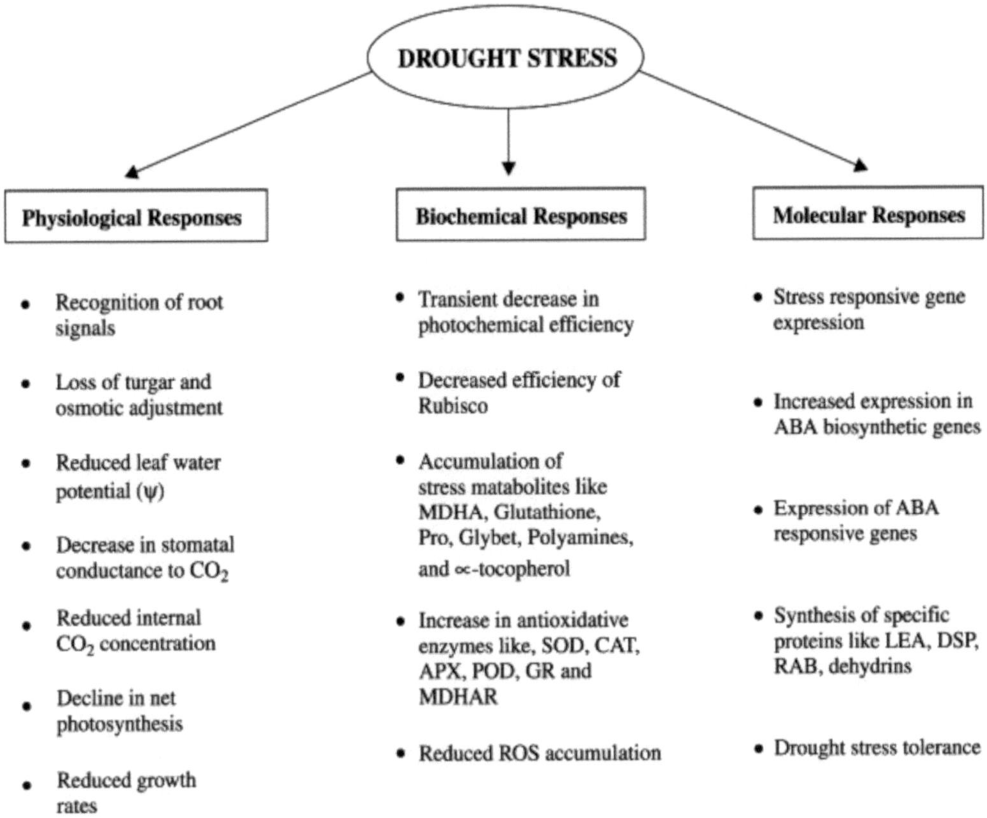

Fig. 1. Physiological, biochemical and molecular basis of drought stress tolerance in plants [Reprinted with permission from 'Shao *et al.* (2008). *Comptes Rendus Biologies*, 331:215-225.' ©2008, Elsevier Limited, UK]

Genetic engineering has allowed the introduction of new pathways for the biosynthesis of various compatible solutes into plants, resulting in the production of transgenic plants with improved stress tolerance. Overexpression of compatible solutes in transgenic plants also resulted in improved stress tolerance. Proline and betaine are well known compatible solutes, synthesized in plants. Genetic engineering of the proline biosynthesis pathways led to increased osmotolerance and salinity stress tolerance in transgenic plant. Proline is synthesized from glutamate via glutamic-ɣ-semialdehyde (GSA) and Δ^1-pyrroline-5-carboxylate (P5C). P5C synthase (P5CS) catalyzes the conversion of glutamate to P5C, followed by P5C reductase (P5CR), which reduces P5C to proline. Transgenic tobacco (*Nicotiana tabacum*) and rice plants overexpressing the p5cs gene that encodes P5CS produced more proline and exhibited better performance, reduced free radical levels and higher biomass under osmotic stress.

Betaine occurs naturally in a wide variety of plants and other organisms to help them better respond to water stress. Glycinebetaine is synthesized in the chloroplast. Choline is converted to betaine aldehyde by the choline mono oxygenase (CMO) under drought, and salinity stresses. Betaine aldehyde is then catalyzed into glycine betaine by betaine aldehyde dehydrogenase. Many important crops, such as rice, potato and tomato, do not naturally accumulate glycinebetaine and are therefore, potential candidates for the engineering of betaine biosynthesis to make them perform better under the drought conditions. Transformed plants like Arabidopsis, tobacco and Brassica that overexpressing choline oxidase (involved in glycinebetaine synthesis) gene (CodA/cox), showed higher tolerance to drought and salt stress. Arabidopsis that expressed N-methyltransferase genes, accumulated glycinebetaine in roots, stems, leaves, and flowers and showed improved seed yield under stress conditions. Transgenic cyanobacteria and other plants engineered to synthesize higher glycinebetaine have shown better resistance to drought, salinity and cold.

On the other hand, polyamines accumulate in plants under a variety of abiotic stresses including drought. Over expression of spermidine synthase in Arabidopsis enhanced its tolerance to multiple stresses including the water deficit. Besides, a number of sugar alcohols like mannitol, trehalose, *myo*-inositol, D-ononitol and sorbitol have been targeted for the engineering of compatible solute overproduction. For example, transgenic tobacco plants carrying a cDNA encoding *myo*-inositol O-methyltransferase (IMT1) accumulated D-ononitol and, as a result, acquired enhanced photosynthesis protection and increased recovery under drought and salt stress. Trehalose overproducing transgenic rice plants showed high tolerance to different abiotic stresses and maintained optimal K^+/Na^+ ratios necessary for cellular functions. Transgenic tobacco plants expressing alfalfa aldose aldehyde reductase, a stress-activated enzyme, showed reduced damage when exposed to oxidative stress and increased tolerance to some stresses like dehydration.

Ion homeostasis in cells is critical for plant resistance and growth under many stress conditions like drought and salinity. Proton pumps such as P-ATPase, V-ATPase and H^+-PPase have the more important role in this function. Arabidopsis plants, transformed with a vacuolar H^+-PPase pump (*AVP1* gene), expressed higher levels of *AVP1* and were more resistant to salt and drought than wild-type plants.

LEA protein accumulates in plants in response to water stress and have several functions in plant resistance to drought. Constitutive over expression of ABF3 (ABRE binding factor) in

Arabidopsis enhances expression levels of target LEA- type genes (RAB18 and RD29) and these transgenic plants are drought tolerant during seedling stage. Over expression of a barley group 3 LEA protein gene, HAV1 in transgenic rice, showed better stress tolerance under salt and drought stress than wild-type plants.

8. Conclusion

The changes in the climatic condition all over the world under the influence of global warming is creating unusual weather phenomena often in the form of water deficit or in the form of floods and waterlogging. Drought is severe of these two due to the prolonged exposure of plants to a water deficient condition. Through the evolutionary mechanism, plants have developed their innate mechanism to combat water stress. All the plants are not equally capable in withstanding water stress and their response to the stress also varies. Even in the highly tolerant species of plants, tolerance comes through changes in the molecular and physiological mechanisms that make plants morphologically adaptable to water deficits. However, they have to pay the price of such tolerance in the form of reduced photosynthesis and resulting in lower biomass yields often caused by the conservative water management scheme adopted by plants. It helps plants to reduce water loss and to maximize available water uptake while making sure the maxim utilization of physiologically available water. The adaptation in this form came from genetic machineries that help plants to produce enzymes, proteins and synthesize molecules suitable in various means to combat water shortages. The drought related oxidative stress is also addresses by biomolecules synthesize by plants. Our knowledge on the causes and effects of water stress on plants along with the understanding of plant's responses in different forms to become tolerant to such stress has already created biotechnological intervention to enhance the drought adaptability of fewer adaptive plants. However, our knowledge about causes and consequences of the water stress in plants still has many dark areas, and we need to enhance our efforts in furthering our appreciation of the issue.

9. References

[1] M. Ashraf, M. Ashfaq, and M. Y. Ashraf, "Effects of increased supply of potassium on growth and nutrient content in pearl millet under water stress," *Biologia Plantarum,* vol. 45, pp. 141-144, 2002.

[2] G. Bernacchia and A. Furini, "Biochemical and molecular responses to water stress in resurrection plants," *Physiologia Plantarum,* vol. 121, pp. 175-181, 2004.

[3] M. M. Chaves, J. S. Pereira, J. Maroco, M. L. Rodrigues, C. P. P. Ricardo, M. L. Osorio, I. Carvalho, T. Faria, and C. Pinheiro, "How plants cope with water stress in the field? Photosynthesis and growth," *Annals of Botany,* vol. 89, pp. 907-916, 2002.

[4] S. Cherian, M. Reddy, and R. Ferreira, "Transgenic plants with improved dehydration-stress tolerance: Progress and future prospects," *Biologia Plantarum,* vol. 50, pp. 481-495, 2006.

[5] I. I. Chernyad'ev, "Effect of water stress on the photosynthetic apparatus of plants and the protective role of cytokinins: A review," *Applied Biochemistry and Microbiology,* vol. 41, pp. 115-128, 2005.
[6] F. M. DaMatta, "Exploring drought tolerance in coffee: a physiological approach with some insights for plant breeding," *Brazilian Journal of Plant Physiology,* vol. 16, pp. 1-6, 2004.
[7] O. Ghannoum, "C_4 photosynthesis and water stress," *Annals of Botany,* vol. 103, pp. 635-644, 2009.
[8] H. Hirt and K. Shinozaki, *Plant Responses to Abiotic Stress.* Berlin; New York: Springer, 2003.
[9] A. Iannucci, A. Rascio, M. Russo, N. Di Fonzo, and P. Martiniello, "Physiological responses to water stress following a conditioning period in berseem clover," *Plant and Soil,* vol. 223, pp. 219-229, 2000.
[10] M. A. Jenks and P. M. Hasegawa, *Plant Abiotic Stress.* Chichester: John Wiley & Sons, 2007.
[11] D. W. Lawlor, "Limitation to photosynthesis in water-stressed leaves: Stomata vs. metabolism and the role of ATP," *Annals of Botany,* vol. 89, pp. 871-885, 2002.
[12] H. Nayyar and D. Gupta, "Differential sensitivity of C_3 and C_4 plants to water deficit stress: Association with oxidative stress and antioxidants," *Environmental and Experimental Botany,* vol. 58, pp. 106-113, 2006.
[13] K. V. M. Rao, A. S. Raghavendra, and K. J. Reddy, Eds., *Physiology and Molecular Biology of Stress Tolerance in Plants.* Dordrecht: Springer, 2006.
[14] A. R. Reddy, K. V. Chaitanya, and M. Vivekanandan, "Drought-induced responses of photosynthesis and antioxidant metabolism in higher plants," *Journal of Plant Physiology,* vol. 161, pp. 1189-1202, 2004.
[15] T. Y. Reddy, V. R. Reddy, and V. Anbumozhi, "Physiological responses of groundnut (*Arachis hypogea* L.) to drought stress and its amelioration: a critical review," *Plant Growth Regulation,* vol. 41, pp. 75-88, 2003.
[16] E. Sacala, "Role of silicon in plant resistance to water stress," *Journal of Elementology,* vol. 14, pp. 619-630, Sep 2009.
[17] H.-B. Shao, L.-Y. Chu, C. A. Jaleel, P. Manivannan, R. Panneerselvam, and M.-A. Shao, "Understanding water deficit stress-induced changes in the basic metabolism of higher plants - biotechnologically and sustainably improving agriculture and the ecoenvironment in arid regions of the globe," *Critical Reviews in Biotechnology,* vol. 29, pp. 131-151, 2009.
[18] H.-B. Shao, L.-Y. Chu, C. A. Jaleel, and C.-X. Zhao, "Water-deficit stress-induced anatomical changes in higher plants," *Comptes Rendus Biologies,* vol. 331, pp. 215-225, 2008.
[19] K. Shinozaki and K. Yamaguchi-Shinozaki, "Molecular responses to dehydration and low temperature: differences and cross-talk between two stress signaling pathways," *Current Opinion in Plant Biology,* vol. 3, pp. 217-223, 2000.
[20] L. Taiz and E. Zeiger, *Plant Physiology.* Sunderland, Mass.: Sinauer Associates, 2002.

[21] W. Wang, B. Vinocur, and A. Altman, "Plant responses to drought, salinity and extreme temperatures: towards genetic engineering for stress tolerance," *Planta,* vol. 218, pp. 1-14, 2003.

[22] I. Yordanov, V. Velikova, and T. Tsonev, "Plant Responses to Drought, Acclimation, and Stress Tolerance," *Photosynthetica,* vol. 38, pp. 171-186, 2000.

1

Plant Water-Stress Response Mechanisms

Şener Akıncı[1,*] and Dorothy M. Lösel[2]
[1]*Department of Biology, Faculty of Arts and Sciences,*
Marmara University, İstanbul,
[2]*Department of Animal and Plant Sciences,*
University of Sheffield
[1]*Turkey*
[2]*U.K.*

1. Introduction

Drought (water stress) is one the major abiotic stress factors that affect all organisms lives including human in terms of health and food. Water absence from the soil solutions affect the natural evaporative cycle between earth and atmosphere that contribute amount of rainfall. Drought occurs when soil moisture level and relative humidity in air is low while temperature is also high. UN reports (2006) [1] estimate that one third of world population has been living in areas where the water sources are poor. Water stress resulting from the withholding of water, also changes the physical environment for plant growth as well as crop physiology [2]. Almost every plant process is affected directly or indirectly by water supply [3]. Plants, as one of basic food sources, either in nature or cultivations, in their growing period, require water or at least moisture for germination. Certainly, most land plants are exposed to short or long term water stress at some times in their life cycle and have tended to develop some adaptive mechanisms for adapting to changing environmental conditions. Some plants may adapt to changing environment more easily than others giving them an advantage over competitors. Water stress may range from moderate, and of short duration, to extremely severe and prolonged summer drought that has strongly influenced evolution and plant life. [4-6]. Crop yields are restricted by water shortages in many parts of the world [7]. The physiological responses of plants to water stress and their relative importance for crop productivity vary with species, soil type, nutrients and climate. On a global basis, about one-third of potential arable land suffers from inadequate water supply, and the yields of much of the remainder are periodically reduced by drought [2]. It is estimated that 10 billion people in the world will be hungry and malnourished by the end of this century [8]. One of the aims of the researches is to gain an understanding of survival mechanisms which may be used for improving drought tolerant cultivars for areas where proper irrigation sources are scarce or drought conditions are common.

In research aimed at improvements of crop productivity, the development of high-yielding genotypes, which can survive unexpected environmental changes, particularly in regions dominated by water deficits, has become an important subject.

---

*Corresponding Author

2. As a major plant growth inhibitor: Drought

2.1 Water and whole plant responses

The amount of water available to plants is important, since water accounts for 80-90% of the fresh weight of most herbaceous plant structures and over 50% of the fresh weight of woody plants [9]. Water supply is restricted in many parts of the world and productivity in these environments can only be increased by the development of crops that are well adapted to dry conditions [10].

Data of Raheja (1966, cited in Hurd, 1976) [11] show that 36% of world's land area is under semi-arid conditions, receiving only 5 to 30 in. of rainfall annually and the remaining (64%) is exposed to temporary drought during the crop season. On a global basis, about one-third of potential arable land suffers from inadequate water supply, and the yields of much of the remainder are periodically reduced by drought [2]. Moreover, water deficits may occur during a plant's life cycle outside of arid and semi-arid regions [12,13] even in tropical rainforests [14]. Water is progressively lost from a fully "**saturated soil**", firstly by draining freely, under the influence of gravity, and the rate of loss gradually slows down until no further water drains away, when the soil is said to be at "**field capacity**". Further loss of water by evaporation or by absorption by plant roots reduces the moisture content still further, until no further loss from these causes can occur, a stage known as the "**wilting point**" at which plants can no longer obtain the water necessary to meet their needs and they therefore wilt and die from moisture starvation.

Initially, stress conditions occur transiently as "cyclic water stress" even under adequate soil moisture conditions and may prevail certain time in the daytime and normalized after reduction of transpiration rate by the night [15].

Crop yields are restricted by water shortages in many parts of the world and the total losses due to this cannot be estimated with confidence [2,7]. According to Rambal and Debussche (1995) [16], changes in plant conductance under water stress are attributable to effects on the roots and xylem. As the soil dries, decreased permeability, due to root suberization and/or increased loss of fine roots, can reduce the balance between water extraction capacity and transpiring leaf area. Roots of unwatered plants often grow deeper into the soil than roots of plants that are watered regularly.

Plants exposed to stress due to decreasing supply of water or other resources, or because of climatic changes, show different responses according to species and the nature and severity of the stress. By altering the chemical and physical composition of tissues, water deficits also modify various aspects of plant quality, such as the taste of fruits and the density of wood [17]. Water shortage significantly affects extension growth and the root-shoot ratio at the whole plant level [18,19]. Although plant growth rates are generally reduced when soil water supply is limited, shoot growth is often more inhibited than root growth and, in some cases; the absolute root biomass of plants in drying soil may increase water use efficiency relative to that of well-watered controls [15,20,21]. Almost every plant process is affected directly or indirectly by water supply. When soil dries, the reduction in water content is accompanied by other changes such as increase in salt concentration and increasing mechanical impedance. The growth of plants is controlled by rates of cell division and enlargement, as well as by the supply of organic and inorganic compounds required for the synthesis of new protoplasm and cell walls.

It is well known that water stress not only affects morphological appearance but also changes bio-mass ratio. Bradford and Hsiao (1982) [22] and Sharp and Davies, (1979) [20] reported that water stress drastically decreases root elongation and leaf area expansion but that these two processes are not equally affected. Leaf growth is usually decreased to a greater extent than root growth, and photosynthate partitioning is altered to increase root/shoot ratio [23,24]. Timpa *et al.*, (1986) [25]; Akıncı and Lösel (2009,2010) [26,27] reported that the water stress caused major reductions in height, leaf number, leaf area index, fresh and dry weight of cotton plants and some *Cucurbitaceae* members.

2.2 Classification attempts of the survival mechanisms of plants response to water stress

Physiological and ecological strategies that plants evolved to cope with water shortages by either avoidance or tolerance to stress. On the nature, since plants are subjected to unavailability of water varying in length from hours to days from the water sources, therefore stress is determined by the extent and duration of the deprivation from water. Plant responses roughly may be classified as; i) short term changes related to mainly physiological responses (linked to stomatal regulation); ii) acclimation to availability of certain level of water (solute accumulation resulted with adjustment of osmotic potential and morphological changes); iii) adaptation to water stress conditions (sophisticated physiological mechanisms and specifically modifications in anatomy) [28-30]. Many processes affect the "fitness" of a plant in water-limited situations but those, such as survival, that may be appropriate in natural ecosystems are often of less interest in some agricultural crops, where productivity is usually of the greatest importance. It is not easy to define "*drought tolerance*", as stability of yield may be the biggest consideration in some situations. However, Jones (1993) [31] has pointed that out drought-tolerant genotypes of most crop plants are those giving some yield in a particular water-limited environment. Kramer (1980) [2] classified as "*drought avoidance*", the adaptations by which plants survive in regions subject to drought, in addition to drought tolerance, since this name fitted the actual situation more accurately than Levitt's term "*drought escape*".

Plants showing improved growth with limited water are considered to tolerate drought, regardless of how the improvement occurs. Kramer and Boyer (1995) [9] have reviewed strategies of drought tolerance, including (1) rapid maturation before onset of drought, or reproduction only after rain, (2) postponement of dehydration by having deep roots, (3) protection against transpiration or storing water in fleshy tissues, (4) allowing dehydration of the tissues and simply tolerating water stress by continuing to grow when dehydrated or surviving severe dehydration. These effects are generally distinct from the factors controlling water use efficiency. *Drought avoiders* often reproduce rapidly after only brief minimal accumulation of dry weight, ensuring that they are represented in the next generation. *Dehydration postponers,* with deep roots, may have a water use efficiency identical to that of other species but will accumulate more dry weight because they can reach a larger amount of water than shallow rooted types, although their water use efficiency may be similar to other spp. *Dehydration tolerators* may have the same water use efficiency as dehydration sensitive species when water is available, but can also grow at lower tissue hydration levels than the other species [9].

The physiology of crop plant responses to drought stress has been classified by Blum (1989) [32] into two domains: (1) a positive carbon balance is maintained by the plant under moderate stress, so that resistant genotypes achieve a greater net gain of carbon than susceptible ones, and have a correspondingly better yield, (2) a net loss of carbon takes place under severe stress, so that growth stops and plants are merely surviving stress. Resistant genotypes generally survive and recover better upon rehydration than susceptible ones, depending upon the degree of stress.

2.3 Plant strategies under water depletion

Major efforts of plant physiologists and breeders during the past 30 years, have concentrated on improving the drought tolerance of many agricultural and horticultural crops. It is clear that, with the increasing world requirement for food, there is an urgent need for research to improve the stress tolerance of crop plants and to develop better management techniques to keep food production at levels near to demand, in spite of limited availability of land and water [33]. According to Borlaug and Dowswell (2005) [34] crop production will have to be doubled achieved by expanding land area for cultivation or increase crop productivity from per hectare. As pointed out earlier by Kozlowski (1968) [17] there is a need to increase crop production, in the face of mounting food shortages, and water conservation is an important factor in overcoming food deficiencies.

Land plants adapted to a moderate water supply are termed **mesophytes** while those adapted to arid zones are **xerophytes**. There are, of course, all gradations between these groups and it is, therefore, not always easy to place a plant in one or other group. It is even possible for a plant to fit into more than one group (Levitt, 1972) [35]. Plants under severe drought conditions tend to develop xeromorphic characteristics including those listed by Walter (1949, cited in Parker, 1968) [36], namely increases in proportion of leaf vein tissue compared to leaf surface, increased stomatal number per unit leaf area, smaller sizes of stomata, epidermal and mesophyll cells, greater density of leaf hairs but smaller hairs, thicker outer epidermal walls and cuticle.

Fresnillo Fedorenko *et al.*, (1995) [37] found similar trends for live and total leaf production, total length per plant of the central leaflet in leaves, and branch and root segment production, all of which decreased proportionally with increasing water stress in *Medicago minima*. Schulze (1986) [38] also reported that water shortage significantly affects extension growth and the root-shoot ratio at the whole-plant level.

Leaf adaptations are among the main factors favouring the success of a species in a water-stressed environment [39]. Morgan (1980) [40] pointed out that, in some species, reduction in leaf area by rolling may also be important in controlling water loss and reflects changes in leaf turgor. Fitter and Hay (1987) [41] pointed out that any reduction in cell size of mesophytes or xerophytes, due to loss of turgor during expansion, will lead to a higher stomatal frequency than in unstressed leaves, since the number of potential guard cells is unchanged.

A few reports discuss changing (reducing) of stomatal index by water stress [42,43]. It may also relate to reduction in leaf growth and production of smaller cells [42,44,45]. Decreasing water content is accompanied by loss of turgor and wilting, cessation of cell enlargement, closure of stomata, reduction in photosynthesis, and interference with many other basic metabolic processes [9].

Larcher (1995) [46] also stated that leaves growing under conditions of water deficiency develop smaller, but more densely distributed, stomata, enabling the leaf to reduce transpiration by a quicker onset of stomatal regulation. In addition, leaves of genotypically adapted plants tend to have more densely cutinized epidermal surfaces, covered with thicker layers of wax.

Water deficit increases wax deposition on the leaf surface, and results in a thicker cuticle that reduces water loss at the epidermis. This reduces CO_2 uptake, but without affecting leaf photosynthesis, because the epidermal cells underneath the cuticle are nonphotosynthetic [47].

3. Physiological and morphological responses to water stress

3.1 Drought effect on photosynthesis

Water stress reduces photosynthesis by decreasing both leaf area and photosynthetic rate per unit leaf area [48]. Photosynthesis by crops is severely inhibited and may cease altogether as water deficits increase. The decrease in leaf growth, or increasing senescence of leaves under drought conditions, may also inhibit photosynthesis in existing leaves [49]. Decreasing water content is accompanied by loss of turgor and wilting, cessation of cell enlargement, closure of stomata, reduction in photosynthesis, and interference with many other basic metabolic processes [9]. Photosynthesis by crops is severely inhibited and may cease altogether as water deficits increase. The decrease in leaf growth, or increasing senescence of leaves under drought conditions, may also inhibit photosynthesis in existing leaves [49]. Ehleringer (1980) [50] pointed out that leaf pubescence, which increases under water stress, can decrease the photosynthesis by reflecting quanta that might have been used in photosynthesis.

In the field, plants are normally not deprived of water rapidly. During slowly increasing water stress photosynthesis and transpiration usually decrease at similar rates [51]. The two main factors causing stomatal closure are usually an increase in the concentration of gaseous carbon within leaves and a decrease in water potential of leaf cells [52,53].

The simplest explanation for the inhibition of photosynthesis during water stress would be that the stomata close and the internal CO_2 concentration decreases [54,55], since stomatal limitation is more severe when a plant is stressed than when it is not [54]. Therefore, it is rather surprising that photosynthesis often decreases in parallel with, or more than, stomatal conductance [56-59]. The photosynthetic rate in higher plants decreases more rapidly than respiration rate with increased water stress, since an early effect of water reduction in leaves is usually a partial or complete stomatal closure, markedly decreasing the movement of carbon dioxide into the assimilating leaves and reducing the photosynthetic rate up to ten times, according to the amount of water removal and the sensitivity of the plant [35]. In terms of the relationship between photosynthesis and leaf water status, Quick et al. (1992) [60] reported that, in field conditions, photosynthesis in ambient CO_2 reached a maximum value in the morning and declined later in the day when water potential decreased and leaf-to-air water vapour pressure deficits increased. In non-watered plants the decline was larger, and occurred earlier. In most cases stomatal conductance followed a diurnal pattern similar to that of photosynthesis.

3.2 Osmotic adjustment mechanisms under water stress

Water is essential in the maintenance of the turgor which is essential for cell enlargement and growth and for maintaining the form of herbaceous plants. Turgor is also important in the opening of stomata and the movements of leaves, flower petals, and various specialised plant structures [9]. Although turgor measurements on segments the non-growing lamina have often appeared to show declining rates of leaf growth with decreasing turgor, turgor measurement in regions of leaves and stems, where cell enlargement usually occurs, often show little or no decrease, even when cell enlargement is largely inhibited due to soil drying [9,61-63]. This is believed to be due to osmotic adjustment, the process in which solutes accumulate in growing cells as their water potential falls [64,65] of osmotic potential arising from the net accumulation of solutes in response to by maintaining turgor in tissues, osmotic adjustment may allow growth to continue at low water potential. Turner and Jones (1980) [64] have defined osmotic adjustment as "the lowering water deficits or salinity".

Osmotic adjustment usually depends mainly on photosynthesis to supply compatible solute. As dehydration becomes more severe, photosynthesis is inhibited, resulting in a smaller solute supply for osmotic adjustment. With continued water limitation, osmotic adjustment delays, but cannot completely prevent, dehydration [9]. In leaves and stems at least, solute accumulation does not fully compensate for the effects of limited water supply on cell enlargement. Turner and Jones (1980) [64] stated that the rate of development of stress has a major effect on the degree of osmotic adjustment. Oosterhuis and Wullschleger (1987) [66] pointed out that increasing the number of stress cycles increased the amount of osmotic adjustment in cotton. Turner (1986) [67] noted that leaf expansion can decrease without change in leaf turgor pressure.

Osmotic adjustment has been found in many species [64,65], and has been implicated in the maintenance of stomatal conductance, photosynthesis, leaf water volume and growth [64,65,67].

Wheat and other cereals show other additional strategies: turgor loss and stomatal closure may occur at different relative water contents, while osmotic adjustment leads to rapid responses decreasing the effect of water stress [68].When soil dries, the reduction in water content is accompanied by other changes such as increase in salt concentration and increasing mechanical impedance [69]. The growth of plants is controlled by rates of cell division and enlargement, as well as by the supply of organic and inorganic compounds required for the synthesis of new protoplasm and cell walls [9].

Wheat and other cereals show other additional strategies: turgor loss and stomatal closure may occur at different relative water contents, while osmotic adjustment leads to rapid responses decreasing the effect of water stress [68]. Russel (1976) [70] pointed out that water stress increases the osmotic pressure of the cell sap, increasing the percentage of sugar in sugar-cane and often in sugar beet, although the yield per acre may be reduced.

Solutes known to accumulate with water stress and to contribute to osmotic adjustment in non-halophytes, include inorganic cations, organic acids, carbohydrates and free amino acids. In some plants potassium is the primary inorganic cation accumulating during water stress and it is often the most abundant solute in a leaf [71,72]. Osmotic adjustment is usually not permanent and plants often respond rapidly to increased availability of water. Loss of osmotic adjustment can occur in less than 2d in durum wheat [73], and both osmotic potentials and concentrations of some individual solutes have been shown to return to pre-stress levels within 10d after watering [64,74].

Studies by Blum and co-workers (reviewed by Blum, 1989) [34] and Kameli (1990) [75] have suggested that drought-resistant wheat varieties, with long-term yield stability under drought stress, were characterised by a greater capacity for osmoregulation than less resistant varieties. Landraces of sorghum and millet from dry regions in India and Africa were found to be more drought resistant (in terms of plant growth and delayed leaf senescence) than those from humid regions [32]. Munns *et al.*, 1979 [76] found that the change in osmotic potential in the apex and enclosed developing leaves of wheat seedlings under rapidly developing water stress, was due mainly to the accumulation of soluble sugars, amino acids (particularly asparagine and proline) and K+ ions.

Morgan and Condon (1986) [77] showed that such increase in solute concentration gives tissues a temporary advantage, enabling turgor to be maintained at low water potentials by decreasing their osmotic potentials. Westgate and Boyer (1985) [78] pointed out that, when dehydration occurs in the absence of high external salinities, there can also be rapid increases in solute content of cells. The growing tissues throughout the plant may show osmotic adjustment when the soil loses water.

In less severe stress, the elongating regions of wheat leaves were found to adjust osmotically by the accumulation of sugars, principally glucose [73,79]. Osmoregulation was very active in races from dry regions. Osmoregulation and turgor maintenance permit continued root growth and efficient uptake of soil moisture [20]. However, despite the accumulation of ions and organic solutes, allowing osmotic adjustment in the meristematic and expanding regions, growth of the shoot may still be inhibited by stress, either because osmotic adjustment may not be sufficiently rapid to compensate for growth or due to a stress-induced fall in turgor.

4. Plant metabolic response to water scarcity

One of the gains an understanding of survival mechanisms which may be used for improving drought tolerant cultivars for areas where proper irrigation sources are scarce or drought conditions are common. Plants adaptations to dry environments can be expressed at four levels: phenological or developmental, morphological, physiological, and metabolic the least known and understood of which are the metabolic or biochemical adaptations involved [80]. Physiological and biochemical changes including carbohydrates, proteins and lipids observed in many plant species under various water stress levels, which may help in better understanding survival mechanisms in drought.

4.1 Carbohydrates changes under water stress

The available reports (listed in Table 1) stated that the content of soluble sugars and other carbohydrates in the leaves of various water stressed plants are altered and may act as a metabolic signal in the response to drought [26,27,81-83] however, accumulation or decrease of sugars depending on stress intensity and role of sugar signalling in these processes is not totally clear yet [84].

Among the major effects are those involving carbohydrate metabolism, with the accumulation of sugars and a number of other organic solutes [75]. Munns *et al.*, (1979) [76] and Quick *et al.*, (1992) [60] showed that sugars are major contributors to osmotic adjustment in expanding wheat leaves. Moreover, short-term water stress inhibited starch synthesis more strongly than sucrose synthesis, in both ambient CO_2 and in saturating CO_2. Short-term water stress was earlier also reported to stimulate the conversion of starch to sucrose in bean leaves [85,86]. The increase of sugar in various plant tissues response to water stress are supported the idea of contribution of solutes while the plants exposed to different stress levels. The studies have shown that soluble sugars accumulate in leaves during water stress [60,71,79,87-90], and have suggested that these sugars might contribute to osmoregulation [65], at least under moderate stress.

Quick *et al.*, (1992) [60] compared the effect of water stress on the rate of photosynthesis and the partitioning of photosynthate in four different species, including two annuals (*Lupinus albus* L. and *Helianthus annuus* L.), and two woody perennials (*Vitis vinifera* cv. Rosaki and

Eucalyptus globulus Labill.) and concluded that, when water stress develops under field conditions, there is an alteration in the balance between sucrose synthesis and translocation, which allows many species to maintain or increase the pool of soluble sugars in their leaves. In *Eucalyptus* soluble sugars were low compared to starch and non-watered plants contained higher levels of soluble sugars in their leaves than watered plants, but much less starch. Similarly, leaves of non-watered sunflower plants contained almost twice as much soluble sugar those of watered plants. Hodges and Lorio (1969, cited in Levitt, 1972) [35] detected a marked increase in reducing sugars, nonreducing sugars, and total carbohydrates, with an approximately equivalent decrease in starch.

Carbohydrates changes	**References**
Increasing total carbohydrates	cotton (Timpa *et al.*, 1986)
Total soluble sugars increasing	durum wheat (Kameli and Lösel, 1996) Nodulated alfalfa (Irigoyen *et al.*, 1992)
Increasing soluble sugars	South African grasses (Westgate *et al.*, 1989)
Increasing sucrose	lupins and *Eucalyptus* (Quick *et al.*, 1992), Alfalfa (Al-Suhaibani, 1996), embryos from Soybean (Westgate *et al.*, 1989), wheat (Drossopoulos *et al.*, 1987), wilted bean (Steward, 1971), durum wheat (Kameli and Lösel, 1993), -in leaves under severe stress-*Cucumis sativus* L., *C. melo* L. (snake cucumber), *Cucurbita pepo* L., *Ecballium elaterium* (L.) A. Rich. (Akıncı and Lösel, 2009).
Fructose, glucose accumulation	durum wheat (Kameli and Lösel, 1993)
Starch accumulation	cotton (Ackerson and Hebert, 1981)
Fructans enhancing resistance	tobacco (Pilon-Smiths *et al.*, 1995)
Carbohydrate unchanged	*Artemisia tridentata* (Evans *et al.*, 1992)
Sucrose content decreasing	soybean cotyledon (Westgate *et al.*, 1989)
Sucrose and starch decreasing	grapevine (Rodriguez *et al.*, 1993)
Raffinose utilisation prevented by water stress	*Citrullus lanatus* seeds (Botha and Small, 1985)
Starch depletion	*Lupinus, Helianthus, Vitis, Eucalyptus* (Quick *et al.*, 1992), wilted bean leaves (Steward, 1971), South African grasses (Schwab and Gaff, 1986), cucumber (Akıncı and Lösel, 2010)

Table 1. Changes in plant metabolics (Carbohydrates)

Drossopoulos *et al.*, (1987) [91] concluded that, in two wheat cultivars, sucrose generally formed the major portion of the ethanol soluble carbohydrates. High concentrations of glucose and fructose were observed in the stems of the water-stressed plants towards maturation as well as in the ears, immediately after heading. The major differences between cultivars were in the sucrose levels of leaves and roots before heading. There have been many reports that water stress leads to a general depletion of soluble sugars and starch in leaves. Hanson and Hitz (1982) [80] and Huber *et al.*, (1984) [57] have concluded that water stress has a larger effect on carbon assimilation than on translocation and use of photosynthate.

Barlow (1986) [92] showed that much of the sugar accumulation, which began with the first indication of suppression in leaf elongation under water stress in wheat, was due to glucose, fructose and sucrose. The inhibition of germination of *Citrullus lanatus* seeds by water stress was investigated by Botha and Small (1985) [93], who observed a marked effect on carbohydrate metabolism. Smaller changes in glucose content and in the reducing substance content of the control seeds occurred during germination, coinciding with a decrease in sucrose. However, this decrease did not entirely account for the observed increase in glucose content. Fructose decreased in control seeds, over the first 30 h of incubation, and then increased again, whereas the glucose content of stressed seeds tended to increase throughout the 48 h incubation period, with fructose remaining fairly constant. On the other hand, Pattanagul and Madore, (1999) [94] also reported various sugars depletion in variegated coleus (*Coleus blumei* Benth.). In the green leaf tissues the diurnal - light period levels of the raffinose family oligosaccaharides stachyose and raffinose and non - structural carbohydrates (galactinol, sucrose, hexoses and starch) decreased whereas drought had little effect on soluble carbohydrate content in the other part of non - photosynthetic white leaf tissues. There was no difference in glucose and fructose levels between the wilted (incubated) and turgid bean leaves as well as depletion of starch concentrations was observed in plants of bean exposed to drought stress [60,85].

4.2 Plant proteins: Responses to drought

Many specified protein synthesized under water scarcity have been isolated and characterized by researches [95-98]. The water stress-specific proteins (stress induced) have been described by different groups such as dehydrins (polypeptide), LEAs (late embryogenesis abundant), RABs (responsive to ABA), storage proteins (in vegetative tissues) [99]. LEAs proteins are also subdivided into several groups and expect to be located in cytosol and with hydrophylic and soluble on boiling featured [100].

Under water stress conditions plants synthesize alcohols, sugars, proline, glycine, betaine and putrescine and accumulate that of those molecular weights are low [101,102]. Dehydrins have been the most observed group among the accumulated proteins in response to loss of water and increased in barley, maize, pea, and *Arabidopsis* and under water stress LEA proteins plays important role as protection of plants. Osmotin is also accumulated protein under water stress in several plant species such as tobacco, triplex, tomato and maize [103].

Changes of amino acids and protein have been mentioned in many reports which have stated that water stress caused different responses depending on the level of stress and plant type and listed in Table 2. For instance, in *Avena* coleoptiles water stress clearly caused a significant reduction in rate of protein synthesis [104]. Water stress has a profound effect upon plant metabolism, and results in a reduction in protein synthesis. Several proteins were reduced by stress in maize mesocotyls [105,106]. Dasgupta and Bewley (1984) [107] pointed out water stress reduced protein synthesis in all regions of barley leaf. Vartanian *et al.*, (1987) [108] mentioned the presence of drought specific proteins in tap root in *Brassica*. Dasgupta and Bewley (1984) [107] pointed out water stress reduced protein synthesis in all regions of barley leaf.

Protein changes	**References**
Inhibited protein synthesis	*Avena* coleoptiles (Dhindsa and Cleland, 1975)
Increased protein levels	*Cicer arietinum* (Rai *et al.*, 1983)
Increased protein content	*Zea mays* (Rai *et al.*, 1983)
Water stress induced spefisic proteins (Dehydrins, LEAs, RABs, vegetative storage proteins)	cotton (Artlip and Funkhouser, 1995), rice (Xu *et al.*, 1996)
Proline, glycine accumulation, betain, putrescine	tobacco (Chopra *et al.*, 1998), (Galston and Sawhney, 1990)
Dehydrin, LEA group 1 (D19)	cotton, barley, carrot (Ramagopal, 1993)
Dehydrin-like transcripts accumulate	*Lathyrus sativus* (Sinha *et al.*, 1996)
LEA (D7, D29)	desiccating mature cotton embryos, *Craterostigma plantagineum* chloroplast, *Citrus* seedlings exposed to drought (Bray, 1995; Naot *et al.*, 1995)
Osmotin	tobacco, triplex, tomato and maize (Ramagopal, 1993)
87kDa and 85kDa proteins (stress-associated -SAPs-) accumulation	rice varieties (Pareek *et al.*,1997)
Boiling-staple protein (BspA) accumulation	*Populus popularis* (Pelah *et al.*, 1997)
RAB18 protein accumulation	*Arabidopsis thaliana* (Mantyla *et al.*, 1995)
Chloroplastic proteins (CDSP 32 and CDSP 34)	*Solanum tuberosum* (Pruvot, *et al.*, 1996)
Total proteins decrease	sugar beet (Shah and Loomis, 1965)
Soluble protein decrease	Bermuda grass (Barnett and Naylor, 1966)
12.6- k.Da protein (cell wall) synthesis decrease	*Lycopersicon chilense* (Yu *et al.*, 1996)
Soluble protein level decline	*Pisum sativum* L. nodules (Gogorcena *et al.*, 1995)
Total and soluble protein content	*Populus popularis* (Pelah *et al.*, 1997)

Table 2. Changes in plant metabolics (Proteins)

A stress episode which inhibits cell division and expansion, and consequently leaf expansion, will also halt protein synthesis, which is also inhibited by osmotic stress imposed on excised plant parts. The direct significance of the inhibition of protein synthesis by stress to growth and leaf expansion is difficult to assess. Hsiao (1970) [109] concluded that inhibition of cell expansion precedes the decline in polysome content and that changes in polysome profile might be caused by cell growth rather than the reverse. Although water stress may inhibit protein synthesis [104,110] some specific types of proteins and mRNA increase in water stressed plants. For instance, free proline accumulation in response to drought in many plant species tissues is well documented [111-115]. Vartanian *et al.*, (1987) [108] mentioned the presence of drought specific proteins in tap root of *Brassica*.

The functions of many of these proteins have not been established [116]. However, water stress may inhibit the synthesis of different proteins equally whilst inducing the synthesis of a specific stress protein [107]. Changes of amino acids and protein have been mentioned in many reports which have stated that water stress caused different responses depending on the level of stress and plant type. For instance, in *Avena* coleoptiles water stress clearly caused a significant reduction in rate of protein synthesis [104]. Treshow (1970) [117] concluded that water stress inhibited amino acid utilisation and protein synthesis. While amino acid synthesis was not impaired, the cellular protein levels decreased and since utilisation of amino acids was blocked, amino acids accumulated, giving a 10- to 100-fold accumulation of free asparagine. Valine levels increased, and glutamic acid and alanine levels decreased. Barnett and Naylor (1966) [118] found no significant differences in the amino acid and protein metabolism of 2 varieties of Bermuda grass during water stress and reported that amino acids were continually synthesised during the water stress treatments, but protein synthesis was inhibited and protein content decreased. Similarly, water stress did not change protein content uniformly in the different cultivars of Cucumber and *Cucurbita pepo* L., *Cucumis melo* L. (snake cucumber) and *Ecballium elaterium* (L.) A. Rich. (Squirting cucumber) which show differing responses to moderate and severe stress treatment and during recovery [3]. Tully and Hanson (1979) [119] found that water stress slightly increased the amino acid to sugar ratio of the exudate, but did not change the amino acid composition very markedly. Several proteins were reduced by stress in maize mesocotyls [105,106].

4.3 Plant lipids – water stress interactions

The effect of water stress lipid composition on the higher plants have been the subject of considerable research. Phospholipids and glycolipids serve as the primary nonprotein components of plant membranes, while triglycerides (fats and oils) are an efficient storage form of reduced carbon, at various developmental stages and particularly in seeds [47]. The functions of membrane proteins are influenced by the lipid bilayer, in which they are either embedded or bound at the surface. For this reason, a knowledge of the lipid composition of membranes in plant cells is important.

Ideas about the adaptive value of lipid changes induced by environmental conditions are often based upon physical properties of the lipids involved in membrane structure, such as phase separation temperatures and fluidity, which may affect the permeability of bio membranes [120]. About 70% of the total protein and 80% of the total lipid of leaf tissue are present in chloroplasts. Any changes in chloroplast membranes, therefore, will usually be reflected by corresponding alterations to leaf total lipids [121].

Lipids, being one of the major components of the membrane, are likely to be affected by water stress. In plant cell, polar acyl lipids are the main lipids associated with membraneous structures [122,123]. Glycolipids (GL) are found in chloroplasts membranes (more than 60%) and phospholipids (PL) are thought to be the most important mitochondrial and plasma membrane lipids [124]. Many workers have investigated the effect of different levels of water stress on lipid content and composition in different parts of plants [75,90,125-132] and their changes listed in Table 3. However, researches concerning on plant lipids affected by water stress have often contradictory since absence of enough information about the plant water status i.e. description of stress effects [133].

Lipid changes	**References**
PL and GL decline	cotton (Wilson *et al.*, 1987)
GL decrease	cotton (Ferrari Ilou *et al.*, 1984),wheat, barley (Chetal *et al.*, 1981)
Total lipids and PL, GL and diacylglycerols decrease	sunflower (Navari-Izzo *et al.*, 1993)
PL decrease	sunflower (Quartacci and Navari-Izzo, 1992), maize (Navari-Izzo *et al.*, 1989) cotton (Wilson *et al.*, 1987), cotton (El-Hafid *et al.*, 1989), oat (Liljenberg and Kates, 1982)
Diacylglycerol, free fatty acid and polar lipid decrease	maize (Navari-Izzo *et al.*, 1989)
Total lipid content decrease	cucumber Cvs., squash, squirting cucumber (Akıncı, 1997)
Trans-hexadecenoic acid decrease	cotton (Pham Thi *et al.*, 1982)
Linoleic and linolenic acid biosynthesis, galactolipid decrease	cotton (Pham Thi *et al.*, 1985)
Diacylglycerol, triacylglycerol and glycolipid increase	soybean (Navari-Izzo *et al.*, 1990)
Saturation of the fatty acids increase	cotton (Pham Thi *et al.* 1982)
Phospholipid (phosphatidylcholin) increase	wheat (Kameli, 1990)
Total lipid content increase	alfalfa (Al-Suhaibani, 1996)
Triglyceride ands streryl ester levels increase	maize (Douglas and Paleg,1981)
Free fatty acids (FFA) increase	wheat (Quartacci *et al.*, 1994)

Table 3. Changes in plant metabolics (Lipids)

Navari-Izzo *et al.*, (1993) [131] pointed out that, since the plasma membrane has a key position in cell biology, understanding membrane function is a major challenge. The selectivity of membranes and their functioning vary with the types and proportions of lipid and protein components.

Investigations on various crop species record a general decrease in phospholipid, glycolipid and linoleic acid contents and an increase in the triacylglycerol of leaf tissues exposed to long periods of water deficits, although the intensity of the stress applied is not always specified. [126,127,134]. The physical state and composition of the lipid bilayer, in which enzymic proteins are embedded, influence both structural and functional properties of membranes. Enzyme activity and transport capacity are affected by the composition and phase properties of the membrane lipids [120,135,136]. Wilson *et al.*, (1987) [137] observed that water deficit caused a significant decline in the relative degree of acylunsaturation (i.e. FA -unsaturation) in phospholipids and glycolipids in two different drought tolerant cotton plants. Pham Thi *et al.*, (1987) [130] pointed out that changes in oleic and linoleic acid during water stress resulted in desaturation changes in one drought sensitive and another more resistant cotton variety and showed that water stress markedly inhibited the incorporation of the precursors into the leaf lipids.

Navari-Izzo *et al.*, (1993) [131] found that, in plasma membranes isolated from sunflower seedlings grown under water stress, there was a reduction of about 24% and 31% in total lipids and phospholipids, respectively, and also significant decreases in glycolipids and diacylglycerols. There was no change in free fatty acids, but triacylglycerols and free sterols increased. However, diacylglycerol, triacylglycerol and glycolipid content increased in soybean seedling shoots under water stress [129]. On the other hand, total lipid content of leaves tended to decrease in two cucumber cultivars as well as *C. pepo* and *Ecballium* in severe stress [3]. The researches indicated that PL in plant tissues under long time drought have been decreased in various crop species [127,129,137,138].

Navari-Izzo *et al.*, (1989) [127] studying responses of maize seedling to field water deficits, found that the diacylglycerol, free fatty acid and polar lipid contents decrease significantly with stress. In the latter class the dryland conditions induced a decrease of more than 50% in phospholipid levels, whereas they did not cause any change in glycolipid levels; and triacylglycerols increased by about 30% over the control.

Pham Thi *et al.*, (1982) [125] investigated the effect of water stress on the lipid composition of cotton leaves. The most striking effects were a decrease of total fatty-acids, due especially to a decrease of trans-hexadecenoic acid. The fatty acid composition of all acyl lipids changed during stress in the direction of increased saturation of the fatty acids. This increased saturation remained even after 10 days of recovery growth under non-stressed conditions.

Pham Thi *et al.*, (1985) [126] pointed out that water deficits inhibit fatty acid desaturation, resulting in a sharp decrease of linoleic and linolenic acid biosynthesis. The decrease in unsaturated fatty acid biosynthesis occurs in all lipid classes, but is greatest in the galactolipid fractions. Wilson *et al.*, (1987) [137] similarly observed that water deficit caused a significant decline in the relative degree of acylunsaturation (i.e. FA -unsaturation) in phospholipids and glycolipids in two different drought tolerant cotton plants. Navari-Izzo *et al.*, (1993) [131] found that, in plasma membranes isolated from sunflower seedlings grown under water stress, there was a reduction of about 24% and 31% in total lipids and phospholipids, respectively, and also significant decreases in glycolipids and diacylglycerols. There was no change in free fatty acids, but triacylglycerols and free sterols increased. Douglas and Paleg (1981) [128] noted that the fatty acids of triglycerides, of maize seedling were quite responsive to stress and in half of the comparisons were found to differ significantly. Stem triglycerides, in general, responded, whereas the major triglyceride change in the leaf was an increase in linolenic, which is essentially absent from this fraction in stems and roots. Kameli (1990) [75] observed that total leaf phospholipids content and, especially, phosphatidylcholine increased, rose in stressed plants of a relatively water stress resistant cultivar of wheat but did not change significantly in another, less tolerant cultivar.

5. Drought and nutrient uptake

Reduction in photosynthetic activity and increases in leaf senescence are symptomatic of water stress and adversely affect crop growth. Other effects of water stress include a reduction in nutrient uptake, reduced cell growth and enlargement, leaf expansion, assimilation, translocation and transpiration. Water and nutrient availability is one of suboptimal phenomenons like most of the natural environments occur continuously, with respect to one or more environmental parameters. Soils are very important natural source for plant growth where the plants anchored however millions of hectares of land becoming unproductive and affecting plant growth every year. The nutrient uptake of crop plants greatly influenced by including overuse of the land in agricultural activities, climate change, precipitation regimes, root morphology, soil properties, quantity and quality of fertilizers, amount of irrigation [139-141]. The root structures such as root extension rate and length, the means of root radius and root hair density affect the quantity of nutrient uptake by a plant. Nutrient elements availability plays vital role for plant growth, nevertheless these physiological factors in nutrient, in soil, in plant or at the root absorpsion sites may in interact as well as antagonistically and synergistically of the plants [141-143].

Many nutrient elements are actively taken up by plants, however the capacity of plant roots to absorb water and nutrients generally decreases in water stressed plants, presumably because of a decline in the nutrient element demand [141]. It is well documented that essential plant nutrients are known to regulate plant metabolism even the plants exposed to drought by acting as cofactor or enzymes activators [144].

It is rather difficult to identify the effects of water stress on mineral uptake and accumulation in plant organs. Many workers have reported different effects of water stress on nutrient concentrations of different plant species and genotypes, and most studies have reported that mineral uptake can decrease when water stress intensity is increased [145-150]. For instance, nitrogen uptake decreased in soybean plants under water stress conditions [145] and nitrogen deficiency causes cotton plants to be sensitive to stress with a higher water stress [151] and decrease of nutrient presumably because of a decline in the nutrient element demand since the reduced root-absorbing power or capacity absorb water and nutrients generally declines accompanied to decrease in transpiration rates and impaired active transport and membrane permeability of crop plants [152].

Water stress generally favoured increases in nitrogen, K^+, Ca^{2+}, Mg^{2+}, Na^+, and Cl^- but decreases in phosphorus and iron [147]. Although the many report stated that water stress mostly causes reduction in uptake of nutrients [152], for instance phosphorus, K^+, Mg^{2+}, Ca^{2+} in some crops [153-155], Ca^{2+}, Fe^{3+}, Mg^{2+}, nitrogen and phosphorus and potassium in *Spartina alterniflora* [156]; Fe^{3+}, Zn^{2+} and Cu^{2+} in sweet corn [157]; Fe^{3+}, K^+ and Cu^{2+} in *Dalbergia sissoo* leaves [150], Gerakis *et al.*, (1975) [158] and Kidambi *et al.*, (1990) [159] stated that nutrient elements increased in forage plant species and alfalfa and soinfoin (*Onobrychis viciifolia* Scop.) respectively. An increase in some specific elements such as K^+ and Ca^{2+} were reported in maize [145], and K^+ in drought tolerant wheat varieties [160], and in leaves of *Dalbergia sissoo* nitrogen, phosphorus, Ca^{2+}, Mg^{2+}, Zn^{2+} and Mn^{2+} increased with increasing water stress [149].

Under water stress, the uptake of K^+ and Ca^{2+} by maize plants increased [145]. The relative amounts of K^+, Ca^{2+}, and Mg^{2+} increased considerably more in barley than in rye when water stresses were imposed [150]. Potassium contributes to osmotic adjustment as one of the primary osmotic substances in many plant species [161,162] and under water stress conditions, K^+ application is beneficial for plant survival with improved plant growth [163,164]. There are a few reports indicating that water stress favored increases in K^+ [147] in plants such as maize [145], drought-tolerant wheat varieties [160], creeping bentgrass [165] and *Ammopiptanthus mongolicus* (evergreen xerophyte shrub) [166]. Contrary to reports stating that water stress generally favored increases in Ca^{2+} [145,147,167,168]. Kırnak *et al.*, (2003) [148] who stated that water stress can cause Ca^{2+} reduction in bell pepper, and suggested antagonistic affects of Zn^{2+} and Mn^{2+} on Ca^{2+} uptake. In moderate and severe stressed leaves of bean (*Phaseolus vulgaris* L.) Ca^{2+} content was lower than the amount of potassium with a Ca/K ratio of 0.12, 0.15 and 0.16 in the control, and in both stress levels

[168]. The reason for total Ca^{2+} content being lower than K^{+} was considered to be directly related to antagonistic effects of Ca^{2+} on K^{+}[169]. According to Kuchenbuch *et al.*, (1986) [170], a reduction in leaf area of onion plants can be explained by declining amount of K^{+} caused by decreasing water content in the soil.

Unlike previous reports which have stated that water stress causes a reduction in nutrients uptake [152-155] as well as Mn^{2+} [150], Mn^{2+} content in bean leaves tended to increase with increased in water stress levels [168]. Nambiar (1977) [150] pointed out that drying the upper layer of a siliceous soil profile strongly reduced the absorption of Mn^{2+} by rye grass, but Cu^{2+} and Zn^{2+} uptake were not relatively affected. For several grassland plants, total nutrients generally decreased with increasing water stress [158].

It is generally accepted that the uptake of phosphorus by crop plants is reduced in dry soil conditions [171,172]. The studies carried out before the mid 1950s, 12 of the 21 papers reported that P concentration decreased, and 9 papers stated that P status was not changed in plants [158]. Although Fawcett and Quirk (1962) [173] reported that only severe water stress reduced plant phosphorus absorption, Nuttall (1976), [174] stated that increased soil moisture resulted in increased phosphorus but decreased sulphur in alfalfa. It is believed that, P uptake by plants increased with increased P levels in the soil ignoring water stress. Olsen (1961) [175] highlighted that the correlations among the soil P levels and monovalent phosphate uptake by plant and magnitude of water stress. In alfalfa (*Medicago sativa* L.) P and that of Ca^{2+}, Mg^{2+}, and Zn^{2+} in alfalfa and soinfoin (*Onobrychis viciifolia* Scop.) increased with decreased soil moisture supply [159]. On the other hand, there was no effect on moisture stress on the concentrations of P, N, K [176].

Magnesium has an inverse relationship with calcium, phosphorus, iron, manganese and potassium with Ca^{2+} and Mg^{2+} having antagonistic effects on Mn^{2+} of a complex nature [47,177] Although some studies have found that Mg^{2+} absorption is increased by water stress in many crops [147,158], in bean leaves Mg^{2+} content decreased by 18% and 45% respectively in two increased water stress levels [168].

In particularly, the presence of Ca^{2+} is of great importance since zinc absorption is closely related with nutrient concentrations, with Zn^{2+} solubility and availability negatively correlated with Ca^{2+} saturation in soils [177]. The increase in Zn^{2+}, particularly in severely stressed plants, seemed to show a competing relationship between Zn^{2+} and Ca^{2+}, with Ca^{2+} appearing at a lower level in the S2 treatment. Dogan and Akıncı (2011) [168] stated that Zn^{2+} supply is expected to decrease the uptake of most nutrients, K^{+} and Mg^{2+} suppressed, while Ca^{2+}, Fe^{3+} only slightly decreased in bean leaves.

According to Singh and Singh (2004) [149], availability of soil nutrients decreases with increasing soil drying, with K^{+}, Ca^{2+}, Mg^{2+}, Zn^{2+}, Fe^{3+} and Mn^{2+} decreasing by 24%, 6%, 12%, 15%, 25% and 18%, respectively. Nambiar (1977) [150] pointed out that drying the upper layer of a siliceous soil profile strongly reduced the absorption of Mn^{2+} by rye grass, but Cu^{2+} and Zn^{2+} uptake were not relatively affected. In herbage plants, the uptake and solubility of nutrient elements depressed but Ca/K and Ca/P ratios increased under water stress conditions. In dried soil, older roots lost their ability to function and nutrients are absorbed by the more active root tips. Most of the studies revealed that water stress restricted uptake of nutrient elements by crops, active transport systems were impaired or destroyed by severe water stress while the presence of various ions responded differently in growth conditions.

6. Conclusion

Wherever they grow, plants are subject to stresses, which tend to restrict their development and survival. Moisture limitation can affect almost every plant process, from membrane conformation, chloroplast organisation and enzyme activity, at a cellular level, to growth and yield reduction in the whole plant and increased susceptibility to other stresses [178]. Reduction in photosynthetic activity and increases in leaf senescence are symptomatic of water stress and adversely affect crop growth. Other effects of water stress include a reduction in nutrient uptake, reduced cell growth and enlargement, leaf expansion, assimilation, translocation and transpiration. In research aimed at improvements of crop productivity, the development of high-yielding genotypes, which can survive unexpected environmental changes, particularly in regions dominated by water deficits, has become an important subject. As pointed out earlier by Kozlowski (1968) [17] there is a need to increase crop production, in the face of mounting food shortages, and water conservation is an important factor in overcoming food deficiencies. From the above survey, it is clear that a wide range of morphological, physiological and biochemical responses have been correlated with differences in drought tolerance in various crop plants.

7. References

[1] UN Human Development Report (2006) Beyond scarcity: Power, poverty and the global water crisis. Accessed: 8 August 2011.

[2] Kramer, P. J. (1980) Drought, stress, and the origin of adaptations. Adaptations of plants to water and high temperature stress. (ed. by Neil C. Turner, Paul J. Kramer) pp. 7-20. John-Wiley & Sons, New York.

[3] Akıncı, S. (1997) physiological responses to water stress by *Cucumis sativus* L. and related species. Ph. D. Thesis, University of Sheffield. U. K.

[4] Pereira, J. S. and Chaves, M. M. (1993) Plant water deficits in Mediterranean ecosystems. Water Deficits plant responses from cell to community. (ed. by J. A. . Smith, H. Griffiths). pp. 237-251. BIOS Sci. Ltd. Oxford.

[5] Pereira, J. S. and Chaves, M. M. (1995) Plant responses to drought under climate change in mediterranean-type ecosystems. Global change and Mediterranean-type ecosytems. Ecological studies, Vol. 117. (ed. by Jose M. Moreno, Walter C. Oechel), pp. 140-160. Springer-Verlag, New York.

[6] Bottner, P., Couteaux, M. M. and Vallejo, V. R. (1995) Soil organic matter in mediterranean-type ecosystems and global climatic changes: A case study-the soils of the mediterranean basin. Global change and Mediterranean-type ecosytems. Ecological studies, Vol. 117. (ed. by Jose M. Moreno, Walter C. Oechel), pp. 306-325. Springer-Verlag, New York.

[7] Austin, R. B. (1989) Prospect for improving crop production in stressful environments. Plants under stress. Biochemistry, physilogy and ecology and their application to plant improvement. (ed. by Hamyln G. Jones, T.J. Flowers, M.B. Jones). pp. 235-248. Cambridge University Press, Cambridge

[8] FAO (Food and Agriculture Organization, United Nations) (2003) Unlocking the water potential of agriculture. www.fao.org. Accessed: 8 August 2011.

[9] Kramer, P. J. and Boyer, J. S. (1995) Water relations of plants and soils. Academic Press. San Diego.

[10] Sharp, R. E. and Davies, W. J. (1989) Regulation of growth and development of plants growing with a restricted supply of water. Plants under stress. Biochemistry, physiology and ecology and their application to plant improvement. (ed. by Hamyln, G. Jones, T. J. Flowers, M. B. Jones). pp. 71-93. Cambridge University Press, Cambridge.

[11] Hurd, E. A. (1976) Plant breeding for drought resistance. Water deficits and plant growth. (ed. by T.T. Kozlowski). Vol IV. pp. 317-353. Academic Press, U.S.A

[12] LawBE, Williams, M, Anthoni P.M., Baldochi, D. D. and Unsworth, M. H. (2000) Measuring and modelling seasonal variation of carbon dioxide and water vapour exchange of a *Pinus ponderosa* forest subject to soil water deficit. Global Change Biology, 6: 613-630.

[13] Wilson, K. B., Baldocchi, D. D. and Hanson, P. J. (2001) Leaf age affects the seasonal pattern of photosynthetic capacity and net ecosystem exchange of carbon in a deciduous forest. Plant Cell and Environment, 24: 571-583.

[14] Grace, J. (1999) Environmental controls of gas exchange in tropical rain forests. In: Press M. C, Scholes J. D., Barker, M. G. eds. Physiological plant ecology. London, UK: British Ecological Society.

[15] Nagarajan, S. and Nagarajan, S. (2010) Abiotic stress adaptation in plants. Physiological, molecular and genomic foundation (Eds. Pareek, A., Sopory, S. K., Bohnert, H. I, Govindjee). pp. 1-11. Springer, The Netherlands.

[16] Rambal, S. and Debussche, G. (1995) Water balance of Mediterranean ecosystems under a changing climate. Global change and Mediterranean-type ecosytems. Ecological studies, Vol. 117. (ed. by Jose M. Moreno, Walter C. Oechel), pp. 386-407. Springer-Verlag, New York.

[17] Kozlowski, T. T. (1968) Water deficits and plant growth. Vol. I (ed. by T.T. Kozlowski). pp. 1-21. Academic press. New York.

[18] Passioura, J. B., Condon, A. G. and Richards, R. A. (1993) Water deficits, the development of leaf area and crop productivity. Water Deficits Plant responses from cell to community.(ed. by J.A.C. Smith, H. Griffiths). pp. 253-264. BIOS Sci. Ltd. Oxford.

[19] Begg, J. E. (1980) Morphological adaptations of leaves to water stress. Adaptations of plants to water and high temperature stress. (ed. by Neil C. Turner, Paul J. Kramer) pp. 33-42. John-Wiley & Sons, New York.

[20] Sharp, R. E. and Davies, W. J. (1979) Solute regulation and growth by roots and shoots of water-stressed maize plants. Planta, 147: 43-49.

[21] Malik, R. S., Dhankar, J. S. and Turner, N. C. (1979) Influence of soil water deficits on root growth of cotton seedlings. Plant and Soil 53: 109-115.

[22] Bradford, K. J. and Hsiao, T. C. (1982) Physiological responses to moderate water stress. In: Encyclopedia of plant physiology, new series, vol. 12B, Physiological plant ecology II, Water relations and carbon assimilation (ed. by O.L. Lange, P.S. Nobel, C.B. Osmond and H. Ziegler), pp. 263-324. Springer-Verlag, Berlin.

[23] Setter, T. L. (1990) Transport/harvest index: Photosynthate partitioning in stressed plants. Plant Biology. Vol. 12. Stress responses in plants: Adaptation and acclimation mechanisms. (ed. by. Ruth G. Alscher, Jonathan R. Cumming). pp. 17-36. Wiley-Liss, U.S.A.

[24] Crawford, R. M. M. (1989) Studies in plant survival. Ecological case histories of plant adaptation to adversity. Studies in Ecology, Vol. 11. pp.177-202. Blackwell Scientific publications, Oxford.
[25] Timpa, J. D., Burke, J. B., Quisenberry, J. E. and Wendt, C. W. (1986) Effects of water stress on the organic acid and carbohydrate compositions of cotton plants. Plant Physiol., 82: 724-728.
[26] Akıncı, S. and Lösel, D. M. (2009) The soluble sugars determination in *Cucurbitaceae* species under water stress and recovery periods. Adv. Environ. Biol., 3(2): 175-183.
[27] Akıncı, S. and Lösel, D. M. (2010) The effects of water stress and recovery periods on soluble sugars and starch content in cucumber cultivars. Fresen. Environ. Bull., 19(2): 164-171.
[28] Schulze, E. D. (1991) Water and nutrient interactions with plant water stress: In: H.A. Mooney, W.E. Winner and E. J. Pell. Eds. Response of plants to multiple stresses. San Diego: Academic Press.
[29] Kozlowski, T. T., Kramer, P. J. and Pallardy, S. G. (1991) The physiological ecology of woody plants. San Diego: Academic Press.
[30] Pugnaire, F. I., Serrano, L. and Pardos, J. (1999) Constraints by water stress on plant growth. In Handbook of Plant and Crop Stress (M. Pessarakli, ed.), 2nd Edition, Marcel Dekker, Inc., New York. pp. 271-283.
[31] Jones, H. G. (1993) Drought tolerance and water-use efficiency. Water deficits plant responses from cell to community. (ed. by J.A.C. Smith, H. Griffiths). pp. 193-203. BIOS Sci. Ltd. Oxford.
[32] Blum, A. (1989) Breeding methods for drought resistance. Plants under stress. Biochemistry, physiology and ecology and their application to plant improvement. (ed. by Hamyln G. Jones, T.J. Flowers, M.B. Jones). pp. 197-215. Cambridge University Press, Cambridge.
[33] Ritchie, J. T. (1980) Plant stress research and crop production: The challenge ahead. Adaptations of plants to water and high temperature stress. (ed. by Neil C. Turner, Paul J. Kramer) pp. 21-29. John-Wiley & Sons, New York.
[34] Borlaug, N. E. and Dowswell, C. R. (2005) Feeding a world of ten billion people: a 21st century challenge. In R Tuberosa, RL Phillips, M Gale, eds, Proceedings of the international congress in the wake of the double helix: From the green revolution to the gene revolution, 27–31 May 2003, Bologna, Italy. pp 3-23.
[35] Levitt, J. (1972) Responses of plants to environmental stresses. Academic Press, New York.
[36] Parker, J. (1968) Drought-resistance mechanisms. Water deficits and plant growth. Vol. I (ed by.T.T. Kozlowski). pp. 195-234. Academic press, New York.
[37] Fresnillo Fedorenko, D. E., Fernandez, O. A. and Busso, C. A. (1995) The effect of water stress on top and root growth in *Medicago minima.* Journal of Arid Environments, 29: 47-54.
[38] Schulze, E. D. (1986) Whole-plant responses to drought. Aust. J. Plant Physiol., 13: 127-141.
[39] Kummerow, J. (1980) Adaptation of roots in water-stressed native vegetation. Adaptations of plants to water and high temperature stress. (ed. by Neil C. Turner, Paul J. Kramer) pp. 57-73. John-Wiley & Sons, New York.

[40] Morgan, J. M. (1980) Differences in adaptation to water stress within crop species. Adaptation of plants to water and high temperature stress. (ed. by Neil C. Turner and Paul J. Kramer). pp. 369-382. John Wiley & Sons, New York.
[41] Fitter, A. H. and Hay, R. K. M. (1987) Environmental Physiology of Plants. Academic Press, London.
[42] Quarrie, S. A and Jones, H. G. (1977) Effects of abscisic acid and water stress on development and morphology of wheat. J. Exp. Bot., 28 (102): 192-203.
[43] Davies, W. J., Metcalfe, J., Lodge, T.A. and da Costa Alexandra, R. (1986) Plant growth substances and the regulation of growth under drought. Aust. J. Plant Physiol., 13: 105-125.
[44] Hall, H. K. and McWha, J. A. (1981) Effects of abscisic acid on growth of wheat (*Triticum aestivum* L.). Ann. Bot., 47: 427-433.
[45] Van Volkenburgh, E. and Davies, W. J. (1983) Inhibition of light-stimulated leaf expansion by abscisic acid. J. Exp. Bot., 34 (144): 835-845.
[46] Larcher, W. (1995) Physiological plant ecology. Ecophysiology and stress physiology of functional groups. Springer, Berlin.
[47] Taiz, L. and Zeiger, E. (1991) Plant Physiology. pp. 265-291. The Benjamin/Cummings Publishing Company, California
[48] McCree, K. J. (1986) Whole-plant carbon balance during osmotic adjustment to drought and salinity stress. Aust. J. Plant Physiol., 13: 33-43.
[49] Boyer, J. S. (1976) Water deficits and photosynthesis. Water deficits and plant growth. (ed. by T.T. Kozlowski). Vol. IV. pp. 153-190. Academic Press, New York.
[50] Ehleringer, J. (1980) Leaf morphology and reflectance in relation to water and temperature stress. Adaptation of plants to water and high temperature stress. (ed. by Neil C. Turner and Paul J. Kramer). pp. 295-308. John Wiley & Sons, New York.
[51] Farquhar, G. D., Wong, S. C., Evans, J. R. and Hubick, K. T. (1989) Photosynthesis and gas exchange. Plants under stress. Biochemistry, physiology and ecology and their application to plant improvement. (ed. by Hamyln G. Jones, T.J. Flowers, M.B. Jones). pp. 47-69. Cambridge University Press, Cambridge.
[52] Jordan, W. R. and Ritchie, J. T. (1971) Influence of soil water stress on evaporation, root absorption, and internal water status of cotton. Plant Physiol., 48: 783-788.
[53] Russel, R. C. (1977) Plant root systems: Their function and interaction with the soil. MacGraw-Hill, London.
[54] Farquhar, G. D. and Sharkey, T. D. (1982) Stomatal conductance and photosynthesis. Annu. Rev. Plant Physiol., 33: 317-345.
[55] Schulze, E. D. (1986) Carbon dioxide and water vapor exchange in response to drought in the atmosphere and in the soil. Ann. Rev. Plant Physiol., 37: 247-274.
[56] Wong, S. C., Cowan, I. R. and Farquhar, G. D. (1985) Leaf conductance in relation to rate of CO_2 Assimilation. II. Effects of short-term exposures to different photon flux densities. Plant Physiol., 78: 826-829.
[57] Huber, S. C., Rogers, H. H. and Mowry, F. L. (1984) Effects of water stress on photosynthesis and carbon partitioning in Soybean (*Glycine max* [L.] Merr.) plants grown in the field at different CO_2 levels. Plant Physiol., 76: 244-249.
[58] Raschke, K. and Resemann, A. (1986) The midday depression of CO_2 assimilation in leaves of *Arbutus unedo* L.: diurnal changes in photosynthetic capacity related to changes in temperature and humidity. Planta, 168: 546-558.

[59] Cornic, G. Le Gouallec, J.-L., Briantais, J. M. and Hodges, M. (1989) Effect of dehydration and high light on photosynthesis of two C_3 plants (*Phaseolus vulgaris* L. and *Elatostema repens* (Lour.) Hall f.). Planta, 177: 84-90.
[60] Quick, W. P., Chaves, M. M., Wendler, R., David, M., Rodrigues, M. L., Passaharinho, J. A., Pereira, J. S., Adcock, M. D., Leegood, R. C. and Stitt, M. (1992) The effect of water stress on photosynthetic carbon metabolism in four species grown under field conditions. Plant, Cell and Environment, 15: 25-35.
[61] Meyer, R. F. and Boyer, J. S. (1972) Sensitivity of cell division and cell elongation to low water potentials in soybean hypocotyls. Planta, 108: 77-87.
[62] Michelena, V. A. and Boyer, J. S. (1982) Complete turgor maintenance at low water potentials in the elongating region of maize leaves. Plant Physiol., 69: 1145-1149.
[63] Westgate, M. E. and Boyer, J. S. (1985) Osmotic adjustment and the inhibition of leaf, root, stem and silk growth at low water potentials in maize. Planta, 164: 540-549.
[64] Turner, N. C. and Jones, M. M. (1980) Turgor maintenance by osmotic adjustment: A review and evaluation. Adaptations of plants to water and high temperature stress. (ed. by Neil C. Turner, Paul J. Kramer) pp. 87-103. John-Wiley & Sons, New York.
[65] Morgan, J. M. (1984) Osmoregulation and water stress in higher plants. Annu. Rev. Plant Physiol., 35: 299-319.
[66] Oosterhuis, D. M. and Wullschleger, S. D. (1987) Osmotic adjustment in cotton (*Gossypium hirsutum* L.) leaves and roots in response to water stress. Plant Physiol., 84: 1154-1157.
[67] Turner, N. C. and Jones, M. M. (1980) Turgor maintenance by osmotic adjustment: A review and evaluation. Adaptations of plants to water and high temperature stress. (ed. by Neil C. Turner, Paul J. Kramer) pp. 87-103. John-Wiley & Sons, New York.
[68] Richter, H. and Wagner, S. B. (1982) Water stress resistance of photosynthesis: Some aspects of osmotic relations. Effects of stress on photosynthesis. Proceedings of a conference held at the "Limburgs Universtair Centrum" Diepenbeek, Belgium, 22 - 27 August 1982. (ed. by R. Marcelle, H. Clijsters, and M Van Poucke), pp. 45-53. Martinus Nijhoff / Dr W. Junk publishers, The Hague.
[69] Hartung, W., Zhang, J. and Davies, W. J. (1994) Does abscisic acid play a stress physiological role in maize plants growing in heavily compacted soil? J. Exp. Bot., 45: 221-226.
[70] Russel, E. W. (1976) Water and crop growth. (Soil conditions and plant growth). pp. 448-478. Longman, London.
[71] Jones, M. M., Osmond, C. B. and Turner, N. C. (1980) Accumulation of solutes in leaves of sorghum and sunflower in response to water deficits. Aust. J. Plant Physiol., 7: 193-205.
[72] Ford, C. W. and Wilson, J. R. (1981) Changes in levels of solutes during osmotic adjustment to water stress in leaves of four tropical pasture species. Aust. J. Plant Physiol., 8: 77-91.
[73] Kameli, A. and Losel, D. M. (1995) Contribution of carbohydrates and other solutes to osmotic adjustment in wheat leaves under water stress. Plant Physiol., 145: 363-366.
[74] Evans, R. D., Black, R. A., Loescher, W. H., and Fellows, R. J. (1992) Osmotic relations of the drought-tolerant shrub *Artemisia tridentata* in response to water stress. Plant, Cell and Environment, 15: 49-59.

[75] Kameli, A. (1990) Metabolic responses of durum wheat to water stress and their role in drought resistance. Ph.D thesis, Animal and Plant Sci. Dept., University of Sheffield, U.K.
[76] Munns, R., Brady, C. J. and Barlow, E. W. R. (1979) Solute accumulation in the apex and leaves of wheat during water stress. Aust. J. Plant Physiol. 6: 379-389.
[77] Morgan, J. M. and Condon, A. G. (1986) Water-use, grain yield and osmoregulation in wheat. Aust. J. Plant Physiol., 13: 523-532.
[78] Westgate, M. E. and Boyer, J. S. (1985) Osmotic adjustment and the inhibition of leaf, root, stem and silk growth at low water potentials in maize. Planta, 164: 540-549.
[79] Munns, R. and Weir, R. (1981) Contribution of sugars to osmotic adjustment in elongating and expanded zones of wheat leaves during moderate water deficits at two light levels. Aust. J. Plant Physiol., 8: 93-105.
[80] Hanson, A. D. and Hitz, W. D. (1982) Metabolic responses of mesophytes to plant water deficits. Ann. Rev. Plant Physiol., 33: 163-203.
[81] Chaves, M. M., Maroco J. P. and Pereira J. S. (2003) Understanding plant response to drought: from genes to the whole plant. Functional Plant Biology 30: 239-264.
[82] Koch, K. E. (1996) Carbohydrate-modulated gene expression in plants. Annual Reviews of Plant Physiology and Plant Molecular Biology 47: 509-540.
[83] Jang, J.-C. and Sheen, J. (1997) Sugar sensing in plants. Trends Plant Sci., 2: 208-214.
[84] Chaves, M. M. and Oliveira, M. M. (2004) Mechanisms underlying plant resilience to water deficits: prospects for water-saving agriculture. Journal of Experimental Botany, 55: 2365-2384.
[85] Stewart, C. R. (1971) Effect of wilting on carbohydrates during incubation of excised bean leaves in the dark. Plant Physiol., 48: 792-794.
[86] Fox, T. C. and Geiger, D. R. (1986) Osmotic response of sugar beet source leaves at CO_2 compensation point. Plant Physiol., 80: 239-241.
[87] Ackerson, R. C. (1981) Osmoregulation in cotton in response to water stress. II. Leaf carbohydrate status in relation to osmotic adjustment. Plant Physiol., 67: 489-493.
[88] Kameli, A. and Losel, D. M. (1993) Carbohydrates and water status in wheat plants under water stress. New Phytol., 125: 609-614.
[89] Kameli, A. and Losel, D. M. (1996) Growth and sugar accumulation in durum wheat plants under water stress. New Phytol., 132: 57-62.
[90] Al-Suhaibani, N. A-R. (1996) Physiological studies on the growth and survival of *Medicago sativa* L. (alfalfa) seedlings under low temperature. Ph.D thesis, Animal and Plant Sci. Dept., University of Sheffield, U.K.
[91] Drossopoulos, J. B., Karamanos, A. J. and Niavis, C. A. (1987) Changes in ethanol soluble carbohydrates during the development of two wheat cultivars subjected to different degrees of water stress. Annals of Botany. 59: 173-180.
[92] Barlow, E. W. R. (1986) Water relations of expanding leaves. Aust. J. Plant Physiol., 13: 45-58.
[93] Botha, F. C. and Small, J. G. C. (1985) Effect of water stress on the carbohydrate metabolism of *Citrullus lanatus* seeds during germination. Plant Physiol., 77: 79-82.
[94] Pattanagul, W. and Madore, M. A. (1999) Water deficit effects on raffinose family oligosachharide metabolism in *Coleus*. Plant Physiol., 121: 987-993.

[95] Singh, N. K., Bracker, C. A., Hasegawa, P. M., Handa, A. K., Buckel, S., Hermodson, M. A., Pfankoch, E., Regnier, F. E. and Bressan, R. A. (1987) Characterization of osmotin. Plant Physiol., 85: 529-536.

[96] Close, T. J. (1997) Dehydrins: A commonalty in the response of plants to dehydration and low temperature. Physiol. Plant., 100: 291-296.

[97] Pelah, D., Wang, W., Altman, A., Shoseyov, O. and Bartels, D. (1997). Differential accumulation of water stress-related proteins, sucrose synthase and soluble sugars in *Populus* species that differ in their water stress response. Physiol. Plant., 99: 153-159.

[98] Claes, B., Dekeyser, R., Villarroel, R., Van den Bulcke, M., Bauw, G., Van Montagu, M. and Caplan, A. (1990) A. Characterization of a rice gene showing organ-specific expression in response to salt stress and drought. Plant Cell, 2: 19-27.

[99] Artlip, T. S. and Funkhouser, E. A. (1995) Protein synthetic responses to environmental stresses. In: M. Pessarakli, ed. Handbook of plant and crop physiology. New York: Marcel Dekker. pp. 627-644.

[100] Dubey, R. S. (1999) Protein synthesis by plants under stressful conditions. In Handbook of Plant and Crop Stress (M. Pessarakli, ed.), 2nd Edition, Marcel Dekker, Inc., New York, pp. 365-397.

[101] Chopra, R.K., Sinha, S. K. (1998) Prospects of success of biotechnological approaches for improving tolerance to drought stress in crop plants. Curr. Sci., 74 (1): 25-34.

[102] Galston, A. W. and Sawhney, R. K. (1990) Polyamines in plant physiology. Plant Physiol., 94: 406-410.

[103] Ramagopal, S. (1993) Advances in understanding the molecular biology of drought and salinity tolerance in plants- the first decade. Adv. Pl. Biotech. Biochem., (Eds.) M.L. Lodha, S.L. Mehta, S. Ramagopal and G.P. Srivastava, Indian Society of Agriculture Biochemists, Kanpur, India, pp. 39-48.

[104] Dhindsa, R. S. and Cleland, R. E. (1975) Water stress and protein synthesis. II. Interaction between water stress, hydrostatic pressure, and abscisic acid on the pattern of protein synthesis in *Avena* coleoptiles. Plant Physiol., 55: 782-785.

[105] Bewley, J. D. and Larsen, K. M. (1982) Differences in the responses to water stress of growing and non-growing regions of maize mesocotyls: Protein synthesis on total, free and membrane-bound polyribosome fractions. J. Exp. Bot., 33(134): 406-415.

[106] Bewley, J. D., Larsen, K. M. and Papp, J. E. E. (1983) Water-stress-induced changes in the pattern of protein synthesis in maize seedling mesocotyls: A comparison with the effects of heat shock. J. Exp. Bot., 34(146): 1126-1133.

[107] Dasgupta, J. and Bewley, D. (1984) Variations in protein synthesis in different regions of greening leaves of barley seedlings and effects of imposed water stress. J. Exp. Bot., 35(159): 1450-1459.

[108] Vartanian, N., Damerval, C. and De Vienne, D. (1987) Drought-induced changes in protein patterns of *Brassica napus* var. *oleifera* roots. Plant Physiol. 84: 989-992.

[109] Hsiao, T.C. (1970) Rapid changes in levels of polyribosomes in *Zea mays* in response to water stress. Plant Physiol., 46: 281-285.

[110] Ho, T.-H. D. and Sachs, M. M. (1989) Environmental control of gene expression and stress proteins in plants. Plants under stress. Biochemistry, physiology and ecology and their application to plant improvement. (ed. by Hamyln G. Jones, T.J. Flowers, M.B. Jones). pp. 157-180. Cambridge University Press, Cambridge.

[111] Andrade, J. L, Larque-Saavedra, A. and Trejo, C. L. (1995) Proline accumulation in leaves of four cultivars of *Phaseolus vulgaris* L. with different drought resistance. Phyton, 57: 149-157.
[112] Aspinall, D. and Paleg, L. G. (1981) Proline accumulation: physiological effects. In: The Physiology and biochemistry of drought resistance in plants (Paleg, L. G. and Aspinall, D. ed), Acad. Press, New York. pp. 206-225.
[113] Chandrasekhar, V., Sairam, R.K. and Srivastava, G.C. (2000) Physiological and biochemical responses of hexaploid and tetraploid wheat to drought stress. Agron. Crop. Sci., 185(4): 219-227.
[114] Tholkappian, P., Prakash, M., Sundaram, M. D. (2001) Effect of AM-fungi on proline, nitrogen and pod number of soybean under moisture stress: Indian J. Plant Physiol, 6(1): 98-99.
[115] Nair, A. S., Abraham, T. K. and Jaya, D. S (2006) Morphological and physiological changes in cowpea (*Vigna unguiculata* L.) subjected to water deficit Indian J. Plant Physiol., 11(3): 325-328.
[116] Hughes, S. G., Bryant, J. A. and Smirnoff, N. (1989) Molecular biology: application to studies of stress tolerance. Plants under stress. Biochemistry, physiology and ecology and their application to plant improvement. (ed. by Hamyln G. Jones, T.J. Flowers, M.B. Jones). pp. 131-155. Cambridge University Press, Cambridge.
[117] Treshow, M. (1970) Disorders associated with adverse water relations. Environment & Plant Response. pp. 153-174. McGraw-Hill Book Company, U.S.A.
[118] Barnett, N.M. and Naylor, A.W. (1966). Amino acid protein metabolism in Bermuda grass during water stress. Plant Physiol., 41: 1222-1230.
[119] Tully, R. E. and Hanson, A. D. (1979) Amino acid translocated from turgid and water-stressed barley leaves. Plant Physiol. 64: 460-466.
[120] Kuiper, P. J. C. (1985) Environmental changes and lipid metabolism of higher plants. Physiol. Plant., 64: 118-122.
[121] Harwood, J. L. and Russell, N. J. (1984) Lipids in plants and microbes. George Allen & Unwin, London.
[122] Harwood, J. L. (1979) The synthesis of acyl lipids in plant tissues. Prog Lipid Res., 18: 55-86.
[123] Bishop, D. G. (1983) Functional role of plant membrane lipids. In: Proceeding of the 6 th Annual Symposium in Botany (January 13–15). Riverside, CA: University of California,:81-103.
[124] Harwood, J. L. (1980) Plant acyl lipids. In: P.K. Stumpf, ed. The Biochemistry of Plants. Vol 4. New York: Academic Press, pp.1-55.
[125] Pham Thi, A. T., Flood, C. and Vieira da Silva, J. (1982) Effects of water stress on lipid and fatty acid composition of cotton leaves. Biochemistry and metabolism of plant lipids. (ed. by J.F.G.M. Wintermans, and P.J.C. Kuiper). pp. 451-454. Elsevier Biomedical press, Amsterdam.
[126] Pham Thi, A. T., Borrel-Flood, C., Vieira da Silva, J., Justin, A.M. and Mazliak, P. (1985) Effects of water stress on lipid metabolism in cotton leaves. Phytochemistry, 24 (4):723-727.
[127] Navari-Izzo, F., Quartacci, M. F. and Izzo, R. (1989) Lipid changes in maize seedlings in response to field water deficits. J. Exp. Bot., 40 (215): 675-680.

[128] Douglas, T. J. and Paleg, L. G. (1981) Lipid composition of *Zea mays* seedlings and water stress-induced changes. J. Exp. Bot., 32 (128): 499-508.
[129] Navari-Izzo, F., Vangioni, N. and Quartacci, M. F. (1990) Lipids of soybean and sunflower seedlings grown under drought conditions. Phytochemistry, 29(7): 2119-2123.
[130] Pham Thi, A. T., Borrel-Flood, C., Vieira da Silva, J., Justin, A. M. and Mazliak, P. (1987) Effects of drought on [1–^{14}C]- oleic acid and [1–^{14}C]- linoleic acid desaturation in cotton leaves. Physiol. Plant., 69: 147-150.
[131] Navari-Izzo, F., Quartacci, M.F., Melfi, D. and Izzo, R. (1993). Lipid composition of plasma membranes isolated from sunflower seedlings grown under water-stress. Physiol Plant., 87: 508-514.
[132] Liljenberg C. and Kates, M. (1982). Effect of water stress on lipid composition of oat sedling root cell membranes. Biochemistry and metabolism of plant lipids. (ed. by J.F.G.M. Wintermans, and P.J.C. Kuiper). pp. 441-444. Elsevier Biomedical press, Amsterdam.
[133] Navari-Izzo, F. and Rascio, N. (1999) Plant response to water-deficit conditions. In Handbook of Plant and Crop Stress (M. Pessarakli, ed.), 2nd Edition, Marcel Dekker, Inc., New York. pp. 231-270.
[134] Martin, B. A., Schoper, J. B. and Rinne, R. W. (1986) Changes is soybean (*Glycine max* [L.] Merr.) glycerolipids in respose to water stress. Plant Physiol., 81: 798-801.
[135] Gronewald, J. W., Abou-Khalil, W., Weber, E. J. and Hanson, J. B. (1982) Lipid composition of a plasma membrane enriched fraction of maize roots. Phytochemistry, Vol. 21(4): 859-862.
[136] Whitman, C. E. and Travis, R. L. (1985) Phospholipid composition of a plasma membrane-enriched fraction from developing soybean roots. Plant Physiol., 79: 494-498.
[137] Wilson, R. F., Burke, J. J. and Quisenberry, J. E. (1987) Plant morphological and biochemical responses to field water deficits. II. Responses of leaf glycerolipid composition in cotton. Plant Physiol., 84: 251-254.
[138] Quartacci, M. F. and Navari-Izzo, F. (1992) Water stress and free radical mediated changes in sunflower seedlings. J. Plant Physiol., 139: 621-625.
[139] Barber, S. A. (1995) Soil nutrient bioavailability: A Mechanistic Approach. 2 nd Ed. New York: J. Wiley.
[140] Patel, S. K., Rhoads, F. M., Hanlon, E. A. and Barnett, R. D. (1993) Potassium and magnesium uptake by wheat and soybean roots as influenced by fertilizer rate. Commun. Soil Sci. Plant Anal., 24 (13-14): 1543-1556.
[141] Alam, S. M. (1999). Nutrient uptake by plants under stress conditions.. In M. Pessarakli (ed.) Handbook of plant and crop stress. Second ed. rev. and exp. Marcel Dekker, New York. pp. 285-313.
[142] Alam, S. M. (1996) Allelopathic effects of weeds on the growth and development of wheat and rice under saline conditions. Ph. D dissertation. University of Sindh, Jamshoro, Pakistan.
[143] Sadiq, M. S., Arain, C. R. and Azmi, A. R. (1997) Wheat breeding in a water stressed environment. V. Carbon isotope discrimination as a selection criterion. Cer. Res. Comm., 25 (1): 43-49.

[144] Nicholas, D. J. D. (1975) The functions of trace elements in plants. In: D. J. D. Nicholas, ed. Trace Elements in Soil-Plant-Animal Systems. New York: Academic Pres.

[145] Tanguilig, V. C., Yambao, E. B., O' Toole, J. C. and De Datta, S. K. (1987) Water stress effects on leaf elongation, leaf water potential transpiration and nutrient uptake of rice, maize and soybean. Plant Soil, 103: 155-168.

[146] Viets, Jr. F. G. (1972) Water deficits and nutrient availability. In: T.T. Kozlowsky, ed. Water deficits and plant growth. Vol 13. New York: Academic Press.

[147] Abdel Rahman, A. A., Shalaby, A. F. and El Monayeri, M. O. (1971) Effect of moisture stress on metabolic products and ions accumulation. Plant and Soil, 34: 65-90.

[148] Kirnak, H., Kaya, C., Higgs, D. and Tas, I. (2003) Responses of drip irrigated bell pepper to water stress and different nitrogen levels with or without mulch cover. J. Plant Nutr., 26: 263-277.

[149] Singh, B. and Singh, G. (2004) Influence of soil water regime on nutrient mobility and uptake by *Dalbergia sissoo* seedlings. Tropical Ecology 45(2): 337-340.

[150] Nambiar, E. K. S. (1977) The effects of drying of the topsoil and of micronutrients in the subsoil on micronutrient uptake by an intermittently defoliated ryegrass. Plant Soil, 46(1): 185-193.

[151] Singh, A. K. and Gupta, B. N. (1993) Biomass production and nutrient distribution in some important tree species on Bhatta soil of Raipur (Madhya Pradesh) India. Ann. For., 1(1): 47-53.

[152] Levitt, J. (1980) Responses of plants to environmental stresses. 2nd ed. New York: Academic Press.

[153] Foy, C. D. (1983) Plant adaptation to mineral stress in problem soils. Iowa J. Res., 57: 339-354.

[154] Abdalla, M. M. and El-Khoshiban, N. H. (2007) The influence of water stress on growth, relative water content, photosynthetic pigments, some metabolic and hormonal contents of two *Triticum aestivum* cultivars. J. Appl. Sci. Res., 3(12): 2062-2074.

[155] Bie, Z., Ito, T. and Shinohara, Y. (2004) Effects of sodium sulphate and sodium bicarbonate on the growth, gas exchange and mineral composition of lettuce. Sci. Hortic., 99: 215-224.

[156] Brown, C. E., Pezeshki, S. R. and DeLaune, R. D. (2006) The effects of salinity and soil drying on nutrient uptake and growth of *Spartina alterniflora* in a simulated tidal system. Environ. Exp. Bot., 58(1-3): 140-148.

[157] Oktem, A. (2008) Effect of water shortage on yield, and protein and mineral compositions of drip-irrigated sweet corn in sustainable agricultural systems. Agr. Water Manage., 95(9): 1003-1010.

[158] Gerakis, P. A., Guerrero, F. P. and Williams, W. A. (1975) Growth, water relations and nutrition of three grassland annuals as affected by drought. J. Appl. Ecol., 12, 125-135.

[159] Kidambi, S. P., Matches, A. G. and Bolger, T. P. (1990) Mineral concentrations in alfalfa and sainfoin as influenced by soil moisture level. Agron. J., 82(2): 229-236.

[160] Sinha, S. K. (1978) Influence of potassium on tolerance to stress. In: G. S. Sekhon, ed. Potassium in soils and Crops. New Delhi: Potash research Institute.

[161] Ashraf, M., Ahmad, A. and McNeilly, T. (2001) Growth and photosynthetic characteristics in pearl millet under water stress and different potassium supply. Photosynthetica, 39: 389-394.

[162] Premachandra, G. S., Saneoka, H. and Ogata, S. (1991) Cell membrane stability and leaf water relations as affected by potassium nutrition of water-stressed maize. J. Exp. Bot., 42: 739-745.

[163] Sangakkara, U. R., Frehner, M. and Nosberger, J. (2001) Influence of soil moisture and fertilizer potassium on the vegetative growth of mungbean (*Vigna radiata* L. Wilczek) and cowpea (*Vigna unguiculata* L. Walp). J. Agron. Crop Sci., 186: 73-81.

[164] Umar, S. M. (2002) Genotypic differences in yield and quality of groundnut as affected by potassium nutrition under erratic rainfall conditions. J. Plant Nutr., 25: 1549-1562.

[165] Saneoka, H., Moghaieb, R. E. A., Premachandra, G. S. and Fujita, K. (2004) Nitrogen nutrition and water stress effects on cell membrane stability and leaf water relation in *Agrostis palustris* Huds. Environ. Exp. Bot., 52: 131-138.

[166] Xu, S., An, L., Feng, H., Wang, X. and Li, X. (2002) The seasonal effects of water stress on *Ammopiptanthus mongolicus* in a desert environment. J. Arid Environ., 51: 437-447.

[167] Pessarakli, M. (1999) Response of green beans (*Phaseolus vulgaris* L.) to salt stress. Handbook of plant and crop stress.. In M. Pessarakli (ed.) Handbook of plant and crop stress. Second ed. rev. and exp. Marcel Dekker, New York. pp. 827-842.

[168] Dogan, N. and Akıncı, S. (2011) Effects of water stress on the uptake of nutrients by bean seedlings (*Phaseolus vulgaris* L.) Fresen. Environ. Bull., 20 (8a): 2163-2173.

[169] Mathers, H. (2002) Fertilizer practices: What's most important (Amounts, relative proportions and timing), Submitted to: NMPRO, pp 1-9.

[170] Kuchenbuch, R. Claassen, N. and Jungk, A. (1986) Potassium availability in relation to soil moisture: I effect of soil moisture on potassium diffusion, root growth and potassium uptake of onion plants. Plant and Soil, 95: 221-231.

[171] Pinkerton, A. and Simpson, J. R. (1986) Interactions of surface drying subsurface nutrients affecting plant growth on acidic soil profiles from an old pasture. Aust. J. Exp. Agric., 26(6): 681-689.

[172] Simpson, J. R. and Lipsett, J. (1973) Effects of surface moisture supply on the subsoil nutritional requirements of lucerne (*Medicago sativa* L.). Aust. J. Agric. Res., 24(2): 199-209.

[173] Fawcett, R. G. and Quirk, J. P. (1962) The effects of soil-water stress on absorption of soil phosphorus by wheat plants. Aust J. Agric. Res., 13(2): 193-205.

[174] Nuttall, W. F. (1976) Effect of soil moisture tension and amendments on yields and on herbage N, P, and S concentrations of alfalfa. Agron. J., 68: 741-744.

[175] Olsen, S. R., Watanabe, F. S. and Danielson, R. E.(1961) Phosphorus absorption by corn roots as affected by moisture and phosphorus concentration. Soil Sci. Soc. Amer. Proc. 25: 289-294.

[176] Gomez-Beltranno, J. F. (1982) Effects of moisture status on alfalfa growth, quality and gas exchange. Ph.D dissertation. New Mexico State University, Las Cruces.

[177] Kabata-Pendias, A. and Pendias, H. (2001) Trace elements in soils and plants (3rd ed.). Boca Raton, FL: CRC Press.

[178] Chevone, B. I., Seiler, J. R., Melkonian, J. and Amundson, R. G. (1990) Ozone-water stress interactions. Plant Biology. Vol. 12. Stress responses in plants: Adaptation and acclimation mechanisms. (ed. by Ruth G. Alscher, Jonathan R. Cumming). pp. 311-328. Wiley-Liss, U.S.A.

[179] Irigoyen, J. J., Emerich, D. W. and Sanchez-Diaz, M. (1992). Water stress induced changes in concentrations of proline and total soluble sugars in nodulated alfalfa (*Medicago sativa*) plants. Physiol. Plant., 84: 55-60.

[180] Westgate, M. E., Schussler, J. R., Reicosky, D. C. and Brenner, M. L. (1989). Effect of water deficits on seed development in soybean. II. conservation of seed growth rate. Plant Physiol., 91: 980-985.

[181] Ackerson, R. C. and Hebert, R. R. (1981). Osmoregulation in cotton in response to water stress. I. alterations in photosynthesis, leaf conductance, translocation, and ultrastructure. Plant Physiol., 67: 484-488.

[182] Pilon-Smits, E. A. H., Ebskamp, M. J. M., Paul, M. J., Jeuken, M. J. W., Weisbeek, P.J. and Smeekens, S. C. M. (1995). Improved performance of transgenic fructan-accumulating tobacco under drought stress. Plant Physiol., 107: 125-130.

[183] Rodrigues, M. L., Chaves, M. M., Wendler, R., David, M. M., Quick, W. P., Leegood, R. C., Stitt, M. and Pereira, J. S. (1993). Osmotic adjustment in water stressed grapevine leaves in relation to carbon assimilation. Aust. J. Plant Physiol., 20: 309-321.

[184] Botha, F. C. and Small, J. G. C. (1985). Effect of water stress on the carbohydrate metabolism of *Citrullus lanatus* seeds during germination. Plant Physiol., 77: 79-82.

[185] Schwab, K. B. and Gaff, D. F. (1986) Sugar and ion content in leaf tissues of several drought tolerant plants under water stress. J. Plant Physiol., 125: 257-265.

[186] Rai, V. K., Singh, G., Thakur, P. S. and Banyal, S. (1983) Protein and amino acid relationship during water stress in relation to drought resistance. Plant Physiol., Biochem. 10: 161-167.

[187] Xu, D. P., Duan, X. L., Wang, B. Y., Hong, B. M., Ho, T. H. D. and Wu, R. (1996) Expression of a late embryogenesis abundant protein gene: HVA1. From barley confers tolerance to water deficit and salt stress in transgenic rice. Plant Physiol., 110: 249-257.

[188] Sinha, K. M., Sachdev, A., Johari, R. P. and Mehta, S. L. (1996) Lathyrus dehydrin-a drought inducible cDNA clone: Isolation and characterization. J. Plant Biochem. Biotechnol. 5: 97-101.

[189] Bray, E. A. (1995) Regulation of gene expression during abiotic stresses and the role of the plant hormone abscisic acid. In:M. Pessarakli, ed. Handbook of plant and crop physiology. New York: Marcel Dekker, pp. 733-752.

[190] Naot, D., Ben-Hayyim, G., Eshdat, Y. and Holland, D. (1995) Drought, heat and salt stress induce the expression of a citrus homologue of an atypical late-embryogenesis *lea5* gene. Plant Mol. Biol., 27: 619-622.

[191] Pareek, A., Singla, S. L., Kush, A. K. and Grover, A. (1997) Distribution patterns of HSP 90 protein in rice. Plant Sci., 125 (10): 221-230.

[192] Mantyla, E., Lang, V. and Palva, E. T. (1995) Role of abscisic acid in drought-induced freezing tolerance, cold acclimation, and accumulation of LTI78 and RAB18 proteins in Arabidopsis thaliana. Plant Physiol., 107: 141-148.

[193] Pruvot, G., Massimino, J., Peltier, G. and Rey, P. (1996) Effects of low temperature, high salinity and exogenous ABA on the synthesis of two chloroplastic drought-induced proteins in *Solanum tuberosum*. Physiol. Plant., 97: 123-131.

[194] Shah, C. B. and Loomis, R. S. (1965) Ribonucleic acid and protein metabolism in sugar beet during drought. Physiol. Plant., 18: 240-254.

[195] Barnett, N. M. and Naylor, A. W. (1966). Amino acid protein metabolism in Bermuda grass during water stress. Plant Physiol., 41: 1222-1230.

[196] Yu, L. X., Chamberland, H., Lafontaine, J. G. and Tabaeizadeh, Z. (1996) Negative regulation of gene expression of a novel proline-, threonine-,and glycine- rich protein by water stress in *Lycopersicon chilense*. Genome, 39: 1185-1193.

[197] Gogorcena, Y., Iturbe-Ormaetxe, I., Escuredo, P. R. and Becana, M. (1995) Antioxidant defences against activated oxygen in pea nodules subjected to water stress. Plant Physiol., 108: 753-759.

[198] Ferrari-Iliou, R., Pham Thi, A. T. and Vieira da Silva, J. (1984) Effect of water stress on the lipid and fatty acid composition of cotton (*Gossypium hirsutum* L.) chloroplasts. Physiol. Plant., 62: 219-224.

[199] Chetal, S., Wagle, D. S. and Nainawatee, H. S. (1981) Glycolipid changes in wheat and barley chloroplast under water stress. Plant Sci. Letter, 20: 225-230.

[200] El-Hafid, L., Pham-Thi, A.T., Zuily-Fodil, Y. and Vieira da Silva, J. (1989) Enzymatic breakdown of polar lipids in cotton leaves under water stress. I. Degradation of monogalactosyl-diacylglycerol. Plant Physiol. Biochem., 27: 495-502.

[201] Quartacci, M. F., Sgherri, C. I. M., Pinzino, C. and Navari-Izzo, F. (1994) Superoxide radical production in wheat plants differently sensitive to drought. Proc. R. Soc. Edinburg. 102B: 287-290.

2

Physiological and Biochemical Responses of Semiarid Plants Subjected to Water Stress

Alexandre Bosco de Oliveira[1],
Nara Lídia Mendes Alencar[2] and Enéas Gomes-Filho[2]
[1]*State University of Piauí,*
[2]*Federal University of Ceará and*
National Institute of Science and Technology Salinity/CNPq
Brazil

1. Introduction

Plants are often subjected to periods of soil and atmospheric water deficits during their life cycle. Moreover, the faster-than-predicted change in global climate (Intergovernmental Panel on Climate Change, 2007) and the different available scenarios for climate change suggest an increase in aridity for the semiarid regions of the globe. Together with overpopulation, this will lead to an overexploitation of water resources for agriculture purposes and increased constraints on plant growth and survival and, therefore, on realizing crop yield potential (Chaves et al., 2003; Passioura, 2007).

Water is one of the fundamental resources for the vital processes of vegetation. Plants need to maintain adequate levels of water in their tissues to assure growth and survival and to perform physiological processes, such as photosynthesis and nutrient uptake (Kramer & Boyer, 1995; Larcher, 1995; Nobel, 1999). In conditions of water deficit, plant cell turgor is reduced, and a series of harmful effects on plant physiology—e.g., reduction of cell growth, cell wall synthesis, protein synthesis, respiration, and sugar accumulation—occur, generating a state of increasing suffering in plants, usually named 'water stress' (Smith & Griffith, 1993; Lauenroth et al., 1994).

Drought is the most important limiting factor for crop production; it is becoming an increasingly severe problem in many regions of the world. In addition to the complexity of drought itself (Passioura, 2007), plant responses to drought are complex, and different mechanisms are adopted by plants when they encounter drought (Jones, 2004). These mechanisms can include: (i) drought escape by rapid development, which allows plants to finish their cycle before severe water stress; (ii) drought avoidance by, for instance, increasing water uptake and reducing transpiration rate by the reduction of stomatal conductance and leaf area; (iii) drought tolerance by maintaining tissue turgor during water stress via osmotic adjustment, which allows plants to maintain growth under water stress; and (iv) resisting severe stress through survival mechanisms (Izanloo et al., 2008). However, this last mechanism is typically not relevant to agriculture (Tardieu, 2005). The maintenance of high plant water status and plant functions at low plant water potential and the recovery of plant function after water stress are the major physiological processes that contribute to the maintenance of high yield under cyclic drought periods (Blum, 1996).

Water stress has been identified as a factor that negatively affects the ratio of reproductive to vegetative growth, seed yield and its components (Iannucci & Martinello, 1998). Many of the changes, which occur in plants during drought stress, represent adaptive responses by which plants cope with water stress. Plants subjected to periods of water stress show an acclimation or hardening and are able to survive subsequent drought periods with less damage compared to plants not previously stressed. The mechanisms developed as survival strategies include tolerance and avoidance of water stress (Zhu et al., 1997).

Generally, stress avoidance involves stomatal closure, hydraulic conductance and root growth patterns. Stress tolerance usually includes osmotic adjustment and changes in tissue elasticity (Jones et al., 1981). Osmotic adjustment (i.e., the lowering of the osmotic potential by net solute accumulation in response to dehydration) aids in the maintenance of turgor at lower water potentials and is considered a beneficial drought tolerance mechanism in both the vegetative and reproductive phases of crop growth (Rascio et al., 1994). The maintenance of turgor above a particular threshold is essential for many physiological processes, such as cell expansion, photosynthesis, gas exchange, enzymatic activities, and continuous growth and maintenance. In addition, the acclimation of plants subjected to drought is also indicated by the accumulation of certain new metabolites associated with structural capabilities to improve plant functions under drought stress (Pinheiro et al., 2001).

Understanding how plants respond to drought can play a major role in stabilizing crop performance under water stress conditions and in the protection of natural vegetation. Adequate management techniques and plant genetic breeding are tools for improving resource use efficiency (including water) by plants (Chaves et al., 2009). Thus, using physiological and molecular genetics tools to enhance our understanding of the physiology and genetic control of these mechanisms will assist breeding programs seeking to improve drought resistance in crop plants. Physiological studies will help to establish the precise screening techniques necessary to identify traits related to plant productivity (Izanloo et al., 2008).

Plant responses to drought stress are complex and involve adaptive changes and/or deleterious effects. The decrease in the water potential results in reduced cell growth, root growth and shoot growth and also causes inhibition of cell expansion and reduction in cell wall synthesis (Chaitanya et al., 2003). A relatively mild water potential around -0.8 MPa reduces the cytokinin content in the leaves of several species (Salisbury & Ross, 1986). Low leaf water potentials inhibit the activities of the enzymes of the pentose phosphate pathway (Hay & Walker, 1989). When water stress is reduced from -1.0 to -2.0 MPa, cells become smaller and leaves develop less, resulting in a reduced area for photosynthesis. At these water potentials, ion transport is slowed and may also lead to a decrease in yield (Medrano et al., 2002).

Cell membranes are the primary targets of many plant stresses (Bajji et al., 2002). Osmotic stress induces rapid changes in cell wall conductivity and plasmalemma (Chazen & Neumann, 1994). The regulation of permeability that occurs during drought stress are accomplished by the opening and closing of water channels formed by membrane polypeptide complexes (Maurel, 1997; Chrispeels et al., 1999) and also by the phase transitions of membrane lipids (Crowe et al., 1992). Water stress affects the regular metabolism of the cell, such as the carbon reduction cycle, light reactions, energy charge and proton pumping, and leads to the production of toxic molecules (Noctor & Foyer, 1998; Chaitanya et al., 2003).

Such change(s) in the metabolisms of plants under environmental stress conditions requires an adjustment of metabolic pathways, aimed at achieving a new state of homeostasis, in a process that is usually referred to as acclimation (Mittler 2006; Suzuki & Mittler, 2006). Several different phases are thought to be involved in acclimation. In the initial stages, the change in the environmental condition is sensed by the plant and activates a network of signaling pathways. In later phases, the signal transduction pathways activated in the first phase trigger the production of different proteins and compounds that restore or achieve a new state of homeostasis (Shulaev et al., 2008).

The intensity, duration and rate of progression of the stress will influence factors that will dictate whether mitigation processes associated with acclimation will or will not occur. Acclimation responses under drought, which indirectly affect photosynthesis, include those related to growth inhibition or leaf shedding that, by restricting water expenditure by source tissues, will help to maintain plant water status and, therefore, plant carbon assimilation. Osmotic compounds that build up in response to a slowly imposed dehydration also have a function in sustaining tissue metabolic activity. Acclimation responses also include synthesis of compatible solutes and adjustments in ion transport. These responses will eventually lead to restoration of cellular homeostasis and, therefore, survival under stress (Chaves et al., 2009). Thus, when plants experience the unfavorable environmental conditions associated with high levels of drought, plant cells protect themselves from the stress of high concentrations of intracellular salts by accumulating a variety of small, organic, electrically neutral molecules that are collectively referred to as compatible solutes or osmoprotectants (Tamura et al., 2003). Compatible solutes are defined as small molecules that are highly soluble in water and are also uniformly neutral with respect to the perturbation of cellular functions, even at high concentrations (Yancy et al., 1982). The properties of compatible solutes allow the maintenance of turgor pressure during water stress, which is an intrinsic feature of major forms of abiotic stress. In addition, some compatible solutes can serve as efficient protective agents by stabilizing the structures and functions of certain macromolecules (Papageorgiou & Murata, 1995).

2. Specific examples for semiarid plants under water stress

2.1 Seed germination and vigor in water stress conditions

Light, temperature and water availability are important abiotic factors that determine the germination of dispersed seeds (Baskin & Baskin, 1998). These factors may be extreme, so their effects can be crucial in the germination and establishment of plants inhabiting arid and semiarid environments (Kigel, 1995). Germination and seedling establishment, due to their dependence on these external factors, are considered to be the most vulnerable stages (Natale et al., 2010). Several desert species are able to germinate at relatively low soil water potential; however, germinability decreases with the reduction in water availability (Kigel, 1995).

In a study with cotton cultivar seeds, it was observed that these seeds are relatively tolerant to water stress induced by PEG-6000. The water stress was most effective in reducing cotton seed viability and vigor at osmotic potentials equal to or more negative than - 0.4 MPa (Meneses et al., 2011). Another interesting example is *Sorghum bicolor* L., which is recognized by its moderate tolerance to water stress (Tabosa et al., 2002), and is an alternative crop that can be used under water deficit. Oliveira et al. (2010), studying the germination and vigor of sorghum seeds under water and salt stress, observed that salt and water stress negatively affected the germination process of these seeds, reducing their vigor. The evaluated genotypes showed different responses to water and salt stress conditions: the CSF 18 sorghum seeds showed higher germination and vigor than did the CSF 20 seeds. With the increase in water deficit, the increase in the number of days for stabilizing the germination was verified. The osmotic potential contributed to the slowest germination, and vigor was more affected than germination (which was more affected in the first counting than in the final germination) (Oliveira et al., 2010).

The germination and seedling establishment of *Tamarix ramosissima,* which constitutes one of the most successful groups of invasive plants in desert riparian ecosystems in the United States, was assessed under different conditions of water availability by Natale et al. (2010). This species seems to be more sensitive to water deficit than to the presence of salts in the substrate. The tolerance limits for germination reached values of -0.4 MPa (PEG-6000). The sensitivity of *T. ramosissima* to water deficit was even more evident in the case of vegetative reproduction. Aerial tissue production by cutting was inhibited by an osmotic potential of -0.4 MPa, a value fairly common in soils that are considered to have a good level of water availability, such as those in the sub-humid region of Argentina, where *Tamarix* species have not been reported to grow spontaneously (Natale et al., 2008).

In a study that evaluated the water stress effects on seed germination responses of *Cereus jamacaru* DC. spp., a cactus widely distributed in Caatinga vegetation, a semiarid ecosystem that characterizes northeastern Brazil (Meiado et al., 2008), low seed germination was observed at a solution concentration of -0.8 MPa in water stress induced by PEG under white light. However, *C. jamacaru* seed germinability was reduced by the reduction of water viability, which affected the mean germination time and the synchronization index. Moreover, in this experiment, germination in the -1.0 MPa range was not observed (Meiado et al., 2010). Another study involving a species of cactus widely distributed in northeastern Brazil was performed *with Hylocereus setaceus* (Salm-Dyck ex DC), whose seeds were subjected to reduced water potentials. In this study, germination was sensitive to decreasing values of Ψs in the medium, and both the germinability and the germination rate shifted negatively with the reduction of Ψs, but the rate of reduction changed with temperature (Simão et al., 2010).

The germination of the semi-deciduous heliophytic tree *Caesalpinia peltophoroides,* popularly known as sibipiruna, was studied under different conditions of light, temperature and water stress. In this study, low water potential reduced both germinability and the germination rate. Under water stress, the seeds were inhibited by white light mediated by phytochrome (Ferraz-Grande & Takaki, 2006).

Water stress affected seed germination and seedling vigor of faveira *(Clitoria fairchildiana* R. Howard, Fabaceae), a widely used tree species in reforestation projects that is recommended for the recovery of degraded areas due to its utility as green manure. Water excess reduced the total amount and the speed of seed germination, whereas the lack of water reduced seedling growth (Silva & Carvalho, 2008).

2.2 Growth and productivity analysis of semiarid plants subjected to water stress

Water deficit is one of the most important environmental stresses affecting agricultural productivity around the world (Hessine et al., 2009). Water availability is undoubtedly the main factor affecting plant growth and development. Moreover, high luminosity, high temperatures and low air relative humidity are further problems plants face in dry lands

(Chaves et al., 2003; Silva et al., 2010a). Morphological and biochemical changes in plants under water deficit lead to acclimation, subsequent functional damage and the loss of plant parts as water stress becomes more severe (Chaves et al., 2003; Costa e Silva et al., 2004), resulting first in a slower growth rate (acclimation phase). This occurs due to the inhibition of cell expansion and a reduction in carbon assimilation (Costa e Silva et al., 2004). Growth maintenance depends on the turgor pressure for cell expansion and division, which is affected by drought (Taiz & Zeiger, 2009). Thus, a reduction in growth is considered the principal effect of drought in plants (Larcher, 1995).

Plants capable of surviving and producing under conditions of low water availability are considered drought-tolerant. From an ecophysiological standpoint regarding strategies for plants living in stressed environments, it is not the maximization of productivity that is important, but the establishment of a balance between yield and survival (Larcher, 1995). However, from an agronomic standpoint, yield is the main objective. Plants that stop growth to survive in a stressful environment could be considered tolerant, but such plants produce fewer grains, leaves or fruit, thus reducing their economic worth (DaMatta, 2004). The yields of crop plants under soil and/or atmospheric drought stress will largely depend on the adaptive mechanisms that allow them to maintain growth and high photosynthetic production under prolonged drought conditions (DaMatta, 2007).

Jatropha curcas is a deciduous stem succulent species with a clear drought avoidance strategy in its leaves, a relatively high water use efficiency and most likely a relatively low water footprint (Maes et al., 2009). In a study involving the biomass production and allocation in *J. curcas* L., Achten et al. (2010) observed that drought treatment significantly influenced growth, biomass allocation, allometry and leaf area. The monitoring of growth and biomass allocation showed that, in optimal conditions, *J. curcas* grows fast, produces a lot of biomass and achieves a high leaf area compared to other tropical deciduous woody species. At the threshold of drought stress (40% plant-available water - PAW), *J. curcas* could still maintain considerable growth and biomass production, without changing the form of its stem and its biomass allocation pattern. Under extreme drought, *J. curcas* started shedding its leaves and stopped growing. In such a situation, the biomass allocation exhibited a higher investment in its roots. Well and moderately watered *Jatropha* plants showed medium biomass investment in leaves and low biomass investment in roots, compared to other tropical deciduous woody tree and shrub species (Achten et al., 2010).

In another study of *J. curcas*, Silva et al. (2010b) observed that water-stressed plants, induced by PEG, suffered higher restrictions in leaf growth compared to salt-stressed plants. In parallel, regarding leaf dry matter reduction, the leaf area decreased by 28% (NaCl) and 39% (PEG) in stressed plants compared to the controls. Again, similar to the results for leaf dry matter, PEG- and NaCl-stressed plants showed only a partial recovery in leaf area after 8 days achieving values of approximately 85% and 79% compared to the controls, respectively. This partial recovery in leaf area occurred in parallel to the restoration in the leaf dry weight. The comparative analyses between the effects of salt and PEG treatments in terms of variables associated with leaf stress intensity (i.e., membrane integrity, leaf growth and water potential) strongly suggest that the PEG treatment caused more negative effects on the physic nut young plants than did NaCl .

Erythrina velutina Willd., a deciduous plant found in the semiarid region of northeastern Brazil, was subjected to different levels of water stress to evaluate growth parameters and water relationships. In this study, all growth parameters were reduced due to water deficit, especially in the plants growth to 25% of the field capacity, as determined by number of leaves, stem diameter, plant height, leaf area, specific leaf area and dry mass in several organs. Leaf area and root: shoot ratios were not affected. *E. velutina* seedlings seem to have developed rusticity to overcome intermittent droughts, with no changes in the pattern of dry matter distribution. The maintenance of turgor pressure seems to be more associated with a reduction in the growth ratio than a reduction in leaf water potential (Silva et al., 2010a).

The influence of water stress on grain yield and vegetative growth were evaluated in two cultivars of beans. The Carioca cultivar, an indeterminate Brazilian landrace, appears to be generally stress-tolerant, whereas the Prince cultivar is stress-intolerant. In this study, water stress reduced yield and yield components at both flowering and pod-filling stages. Moreover, seed weight, number of seeds per plant and number of pods per plant and per seed weight were affected (Boutraa & Sanders, 2001). Water stress during both phenological stages reduced other growth parameters (e.g., the number of trifoliate leaves, stem height, the number of main branches and the number of nodes on the main stem) (Boutraa & Sanders, 2001).

Water deficit is one of the major factors limiting the production of sugarcane (*Saccharum officinarum* L.). Silva et al. (2008) investigated the effects of limited water condition on yield components as indicators of drought tolerance in sugarcane. These authors described that under stress, the tolerant control (TCP93-4245) showed higher productivity, stalk number, stalk height and stalk weight than did the susceptible control (TCP87-3388). However, the susceptible control exhibited higher stalk diameter. A linear association was found between productivity and its yield components, but stalk diameter was shown to be fairly unstable among genotypes.

2.3 Osmoregulation (the participation of organic and inorganic solutes)

Osmotic stress is a physiological event often associated with excessive water deficit that can reduce plant growth through mechanisms that are not yet fully understood. Osmotic adjustment is a cellular adaptive mechanism vital for stress-tolerant plants, allowing for plants to continue growing in the case of drought (Silva et al., 2010c). It is usually defined as a decrease in the cell sap osmotic potential, resulting from a net increase (discounting the concentration effect due to drought-induced reduction in cell volume) in intracellular solutes rather than from a loss of cell water (Kusaka et al., 2005). This process has been considered as an important physiological adaptation characteristic associated with drought tolerance, and it has drawn much attention during the last years (Hessine et al., 2009). Osmotic adjustment involves the net accumulation of solutes in plant cells in response to falls in water potential in the root medium. As a consequence, the cell's osmotic potential is diminished; this, in turn, attracts water into the cell by tending to maintain turgor pressure (Pérez-Pérez et al., 2009). According to Martinez et al. (2005), compatible solubles, such as sugars, glycerol, proline or glycinebetaine, can also contribute to this process.

The roles of organic and inorganic solutes in the osmotic adjustment of drought-stressed *Jatropha curcas* plants were evaluated by Silva et al. (2010c). Regarding inorganic solutes, these authors observed that K^+ had greater quantitative participation in osmotic adjustment, followed by Na^+ and Cl^-. Regarding the organic solutes, the total soluble sugars had the highest contribution to osmotic adjustment, mainly in the most severe cases of water stress. Young *J. curcas* plants exhibited osmotic adjustment in roots and leaves in response to drought stress that was linked with mechanisms to prevent water loss by transpiration by means of the participation of inorganic and organic solutes and stomatal closure (Silva et al., 2010c).

Organic solute production in four umbu tree (*Spondias tuberosa*) genotypes under intermittent drought was evaluated by Silva et al. (2009). The authors concluded that umbu trees presented isohydric behavior during intermittent drought, maintaining high leaf water potential and great variability in the production of organic solutes with marked differences among the genotypes. The high values of water potential and the variability in the organic solutes studied suggest that the storage of water in the xylopodium associated with stomatal closure is responsible for maintaining turgor in the leaves of the umbu tree genotypes; additionally, solute accumulation as a drought-tolerance indicator was not evident in umbu plants.

Osmotically active solutes in cassava leaves during water stress and the contribution of solutes to osmotic adjustment were investigated by Alves & Setter (2003). They observed that K^+ inorganic solute was the major contributor to total osmolytes in both mature and expanding leaves, accounting for approximately 60% of the osmotic potential. The concentration of K-salts increased in response to water stress and was positively correlated with the extent of OA. In contrast, total sugars (sucrose + glucose + fructose) decreased during water deficit, exhibiting a negative correlation with OA. Although the concentration of proline in mature leaves increased in response to water stress, its contribution to the total change in osmotic potential was insignificant (Alves & Setter, 2003).

Osmotic adjustment as a tolerance mechanism to water stress in young jackfruit and sugar apple trees was evaluated by Rodrigues et al. (2010). They observed an accumulation of solute sugars, followed by an increase in protein and amino acid concentrations in water-stressed leaves. They also observed that both evaluated species had a high tolerance to the stressful climatic conditions that occur in the semiarid region of northeastern Brazil. Moreover, jackfruit proved to be more tolerant than sugar apple due to the maintenance of a higher leaf water potential under water deficit.

In a study with cowpea (*Vigna unguiculata*) subjected to water stress and recovery treatment, Souza et al. (2004) showed that carbohydrate metabolic changes revealed an accumulation of soluble sugars in water-stressed leaves, which also persisted for one day after re-watering. This finding suggests a transient end-product inhibition of photosynthesis, contributing to a minor non-stomatal limitation during stress and the initial phase of recovery. Moreover, increases in proline levels were small, and their onset was delayed after stress imposition; thus, their increase may be a consequence instead of a stress-induced beneficial response.

The effect of salinity and PEG-induced water stress on water status, gas exchange, and solute accumulation in *Ipomoea pes-caprae* was evaluated by Sucre & Suárez (2011). They observed that under saline conditions, plants accumulated higher Na^+ concentrations; the accumulation increased with increasing NaCl concentration. However, the leaf Na^+ concentration of plants growing under PEG-induced water stress was higher than expected. Regarding proline, an increase of -1.0 MPa was observed by the addition of NaCl or PEG compared to control plants. Proline accumulation is considered the first response of plants exposed to salt stress and water-deficit stress to reduce leaf osmotic potential; the kinetics of accumulation of this solute depend on the intensity and duration of the stress (Ashraf & Foolad, 2007).

The impact of water shortage on leaf osmotic potential and proline accumulation was evaluated in young plants of two Brazilian green dwarf (BGD) coconut ecotypes. Green dwarf coconut palms exhibited low osmotic adjustment (from 0.05 to 0.24 MPa) and a significant accumulation of proline (1.5 to 2.1 times more than the control) in leaflets in response to water deficit. Considering the growth reduction observed in both ecotypes, proline was not associated to with osmoregulation (Gomes et al., 2010).
The impact of salinity and water stress was analyzed in the xero-halophyte *Atriplex ortensis*, which is a C_3 species well adapted to salt and drought conditions. Na^+ accumulated in response to drought and salinity conditions, suggesting that this element could play a physiological role in the stress response of this xero-halophyte species. However, K^+, Ca^{2+} and Mg^{2+} decreased in response to water and salt stress. Cl^- concentration increased in response to salt stress in all tissues, but water stress had no impact on this parameter (Kachout et al., 2011).
Regarding osmoregulation in the coconut palm, proline's contribution to the overall osmotic adjustment in ecotypes of BGD was recently shown to be low (Gomes et al., 2006; Gomes & Prado, 2007) because: (1) proline concentration was reduced to control levels upon re-watering, whereas osmotic potential did not increase and even decreased in re-watered plants; and (2) the patterns of the two coconut ecotypes evaluated for proline accumulation did not reflect their relative behaviors in terms of osmotic adjustment. Indeed, chloride plays important functions in the water balance of coconut palms. First, Cl^- is important for regulating stomatal apertures in coordinated water flow between six neighboring cells (two guard cells and four subsidiary cells) of the coconut stomatal apparatus. During stomatal pore opening, K^+ and Cl^- ions flow from the subsidiary cell to the guard cell. This movement leads to decreases in (water potential) Ψw of the guard cell and, at the same time, increases Ψw of subsidiary cells, driving water flow from the subsidiary to the guard cells (Braconnier & d´Auzac, 1990).

2.4 Water relations in semiarid plants

Water relations in Caatinga trees were evaluated by Dombroski et al. (2011). The species studied were classified into four groups. (I), *Mimosa caesalpiniifolia* had low leaf water potential (Ψw) at predawn and no significant decrease at midday. Stomatal conductance (gs) analyses indicated that plants had reached their lowest Ψw. (II), *Caesalpinia pyramidalis* and *Auxemma oncocalyx* had low Ψw at predawn and significant decreases at midday. For these species, the recuperation of their water statuses at night may have been sufficient for maintaining open stomata during the day. (III), *Caesalpinia ferrea* and *Calliandra spinosa* had relatively high Ψw at predawn and significant decreases at midday. These species might maintain their water statuses similar to individuals in group II, but they might also have deeper root systems. (IV), *Tabebuia caraiba* had the highest Ψw at predawn and no significant decrease at midday, possibly indicating a combination of good stomatal control of water loss and a deeper root system (Dombroski et al., 2011).
Leaf water relations induced by NaCl and PEG-6000 were assessed in *Ipomoea pes-caprae;* the decrease in the nutrient solution water potential (Ψ_{sol}) was initially stressful and caused turgor loss Brazilian landrace, but 2 d after the beginning of the experiment, the plants that experienced a reduction of -0.5 MPa in Ψ_{sol} could adjust osmotically (Sucre & Suaréz, 2011). However, under high salinity (HS) and water deficit (HW) stress, the osmotic adjustment was delayed, with a consequent turgor loss during the first 16 d of treatment. During this

period in HS and HW, ion delivery and/or synthesis of organic solutes were not enough to achieve the osmotic adjustment necessary for turgor maintenance (Lacerda et al., 2003; Akhiyarova et al., 2005). Subsequently, the resumption of the soil-plant water potential gradient may take place due to the accumulation of inorganic and/or organic solutes in the vacuole, allowing water uptake and enhancing turgor-dependent processes (Akhiyarova et al., 2005).

Water relations evaluated in four umbu tree (*Spondia tuberosa*) genotypes under intermittent drought showed high leaf water potential (Silva et al., 2009). Significant differences in leaf water potential (Ψw) were observed among the genotypes. GBU 68 showed the highest value of Ψw, while GBU 50 showed the lowest. Genotypes GBU 44 and GBU 50 showed significantly reduced Brazilian landraceΨw at 8:00 am, while control plants were reduced at 12:00 am (Silva et al., 2009). Recent studies have demonstrated that the water balance of the umbu tree under drought conditions should be maintained through the utilization of the water storage in the roots and low transpiration rates (Lima-Filho, 2001, 2004).

2.5 Photosynthetic and gas exchange

The ecophysiology parameters of four Brazilian Atlantic Forest under drought were evaluated by Souza et al. (2010). A water deficit induced by 7 days of irrigation suspension reduced the Ψw in all the species studied, except for *C. zeylancium*. This finding was not accompanied by the rates of stomatal conductance (*gs*), net photosynthesis (*A*), transpiration (*E*) and intrinsic water use efficiency (IWUE), except for *B. guianensis,* which exhibited an *A* rate of zero. The gas exchange values obtained tended to be lower in plants under stress; however, *T. guianensis* plants under drought exhibited a high value for *gs* and *A*. No significant differences were observed for *E* and IWUE, except for *B. guianensis,* which showed high *gs* and *E* values with a low *A* rate; thus, the IWUE for this species was the lowest determined (Souza et al., 2010).

The processes involved in the susceptibility of sugarcane plants to water deficit were evaluated in drought-tolerant and drought-sensitive cultivars. The water deficit affected the photosynthetic apparatus of all the plants in different ways within and between cultivars. Photosynthetic rates and stomatal conductances decreased significantly in all cultivars subjected to water deficit Graça et al. (2010). In control tolerant cultivar plants (SP83-2847 and CTC15), the photosynthetic rate was higher than in the sensitive cultivar (SP86-155) (Graça et al., 2010).

The function of the photosynthetic apparatus of cotton (*Gossypium hirsutum*) grown during the onset of water limitation was studied by gas-exchange and chlorophyll fluorescence. The onset of drought stress caused an increase in the operating quantum efficiency of PSII photochemistry, which indicated increased photorespiration, as photosynthesis was hardly affected by water limitation. The increase in PSII was caused by an increase in the efficiency of open PSII reaction centers (Fv_0/Fm_0) and by a decrease in basal non-photochemical quenching. The increased rate of photorespiration in plants during the onset of drought stress can be seen as an acclimation process to avoid an over-excitation of PSII under more severe drought conditions (Massaci et al., 2008).

The responses of photosynthetic gas exchange and chlorophyll fluorescence were studied in the cowpea (*Vigna unguiculata*) during water stress and recovery. The reductions in CO_2 assimilation rates in water-stressed cowpea plants were largely dependent on stomatal closure, which decreased available internal CO_2 and restricted water loss through transpiration. This response seemed to be effective in preventing large decreases in leaf water potential; thus, it appeared to be the basis for dehydration avoidance in the cowpea (Souza et al., 2004).

In the coconut palm, the drought-induced photosynthetic reductions were initially attributable to limited CO_2 diffusion from the atmosphere to the intercellular spaces as a result of stomatal closure (Reppellin et al., 1997). Non-stomatal factors have been shown to contribute to the reduction in photosynthesis, both during a period of severe water deficit and during the recovery phase after resuming irrigation (Gomes, 2006; Gomes et al., 2007). Gomes et al. (2007) used net photosynthesis (NP) and internal CO_2 concentration (Ci) to show drought-induced non-stomatal limitation to NP in dwarf coconut, as indicated by reductions in CO_2-saturated and carboxylation efficiency.

NaCl and PEG stress were able to induce significant and reversible alterations in the physiological stress indicators associated with water relations, growth and leaf gas exchange, but they were unable to cause any changes in the photochemical activity of young *J. curcas* plants (Silva et al., 2010b). Both treatments caused similar impairments of the CO_2 assimilation rate, but the PEG-stressed plants showed higher restriction in stomatal conductance and transpiration. Although both stresses caused significant decreases in leaf chlorophyll content, photochemical activity was not affected (Silva et al., 2010b).

3. Conclusion

The physiological and biochemical changes that occur in semiarid plants subjected to water stress represent adaptive responses by which plants cope with the water deficit. Such species, growing under the low water content that predominates in the soils of these regions, demonstrate an acclimation to this abiotic stress and are able to survive subsequent drought periods with less damage compared to plants from other regions.

4. Acknowledgment

The authors would like to thank National Council of Research and Development (CNPq, Brazil) for the fellowships and financial support. The authors would like to thank National Institute of Science and Technology Salinity (CNPq/MCT/Brazil) for the fellowships and financial support.

5. References

Achten, W.M.J.; Maes, W.H.; Reubens, B.; Mathijs, E.; Singh, V.P.; Verchot, L.; Muys, B. (2 0 1 0). Biomass production and allocation in *Jatropha curcas* L. seedlings under different levels of drought stress. *Biomass and Bioenergy*, Vol.34, No.5, (May 2010) pp.667-676, ISSN 0961-9534

Akhiyarova, G.R.; Sabirzhanova, I.B.; Veselov, D.S.; Frike, V. (2005). Participation of plant hormones in growth resumption of wheat shoots following short-term NaCl treatment. *Russian Journal Plant Physiology*. Vol.52, No.6, (May 2005), pp.788–792, ISSN 1021- 4437

Alves, A.A.C.; Setter, T.L. (2004). Abscisic acid accumulation and osmotic adjustment in cassava under water deficit. *Environmental and Experimental Botany*, Vol.51, No.3, (November 2004), pp.259–271, ISSN 0098-8472

Ashraf, M.; Foolad, M.R. (2007). Roles of glycine betaine and proline in improving plant abiotic stress resistance. *Environmental Experimental Botany*, Vol.59, No.2, (March 2007), pp.207-216, ISSN 0098-8472

Bajji, M.; Kinet, J.M.; Luts, S. (2002). The use of electrolyte leakage method for assessing cell membrane stability as a water stress tolerance test in durum wheat. Plant Growth Regulation, Vol.36, No.1, (December 2001), pp.61-70, ISSN 1435-8107Bajji, M.; Kinet, J.M.; Luts, S. (2002). The use of electrolyte leakage method for assessing cell membrane stability as a water stress tolerance test in durum wheat. *Plant Growth Regulation*, Vol.36, No.1, (December 2001), pp.61-70, ISSN 1435-8107

Baskin, C.C.; Baskin, J.M. (1998). Seeds: ecology, biogeography and evolution of dormancy and germination. Academic Press, ISBN 0-12-080260-0, San Diego

Blum, A. (1996). Crop responses to drought and the interpretation of adaptation. *Plant Growth Regulation*, Vol.20, No.2, (February 1996), pp.135-148, ISSN 1435-8107

Boutraa, T.; Sanders, F.E. (2001). Influence of water stress on grain yield and vegetative growth of two cultivars of bean (*Phaseolus vulgaris* L.) *Journal of Agronomy & Crop Science*, Vol.187, No. 4, (August 2001), pp.251-257, ISSN 0931-2250

Braconnier, S.; d'Auzac J (1990). Chloride and stomatal conductance in coconut. *Plant Physiology and Biochemsitry*, Vol. 28, No.1 (April 1990), pp.105-111, ISSN: 0981-9428

Chaitanya, K.Y.; Sundar, D.; Jutur, P.P.; Ramachandra Reddy, A. (2003). Water stress effects on photosynthesis in different mulberry cultivars. *Plant Growth Regulation*, Vol.40, No.1, (May 2003), pp.75-80, ISSN 1435-8107

Chaves, M.M.; Flexas, J.; Pinheiro, C. (2009). Photosynthesis under drought and salt stress: regulation mechanisms from whole plant to cell. *Annals of Botany*, Vol.103, No.4, (April 2009), pp. 551-560, ISSN 1095-8290

Chaves, M.M.; Maroco, J.P.; Pereira, J.S. (2003). Understanding plant responses to drought - from genes to the whole plant. *Functional Plant Biology*, Vol.30, No.3, (March 2003), pp. 239-264, ISSN 1445-4408

Chazen, O., Neumann, P.M. (1994). Hydraulic signals from the rots and rapid cell wall hardening in growing maize (*Zea mays* L.) leaves are primary responses to polyethylene glycol induced water deficits. *Plant Physiology*, Vol.104, No.4, (April 1994), pp.1385-1392, ISSN 1532-2548

Chrispeels, M.J.; Crawford, N.M.; Schroeder, J.I. (1999). Proteins for transport of water and mineral nutrients across the membranes of plant cells. *Plant Cell*, Vol.11, No.4, (April 1999), pp.661-676, ISSN 1531-298X

Costa e Silva, F.; Shvaleva, A.; Maroco, J.P.; Almeida, M.H.; Chaves, M.M.; Pereira, J.S. (2004). Responses to water stress in two *Eucalyptus globulus* clones differing in drought tolerance. *Tree Physiology*, Vol. 24, No. 10, (August 2004), pp. 1165-1172, ISSN 0829-318X

Crowe, J.H.; Hoekstra, F.A.; Crowe, L.M. (1992). Anhydrobiosis. *Annual Review of Plant Physiology and Plant Molecular Biology*, Vol.54, No.1, (January 1992), pp.570-599, ISSN 0066-4294

DaMatta, F.M. (2004). Exploring drought tolerance in coffee: a physiological approach with some insights for plant breeding. *Brazilian Journal of Plant Physiol*, Vol.16, No.1, (January/April 2004), pp.1-6, ISSN 16770420

DaMatta, F.M. (2007). Ecophysiology of tropical tree crops: an introduction. *Brazilian Journal of Plant Physiology*, Vol.19, No.4, (October/December 2007), pp.239-244, ISSN 16770420

Dombroski, J.L.D.; Praxedes, S.C.; Freitas, R.M.O.; Pontes, F.M. (2011). Water relations of Caatinga trees in the dry season. *South African Journal of Botany*, Vol.77, No.2, (April 2011), pp.430–434, ISSN 0254-6299

Ferraz-Grande, F.G.A.; Takaki, M. 2006. Efeitos da luz, temperatura e estresse de água na germinação de sementes de Caesalpinia peltophoroids Benth. (Caesalpinoidea). Bragantia, Vol.65, No.1, pp.37-42,

Gomes, F.P.; Oliva, M.A.; Mielke, M.S.; Almeida, A-AF.; Leite, H.G. (2006). Photosynthetic irradiance-response in leaves of dwarf coconut palm (*Cocos nucifera* L. 'nana', Arecaceae): Comparison of three models. *Scientia Horticulturae*, Vol.109, No.1, (June 2006), pp.101-105, ISSN 0304-4238

Gomes, F.P.; Prado, C.H.B.A. (2007). Ecophysiology of coconut palm under water stress. *Brazilian Journal of Plant Physiol*ogy, Vol.19, No.4, (October/December, 2007), pp.377-391, ISSN 1677-0420

Gomes, F.P.; Olivab, M.A.; Mielkea, M.S.; Almeidaa, A.A.F.; Aquinob, L.A. (2010). Osmotic adjustment, proline accumulation and cell membrane stability in leaves of *Cocos nucifera* submitted to drought stress. *Scientia Horticulturae*, Vol:126, No.3, (September 2010), pp.379–384, ISSN 0304-4238

Graça, J.P.; Rodrigues, F.A.; Farias, J.R.B.; Oliveira, M.C.N.; Hoffmann-Campo, C.B.; Zingaretti, S.M. (2010). Physiological parameters in sugarcane cultivars submitted to water stress. *Brazilian Journal of Plant Physiology*, Vol.22, No.3, (October 2010), pp. 189-197, ISSN 1677-0420

Hay, K.M.H.; Walker, A.J. (1989). Environmental effects on photosynthesis: water stress, In: *An introduction to the physiology of crop yield. Photosynthesis efficiency: photosynthesis and respiration*, Hay, K.M.H.; Walker, A.J. (Eds.), pp. 68-76, Longman Scientific and Technical, ISBN 978-14051-0859-1, Wiley, New York

Hessine, K.; Martínez, J.P.; Gandour, M.; Albouchi, A.; Soltani, A.; Abdelly, C. (2009). Effect of water stress on growth, osmotic adjustment, cell wall elasticity and water-use efficiency in *Spartina alterniflora*. *Environmental and Experimental Botany*, Vol.67, No. 2, (December 2009), pp. 312–319, ISSN 0098-8472

Iannucci, A.; Mattinello, P. (1998). Analysis of seed yield and yield components in four Mediterranean annual clovers. *Field Crops Research*, Vol.55, No.3, (February 1998), pp.235-243, ISSN 0378-4290

Intergovernmental Panel on Climate Change (2007). http://www.ipcc.ch. Accessed 25 October 2007

Izanloo, A.; Condon, A.G.; Langridge, P.; Tester, M.; Schnurbusch, T. (2008). Different mechanisms of adaptation to cyclic water stress in two South Australian bread wheat cultivars. *Journal of Experimental Botany*, Vol. 59, No. 12, (August 2008), pp. 3327–3346, ISSN 1460-2431

Jones, H. (2004). What is water use efficiency?, In: *Water use efficiency in plant biology*, Bacon, M.A. (Ed.), pp. 27-41, Wiley-Blackwell, ISBN 978-1-4051-1434-9, Oxford

Jones, M.M.; Turner, N.C.; Osmond, C.B. (1981). Mechanisms of drought resistance. In: *The physiology and biochemistry of drought resistance in plants*, Paleg, L.G.; Aspinall, D. (Eds.), pp. 20-35, Academic Press, ISBN 0-12-544380-3, Sydney

Kachout, S.S.; Mansoura, A.B.; Hamza, K.J.; Leclerc, J.C.; Rejeb, M.N.; Ouergui, Z. (2011). Leaf-water relations and ion concentrations of the halophyte *Atriplex hortensis* in response to salinity and water stress. *Acta Physiologia Plantarum,* Vol.33, No.2, (March 2011), pp.335-342, ISSN: 1861-1664

Kigel, J. Seed germination in arid and semiarid regions. (1995). *Seed development and germination.* Kigel, J.; Galili, G (Eds), p. 645-699, Marcel Dekker, Inc., ISBN 0-8247-9229, New York

Kramer, P.J.; Boyer, J.S. (1995). *Water Relations of Plants and Soils.* Academic Press, ISBN 0-12-425060-2, San Diego

Kusaka, M.; Lalusin, A.G.; Fujimura, T. 2005. The maintenance of growth and turgor in pearl millet (*Pennisetum glaucum* [L.] Leeke) cultivars with different root structures and osmo-regulation under drought stress. *Plant Science,* Vol. 168, pp. 1–14.

Lacerda, C.F.; Cambraia, J.; Oliva, M.A.; Ruiz, H.A.; Prisco, J.T. (2003). Solute accumulation and distribution during shoot and leaf development in two Sorghum genotypes under salt stress. *Environmental Experimental Botany,* Vol.49, No.2, (April 2003), pp.107–120, ISSN 0098-8472

Larcher, W. (1995). *Physiological Plant Ecology.* Springer, ISBN 9-78-354043516-7, Berlin

Lauenroth, W.K.; Sala, O.E.; Coffin, D.P.; Kirchner, T.B. (1994). The importance of soil-water in the recruitment of Bouteloua-Gracilis in the shortgrass steppe. *Ecological Applications,* Vol.4, No.4, (April 1994), pp.741–749, ISSN 1051-0761

Lima-Filho, J.M.P. (2001). Internal water relations of the umbu tree under semi-arid conditions. *Revista Brasileira de Fruticultura,* Vol. 23, No. 3, (December 2001), pp. 518 -571, ISSN 0100-2945

Lima-Filho, J.M.P. (2004). Gas exchange of the umbu tree under semi arid conditions. *Revista Brasileira de Fruticultura,* Vol.26, No.2, (August 2004), pp.206-208, ISSN 0100-2945

Maes, M.H.; Achten, W.M.J.; Reubens, B.; Samson, R.; Muys, B. (2009). Plant–water relationships and growth strategies of *Jatropha curcas* L. saplings under different levels of drought stress. *Journal of Arid Environments,* Vol.73, No.10, (October 2009), pp.877–84, ISSN 0140-1963

Martinez, J.P.; Kinet, J.M.; Bajji, M. ; Lutts, S. (2005). NaCl alleviates polyethylene glycol induced water stress in the halophyte species *Atriplex halimus* L. *Journal Experimental Botany,* Vol.56, No.419, (September 2005), pp.2421–2431, ISSN 0022-0957

Maurel, C. (1997). Aquaporins and water permeability of plant membranes. *Annual Review of Plant Physiology and Plant Molecular Biology,* Vol.48, No.1, (June 1997), pp.399-429, ISSN 0066-4294

Massaci, A.; Nabiev, S.M.; Pietrosant, L.; Nematov, S.K.; Chernikova, T.N.; Thor, K.; Leipner, J. (2008). Response of the photosynthetic apparatus of cotton (*Gossypium hirsutum*) to the onset of drought stress under field conditions studied by gas-exchange analysis and chlorophyll fluorescence imaging. *Plant Physiology and Biochemistry,* Vol.46, No.2, (February 2008), pp.189-195, ISSN: 0981-9428

Medrano, H.; Escalona, J.M.; Bota, J.; Gulías, J.; Flexas, J. (2002). Regulation of photosynthesis of C3 plants in response to progressive drought: stomatal conductance as a reference parameter. *Annals of Botany,* Vol.89, No.7, (June 2002), pp.895-905, ISSN 0305-7364

Meiado, M.V.; Albuquerque, L.S.C.; Rocha, E.M.; Rójas-Aréchiga, M.; Leal, I.R. (2010). Seed germination responses of *Cereus jamacaru* DC. ssp. *jamacaru* (Cactaceae) to environmental factors. *Plant Species Biology*, Vol.25, No.2, (May 2010), pp.120–128, ISSN 0913-557X

Meiado, M.V.; Rocha E.A.; Rojas-Aréchiga, M.; Leal, I. R. (2008). Comunidad de cactus en la Caatinga: ¿qué influencia la dinámica de semillas en el ambiente semiárido? *Boletín De La Sociedad Latinoamericana Y Del Caribe De Cactáceas Y Otras Suculentas*. Vol.5, No.3, (September/December 2008), pp.4–6, ISSN 1856-4569

Meneses, C.H.S.G.; Bruno, R.L.A.; Fernandes, P.D.; Pereira, W.E.; Lima, L.H.G.M; Lima, M.M.A,; Vidal, M.S. (2011). Germination of cotton cultivar seeds under water stress induced by polyethyleneglycol-6000. *Scientia Agricola*, Vol.68, No.2, (March/April 2001). pp.131-138, ISSN 0103-9016

Mittler, R. (2006). Abiotic stress, the field environment and stress combination. *Trends in Plant Science*, Vol.11, No.1, (January 2006), pp.11-19, ISSN 1360-1385

Natale, E.; Zalba, S.M.; Oggero, A.; Reinoso, H. (2010). Establishment of *Tamarix ramosissima* under different conditions of salinity and water availability: Implications for its management as an invasive species. *Journal of Arid Environments*, Vol.74, No. 11, (November 2010), pp. 1399-1407, ISSN 0140-1963

Natale, E.; Gaskin, J.; Zalba, S.M.; Ceballos, M.; Reinoso, H. (2008). Species of the genus Tamarix (tamarisk) invading natural and semi-natural environments in Argentina. *Boletín de la Sociedad Argentina de Botánica*, Vol.43, No.1-2, (January/June 2008), pp. 137-145, ISSN 0373-580X

Nobel, P.S. (1999). *Physiochemical and Environmental Plant Physiology*. Academic Press, ISBN 978-0-12-520026-4, San Diego

Noctor, G.; Foyer, C.H. (1998). Ascorbate and glutathione: keeping active oxygen under control. *Annual Review of Plant Physiology and Plant Molecular Biology*, Vol.49, No.1, (June 1998), pp.249-279, ISSN 0066-4294

Oliveira, A.B.; Gomes-Filho, E. (2009). Germinação e vigor de sorgo forrageiro sob estresse hídrico e salino. *Revista Brasileira de Sementes*, Vol. 31, N° 3, (August 2009), pp.48-56, ISSN 0101-3122

Papageorgiou, G.C.; Murata, N. (1995). The unusually strong stabilizing effects of glycine betaine on the structure and function of the oxygen-evolving Photosystem II complex. *Photosynthesis Research*, Vol.44, No.3, (September 1995), pp. 243- 252, ISSN 1573-5079,

Passioura, J.B. (2007). The drought environment: physical, biological and agricultural perspectives. *Journal of Experimental Botany*, Vol.58, No.2, (February 2007), pp. 113–117, ISSN 1460-2431

Pérez-Pérez, J.G.; Robles, J.M.; Tovar, J.C. ; Botía, P. (2009). Response to drought and salt stress of lemon 'Fino 49' under field conditions: water relations, osmotic adjustment and gas exchange. *Scientia Horticultura*, Vol. 122, pp. 83–90.

Pinheiro, R.G.; Rao, M.V.; Palyath, G.; Murr, D.P.; Fletcher, R.A. (2001). Changes in the activities of antioxidant enzymes and their relationship to genetic and paclobutrazol - induced chilling tolerance of maize seedlings. *Plant Physiology*, Vol.114, No.2, (February 2001), pp.695-704, ISSN 1532-2548

Rascio, A.; Platani, C.; Scalfati, G.; Tonti, A.; Di Fonzo, N. (1994). The accumulation of solutes and water binding strength in durum wheat. *Physiologia Plantarum*, Vol.90, No.4, (April 1994), pp.715-721, ISSN 0031-9317

Rodrigues, B.M.; Souza, B.D.; Nogueira, R.M.; Santos, M.G. (2010). Tolerance to water deficit in young trees of jackfruit and sugar apple. *Revista Ciência Agronômi*ca, Vol. 41, No. 2, pp. 245-252, (April/June 2010), ISSN 0045-6888

Shulaev, V.; Cortes, D.; Mittler, R. (2008). Metabolomics for plant stress response. *Physiologia Plantarum*, Vol.132, No.2, (January 2008), pp.1027-1040, ISSN 0031-9317

Silva, S.B.M.; Carvalho, N.M. (2008). Efeitos do estresse hídrico sobre o desempenho germinativo da semente de faveira (*Clitoria fairchildiana* R.A. Howard. - FABACEAE) de diferentes tamanhos. *Revista Brasileira de Sementes*, Vol.30, No.1, (February 2008), pp.55-65, ISSN 0101-3122

Silva, M.A.; Silva, J.A.G.; Enciso, J.; Sharma, V.; Jifon, J. (2008). Yield components as indicators of drought tolerance of sugarcane. *Scientia Agricola*, Vol.65, No.6, (November/December 2008), pp.620-627, ISSN 0103-9016

Silva, E.C., Nogueira, R.J.M.C., Vale, F.H.A., Melo, N.F, Araújo, F.P. (2009). Water relations and organic solutes production in four umbu (*Spondias tuberosa*) tree genotypes under intermittent drought. *Brazilian Journal of Plant Physiology*, Vol.21, No.1, (January/March 2009), pp. 43-53, ISSN 1677-0420

Silva, E.C.; Silva, M.F.A.; Nogueira, R.J.M.C.; Albuquerque, M.B. (2010a). Growth evaluation and water relations of *Erythrina velutina* seedlings in response to drought stress. *Brazilian Journal of Plant Physio*logy, Vol.22, No.4, (October/December 2010), pp. 225-233, ISSN 1677-0420

Silva, E.N.; Ribeiro, R.V.; Ferreira-Silva, S.L.; Viégas, R.A.; Silveira, J.A.G. (2010b). Comparative effects of salinity and water stress on photosynthesis, water relations and growth of *Jatropha curcas* plants. *Journal of Arid Environments*, Vol. 74, No. 10, (October 2010), pp. 1-8, ISSN 0140-1963

Silva, E.N.; Ferreira-Silva, S.L.; Viégas, R.A.; Silveira, J.A.G. (2010c). The role of organic and inorganic solutes in the osmotic adjustment of drought-stressed *Jatropha curcas* plants. *Environmental and Experimental Botany*, Vol. 69, No. 3, (December 2010), pp. 279-285, ISSN 0098-8472

Simão, E.A.; Takaki, M.B.; Cardoso, VJM. Germination response of *Hylocereus setaceus* (Salm-Dyck ex DC.) Ralf Bauer (Cactaceae) seeds to temperature and reduced water potentials. *Brazilian Journal Biology*, Vol. 70, No. 1, (February 2010), p. 135-144, ISSN 1519-6984

Smith, J.A.C.; Griffith, H. (1993). *Water Deficits: Plant Responses from Cell to Community*. Bios Scientific Publisher, ISBN 9-78-187274806-1, Oxford

Souza, R.P.; Machado, E.C.; Silva, J.A.B.; Lagôa, A.M.M.A.; Silveira, J.A.G. (2004). Photosynthetic gas exchange, chlorophyll fluorescence and some associated metabolic changes in cowpea (*Vigna unguiculata*) during water stress and recovery. *Environmental and Experimental Botany*, Vol. 51, No.1, (February 2004), pp. 45-56, ISSN 0098-8472

Souza, B.D.; Rodrigues, B.M.; Endres, L.; Santo, M.G. (2010). Ecophysiology parameters of four Brazilian Atlantic Forest species under shade and drought stress. *Acta Physiologia Plantarum*, Vol. 32, No. 4, (January 2010), pp. 729-737, ISSN 0137-5881

Sucre, B.; Suárez, N. (2011). Effect of salinity and PEG-induced water stress on water status, gas exchange, solute accumulation, and leaf growth in *Ipomoea pes-caprae*. *Environmental and Experimental Botany*, Vol.70, No. 2, (February 2011), pp.192–203, ISSN 0098-8472

Suzuki, N.; Mittler, R. (2006). Reactive oxygen species and temperature stresses: a delicate balance between signaling and destruction. *Physiologia Plantarum*, Vol.126, No.1, (November 2005), pp.45-51, ISSN 0031-9317

Tabosa, J.N.; Reis, O.V.; Brito, A.R.M.B.; Monteiro, M.C.D.; Simplício, J.B.; Oliveira, J.A.C.; Silva, F.G.; Azevedo Neto, A.D.; Dias, F.M.; Lira, M.A.; Tavares Filho, J.J.; Nascimento, M.M.A.; Lima, L.E.; Carvalho, H.W.L.; Oliveira, L.R. (2002). Comportamento de cultivares de sorgo forrageiro em diferentes ambientes agroecológicos dos Estados de Pernambuco e Alagoas. *Revista Brasileira de Milho e Sorgo*, Vol.1, No.2, (May/August 2002), pp.47-58, ISSN 1980-6477

Taiz, L.; Zeiger, E. (2009). *Plant Physiology*. Sinauer Associates, ISBN 978-0878938230, Sunderland

Tamura, T.; Hara, K.; Yamaguchi, Y.; Kolzumi, N.; Sano, H. (2003). Osmotic stress tolerance of transgenic tobacco expressing a gene encoding a membrane-located receptor-like protetn from tobacco plants. *Plant Physiology*, Vol.131, No.2, (February 2001), pp.454-462, ISSN 1532-2548

Tardieu, F. (2005). Plant tolerance to water deficit: physical limits and possibilities for progress. *Comptes Rendus Geoscience*, Vol.337, No.1, (January 2005), pp.57-67, ISSN 1631-0713

Yancy, P.H.; Clark, M.E.; Hand, S.C.; Bowlus, R.D.; Somero, G.N. (1982). Living with water stress: evolution of osmolyte systems. *Science*, Vol.21, No.4566, (September 1982), pp. 1214–1222, ISSN 1095-9203

Zhu, J.K.; Hasegawa, P.M.; Bressan, R.A. (1997). Molecular aspects of osmotic stress in plants. *Critical Reviews in Plant Sciences*, Vol.16, No.3, (February 1997), pp.253-277, ISSN 0735-2689

3

Controlled Water Stress to Improve Fruit and Vegetable Postharvest Quality

Leonardo Nora, Gabriel O. Dalmazo,
Fabiana R. Nora and Cesar V. Rombaldi
University Federal of Pelotas
Brazil

1. Introduction

Healthier food produced in a sustainable manner at affordable price is a necessity in the contemporaneous society. In this context, studies to devise the minimum necessary amount of water to crops towards development are required not only to save water and/or energy, but also to improve plant fitness to cope with biotic and abiotic stresses, even after harvest, and to improve nutritional, functional and sensorial food properties. The awareness of the growing impact of environmental stress has lead to worldwide efforts in adapting agricultural production to adverse environmental conditions, focusing on mitigating quantitative yield losses (Godfray et al., 2010). Far less attention has been devoted to the impact of abiotic environmental stresses on crop quality (Wang & Frei, 2011). Limited water supply (LWS) is a serious threat to agriculture, and often cause yield reduction. However, many plant species are not intolerant to that, otherwise the water input concept should be revised. In this review we describe the effect of limiting water supply during plant development on fruit and vegetables postharvest quality, mainly in terms of sensorial attributes (texture, colour, aroma, and taste) and composition (nutrients and bioactive compounds), and also in terms of gene expression, enzyme activity, yield, and storage behaviour. A brief introduction about regulated deficit irrigation, plant growth regulators and secondary metabolites is presented to illustrate common aspects associated with water stress effects on fruit and vegetable quality.

2. Regulated deficit irrigation

Deficit irrigation (i.e. irrigation below optimal crop water requirements) research to improve productivity of horticultural crops began in the 1970's with the aim to control excessive vegetative vigour in high-density orchards. Tree physiology was intensively studied to examine timing of water deficits that would minimize the impact on fruit growth but maximize effects on shoot growth. This strategy later became known as regulated deficit irrigation (RDI). In recent years there has been resurgence in the application of RDI to horticultural crops due to changes in climate resulting in severe drought. However, focus has now switched from controlling excessive vigour to investigating opportunities to stimulate improvements in fruit and vegetable quality so that any yield loss can be compensated by an increase in crop value (Costa et al., 2007; Stefanelli et al., 2010).

3. Plant growth regulators

Organisms need to adapt themselves to changes in fluctuating environmental conditions. The plants, since they are not able to scape from adverse environmental conditions, have to rely entirely on their developmental plasticity to survive (Krouk et al., 2011). These adaptations include the responses to temperature fluctuations, water and nutrients imbalance, UV radiation, pathogens, and insects, among other biotic and abiotic stresses. Plant growth regulators (phytohormones), compounds derived from plant biosynthetic pathways, mediate these responses by acting either at the site of synthesis or following their transport, elsewhere in the plant. Collectively, plant hormones regulate every aspect of plant growth, development and the responses of plants to biotic and abiotic stresses (Peleg & Blumwald, 2011). Classical phytohormones are abscisic acid (ABA), ethylene, cytokinin (CK), auxin, gibberellin, jasmonate, as well as brassinosteroids, salicylic acid, nitric oxide, and strigolactone, and it is likely that additional growth regulators are yet to be discovered (Santner & Estelle, 2009). ABA synthesis is one of the fastest responses of plants to water stress, triggering ABA-inducible gene expression and causing stomatal closure, thereby reducing water loss via transpiration and eventually restricting cellular growth (Wilkinson & Davies, 2010; Yamaguchi-Shinozaki & Shinozaki, 2006). Many ABA-mediated physiological processes induced by water deficit, including closure of the stomata and acceleration of leaf senescence, are counteracted by CKs which increase stomatal aperture and/or delay ABA-induced stomatal closure. It has been suggested that in longer-term responses to stress, hormones such as ABA and CK may function to regulate the production, metabolism and distribution of metabolites essential for stress survival and recovery (Pospíšilová & Dodd, 2005; Stoll et al., 2000).

4. Secondary metabolites

Plants produce a huge variety of secondary metabolites with roles in various biological processes, such as pollination, seed dispersal, and resistance to biotic and abiotic stresses (Wink, 1999). Until recently it was thought that genes for plant metabolic pathways were not clustered, and this is certainly true in many cases. However, five plant secondary metabolic gene clusters have now been discovered, all of them implicated in synthesis of defence compounds, with enzymes for the first committed steps apparently recruited directly or indirectly from primary metabolic pathways involved in hormone synthesis (Chu et al., 2011). The genes and corresponding gene products for the first committed steps in these pathways can be regarded as signature genes/enzymes, as they are required for the synthesis of the skeleton structures of the different classes of secondary metabolite (Osbourn, 2010). The signature genes all share homology with genes from plant primary metabolism, and so are likely to have been recruited directly or indirectly from primary metabolism by gene duplication and acquisition of new functions (Chu et al., 2011). Plant secondary metabolites have long been associated with ecological roles in antagonistic or mutualistic interactions between plants and their herbivores, pathogens, competitors, pollinators or seed dispersers. However, many of these substances have been demonstrated to function in the primary processes of growth and development or resistance to abiotic stresses (Fig. 1). Clearly, more attention should now be devoted to looking for internal roles. In the past decade, several substances that were once considered to be secondary metabolites, such as jasmonic acid, salicylic acid and brassinosteroids, have been shown to be important internal signals (D'Auria & Gershenzon, 2005).

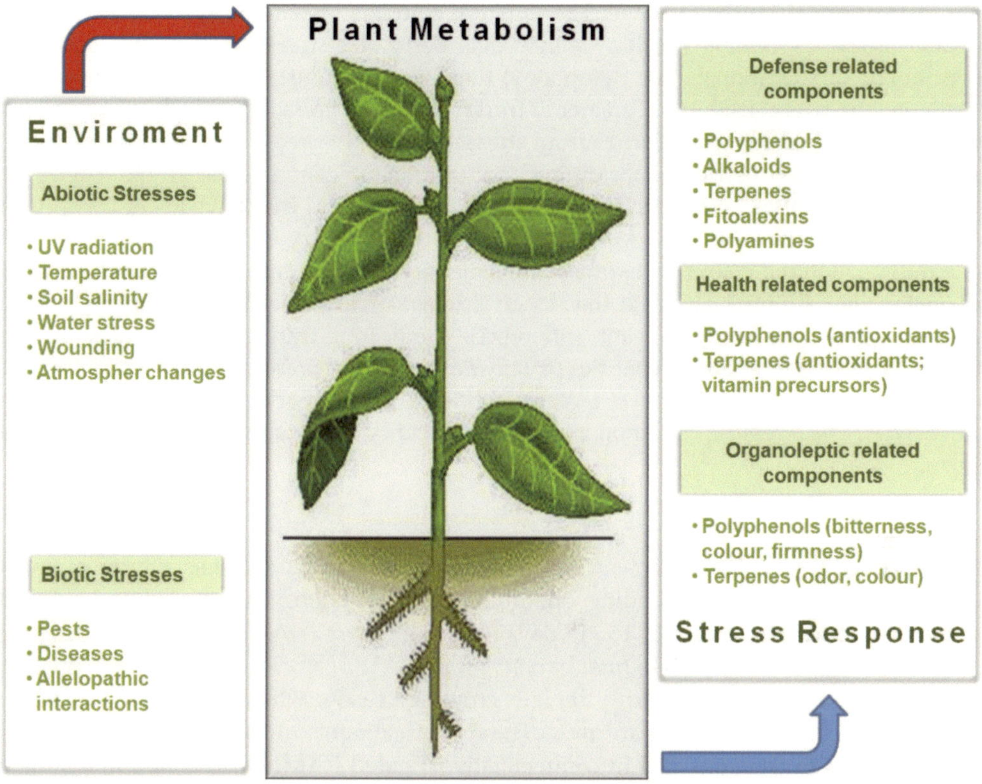

Fig. 1. Schematic representation of plant metabolism response to environmental stress.

5. Water stress effects on fruit quality

5.1 Kiwi

Miller et al. (1998), performed a two year experiment in New Zealand to determine the responses of kiwifruit (*Actinidia deliciosa* cv. Hayward) to water stress conditions. Plants were submitted to a stress regime composed of three different treatments: Control (water received according to culture demands), water stress in early summer and water stress in late summer. They observed a significant loss in fruit weight, especially in plants exposed to stress in early summer (fruit set period). In contrast, an increase in the total soluble solids occurred. Differences in firmness and performance during storage were not detected for stressed kiwifruit. However, in a similar experiment conducted by Reid et al. (1996), kiwifruit harvested from vines exposed to a less severe drought stress were unaffected in size and fruit firmness was retained for 30 days longer in comparison to control (fruit harvested from fully irrigated vines).

5.2 Strawberry

Bordonaba & Terry (2010), testing strawberry (*Fragaria X ananassa* Duch. cvs. Elsanta, Sonata, Symphony, Florence and Christine) response to water deficits obtained promising results. Stressed plants received a quarter of irrigation water compared to control plants (cultivated under field capacity). Water suppress started when most fruit from the primary truss were at flower initiation stage. The different cultivars responded in a specific manner to water stress, expressing specific water usage. Berry size was equivalent (Florence and Christine) or smaller (Sonata and Symphony) than control plants. Dry mater, as proportion of fruit weight, was considerably greater in fruit from water stressed plants than from plants kept at or near field capacity. Berries from stressed plants showed lower redness (higher h^o value) than control berries. Considering that the main components of red colour in strawberries are anthocianins is plausible to presume that these fruit have lower contents of this secondary metabolite. However, in a previous study, Terry et al. (2007) observed the same reduction in red colour, but anthocianins measurements pointed to a higher content of this metabolite. Authors attribute that to an artefact of the objective colorimeter due to smaller berry size. No significant differences were found in sugar contents among treatments. However, monosaccharides (fructose and glucose) were in higher concentrations in stressed plants, thus berries were sweeter. Acids are also important flavor components in strawberries. The stressing condition increased the acidity in all cultivars except Elsanta and Sonata.

5.3 Apple

Mpelasoka et al. (2001b) demonstrated that deficit irrigation (DI) has effects on fruit maturation and ripening depending on timing of application. They tested early deficit irrigation (EDI), applied from 63 to 118 days after full bloom (DAFB), late deficit irrigation (LDI), applied from 118 DAFB to final harvest on 201 DAFB, whole-season deficit irrigation (WDI), irrigated only twice during the late growing season, when volumetric soil water content (θ) declined below 0.15 m^3 m^{-3}. These DI treatments all reduced volumetric soil water content. Control consisted in commercially irrigated (CI) trees, irrigated to maintain soil moisture at or close to field capacity. All DI treatments increased fruit total soluble solids (TSS) and firmness regardless of maturity but had little or no effect on titratable acidity. According to the authors the DI fruit may be harvested over a longer period due to their earlier increased TSS and their higher firmness prior to harvest and for most of the storage period. However, the advanced ripening of the DI fruit is responsible for the loss of advantage by DI regarding firmness after long-term storage. Fruit thinning has been proposed as a feasible strategy to compensate the loss in fruit size caused by water stress (Mpelasoka et al., 2001a). Irrigation treatments did not affect the crop load. Irrespective of fruit thinning treatment, deficit irrigated stress resulted in lower fruit weight, total yield and fresh-market yield at harvest than control. However, under deficit treatment, thinned trees resulted higher fruit weight and equal fresh-market yield. Regarding quality parameters, deficit irrigated plants exhibited higher contents of TSS than fully irrigated plants. This could have a positive impact on fruit taste. An equal increase was observed in fruit firmness. Although this attribute correlates to a smaller fruit size due to dehydration during water restriction. In a similar experiment, with apple, the fruit firmness was higher under water restriction treatments compared to fully irrigated treatments despite of fruit size (Mpelasoka et al., 2000). Regarding postharvest conditions, fruit exposed to water restrictions had lower weight loss during cold storage than those originated from fully irrigated treatments. According to the authors, the reduced weight loss can be explained by the structure and/or composition of the skin or the epicuticular waxes covering the skin. This reduction in weight loss could prolong the cold storage life.

5.4 Pear

Lopez et al. (2011) submitted pear (*Pyrus communis* L. cv. Conference) to water restrictions and fruit thinning at the Stage II (80 and 67 days before harvest in 2008 and 2009, respectively). The experiment was composed of two simple irrigation treatments: fully irrigated plants (100% Evapotranspiration, ET_c) and deficit irrigated (20% ETc, preceded by three weeks of total water deprivation to induce the stress response). Each irrigation treatment was spliced in two thinning treatments: no thinning (approximately 180 fruit per tree) and thinning (approximately 85 fruit per tree). Fruit thinning has been proposed as a feasible strategy to compensate the loss in fruit size caused by water stress (Mpelasoka et al., 2001c). Different Irrigation regimes did not affect the crop load between irrigation treatments. Irrespective of fruit thinning treatment, deficit irrigated trees had lower weight, total yield and fresh-market yield at harvest than control. However, under deficit treatment, thinned trees had higher fruit weight and equal fresh-market yield. Regarding quality parameters, deficit irrigated plants exhibited higher contents of total soluble solids than fully irrigated plants. Deficit irrigated-thinned trees had equivalent fruit weights at harvest compared to full irrigated-non thinned trees, confirming that fruit size can be improved by fruit thinning. Considering that fruit size is an important attribute in pear, deficit irrigation can improve fruit marketability.

5.5 Apricot

Perez-Pastor (2007) evaluated postharvest fruit quality of apricot (*Prunus armeniaca* L. cv. Búlida) harvested from trees exposed to three different treatments: control treatment (100% of evapotranspiration); regulated deficit irrigation (RDI), which consists in fully irrigation during critical periods; and 50% water regime compared to control. At harvest was not observed differences in weight, equatorial diameter and firmness of the fruit among the different treatments. In addition, fruit from water stressed plants had higher values of total soluble solids (TSS), titratable acidity (TA) and h^o value (skin colour). During storage, stressed plants had lower decreases in fruit colour (skin and pulp) compared to fully irrigated plants. During the first 20 days of storage, stressed plants conserved higher values of TSS and TA, after that the differences disappeared between treatments. During a simulated retail sale, fungi of the genera *Rhizopus, Monilinia, Penicillium, Alternaria, Botrytis* and *Cladosporium* caused fruit losses. Interestingly, a lower fungal attack was observed in stressed fruit. The authors link this fact to a thicker cuticle and/or to the absence of microcrackings. Perez Sarmiento et al. (2010) using nine year-old apricot-trees (*Prunus armeniaca* L. cv. 'Búlida') grafted on 'Real Fino' rootstock analyzed the effects of RDI on fruit quality. Two irrigation treatments were established. The first, a control treatment, was irrigated to fully satisfy the crop water requirements (100% ETc) during the critical periods (stage III of fruit growth and two months after harvest period), and the second, a RDI treatment, was subject to water shortage during the non-critical periods of crop development, by reducing the amount of applied irrigation water to: a) 40% of ETc from flowering until the end of the first stage of fruit growth; b) 60% of ETc during the second stage of fruit growth and c) 50% and 25% of ETc during the late postharvest period (that starts 60 days after harvesting), for the first 30 days and until the end of tree defoliation, respectively. They found that some qualitative characteristics such as the level of soluble solids, fruit taste and the colour of the fruit are enhanced.

5.6 Peach

Gelly et al. (2003) evaluated the effects of water deficit on fruit quality of peaches (*Prunus persica* L.). It was shown an increase in soluble solids content and coloration of fruit when during production was applied RDI. Water deficit was applied during either stage II of fruit development (RDI-SII) or during stage II and postharvest (RDI-SII-PH), as compared with non-droughted (control) and postharvest (RDI-PH) treatments. Significant higher concentration of soluble solids and accentuated red colour were observed at harvest in fruit from RDI treatments when compared to control fruit. Accordingly, fruit submitted to RDI during development had greater ethylene production when detached. Ethylene production by RDI-PH fruit did not change, but their quality did in terms of increased soluble solids concentration and improved skin colour, similar to the RDI-SII treatment. Buendía et al. (2008) investigated the influence of regulated deficit irrigation (RDI) on the content of vitamin C, phenolic compounds and carotenoids of peaches. Fruit were harvested from five-year-old early ripening peach trees (cv. "Flordastar", grafted on GF677 rootstock) in Santomera, Murcia, Spain. Two irrigation strategies, fully irrigated (FI) and RDI, were compared at two levels of thinning, commercial and half of the commercial crop load. RDI caused fruit peel stress lowering the content of vitamin C and carotenoids, while increasing the phenolic content, mainly anthocyanins and procyanidins. Fruit weight was the only quality index influenced by the crop load as it increased in FI fruit at low crop load. According to the authors, the decrease in the antioxidant constituents could be due to a higher sunlight exposure of fruit collected from RDI trees as a result of a low vegetative growth of those trees. The increase in the anthocyanin content could be also explained as a response mechanism of peaches against UV irradiation, to mitigate the photooxidative injury of plant tissue.

5.7 Plum

The effects of RDI (RDI) and crop load on Japanese plum (*Prunus salicina*) cv. Black-Gold were investigated by Intrigliolo & Castel (2010). RDI was applied during phase II of fruit growth and postharvest and compared to a control irrigation treatment (full crop evapotranspiration). Plants from each irrigation treatment were thinned to reach a commercial crop load (described as medium) and to approximately 40% less than the commercial practice (described as low). RDI strategy increased the efficiency of water usage, with 30% of water savings, having minimal effect on crop yield and fruit growth. Economic return, calculated from fruit weight distribution by commercial categories, was more affected by RDI than yield. The combination of medium crop load and RDI shifted fruit mass distribution towards the low value categories. This leads to similar or even higher economic returns in the RDI treatment with low crop level than with the medium one. In addition, since both, low crop level and RDI, increased fruit total soluble solids (TSS), fruit produced under RDI and low crop levels had the highest values of TSS. Deficit irrigation improved fruit composition and, in the short-term, increased tree water use efficiency.

5.8 Citrus

Recently García-Tejero (2010) determined the postharvest fruit quality of oranges (*Citrus sinensis* L. Osbeck, cv. Salustiano) exposed to RDI in commercial orchards at the semi-arid region of Andalusia- Spain, in the years 2005, 2006 and 2007. The experiment was composed by four different treatments: Control (irrigation replacing 100% of Evapotranspiration, ET_c), low deficit irrigation (75% of ET_c), moderate deficit irrigation (65% of ET_c) and severe deficit irrigation (50% of ET_c). As a result, fruit quality parameters as TSS and TA increased in all stressed treatments resulting better organoleptic parameters. Significant fruit size reduction was observed in the year 2005 only. However, reduction in fruit size is often associated with an increase in fruit number (Treeby et al., 2007). In the years 2006 and 2007, a significant yield loss was not observed. The low deficit irrigation treatment proportionate a water saving of 93 mm, which consists in lower irrigation costs that might compensate the slightly yield losses in fruit production. In a similar study Velez et al. (2007), using the "maximum daily trunk shrinkage" method, which is used as an indicator of water stress, was able to obtain water savings reaching 18% without significant decreases in average fruit yield, weight and number. In addition, fruit submitted to deficit irrigation had significantly higher TSS and similar TA. According to the authors, a higher accumulation of sugars is a result of an active response to water deficit.

5.9 Grape

Working with grape (*Vitis vinifera* L. cv. Rizamat), table type, Du et al. (2008) performed a two year experiment to investigate the alternate partial root-zone drip irrigation on fruit quality, yield and water use efficiency in the arid region of northwest China. The treatments consisted of a control (both sides of the root zone irrigated); an alternate drip irrigation treatment (both sides of the root zone irrigated, one at a time with half the water used in control) and a fix drip irrigation (only one side of the root zone was irrigated with half the water used in control). As a result, in alternate drip irrigation condition, the photosynthetic rate was similar to control whilst the transpiration rate kept in the same level. The leaf water use efficiency increased indicating that the plants submitted to water stress needed less water to keep hydrated. In stressing condition significant yield losses were not perceived, in addition a higher percentage of edible grapes were achieved, which consequently improve product price. A plausible explanation for such performance relays on changes in balance, between vegetative and reproductive growth, of plants facing limited water supplies, resulting in a higher flow of photoassimilates to berries. Stressed grapes presented higher concentrations of both, ascorbic acid and total soluble solids, and lower titrated acidity, culminating in healthier and sweeter grapes. In a similar study (Santos et al., 2007) compared the effects of partial root zone drying irrigation system (50% ET_c irrigating one side at a time) with the conventional deficit irrigation system (50% ET_c applied on both sides), full irrigation system (100% ET_c applied on both sides) and non irrigated vines. The one-year experiment was performed in mature "Moscatel" grapevines (*Vitis vinifera* L.) in south Portugal. Plants submitted to partial drying regime showed a decreased vegetative growth, expressed by the smaller values of leaf layer number, percentage of water shoots, shoot weight, pruning weight and total leaf area. In synthesis, plants had a higher control over vegetative growth in face of a stressing condition. The partial drying regime and the conventional deficit irrigation treatments, despite receiving the same quantities of water during the experiment, led to different plant responses. The author argue that in the partial drying regime, roots in the dry side produce chemical signals which restrict plant growth. As a consequence, berry temperature increases in response to augmented solar incidence. In response, the concentrations of anthocyanins, phenols and glycosyl-glucose were higher in water stressed plants culminating in higher quality grapes. Yield loss was significant only in non-irrigated treatment.

5.10 Watermelon

Proietti et al. (2008) determined quality parameters of mini-watermelon (*Citrullus lanatus*) cv. Ingrid, ungrafted or grafted onto a squash hybrid rootstock, and grown under different irrigation regimes: 1.0, 0.75, and 0.5 of evapotranspiration (ET_c) rates. The highest fruit yields were observed at 1.0 ET_c and at 0.75 ET_c when compared to 0.5 ET_c, in grafted plants. Grafting of mini-watermelon under irrigation deficit did not modify crop response to water availability, but increased productivity and induced small positive changes in plant quality and nutritional value.

6. Water stress effects on vegetable quality

6.1 Potato

Bejarano et al. (2000) evaluated the content of glycoalkaloids (GAs, α-solanine and α-chaconine) in drought-tolerant potato (*Solanum tuberosum* L.) grown in the Bolivian highlands under drought stress. Under drought stress conditions GAs concentration increased an average of 43% and 50% in the improved and control cultivars, respectively, but never above the recommended food safety limit (200 $mg.kg^{-1}$ fresh tubers). GAs are natural toxins synthesised by plants of the *Solanaceae* family and are believed to be associated with resistance to certain insects. At least 95% of total GAs are in the form of α-solanine and α-chaconine. Both compounds are heat-stable and therefore are not destroyed by common cooking processes such as boiling or frying (Friedman, 2006).

6.2 Tomato

Sánchez-Rodríguez et al. (2011) obtained insightful results regarding the phenolic metabolism in response to water stress. Phenolic compounds are the most abundant type of secondary metabolites in plants, being frequently associated with beneficial effects in human health (Dixon & Paiva, 1995; Hooper & Cassidy, 2006). Tomato plants (*Solanum lycopersicum* L.), cultivars Kosaco, Josefina, Katalina, Salomé and Zarina, were submitted to water stress conditions. The experimental design consisted of two watering treatments: Control (100% of field capacity) and moderate water stress (50% of field capacity). Zarina, characterized as a drought tolerant cultivar, presented the best responses. All the cultivars had significant decreases in phenolic compounds when submitted to water stress, except Zarina, which showed an increase, particularly in flavonoids, when cultivated under moderate water stress. This increase was concomitant with a 33% higher DAHPS (EC 4.1.2.15) activity, the key enzyme controlling the carbon flow towards phenolic metabolism, and equivalent increases were observed for other related enzymes. The enzymes that degrade phenols, PPO (EC 1.10.3.2) and GPX (EC 1.11.1.7), declined in activity, 30% and 47%, respectively. Thipyapong et al. (2004) propose that in tomato plants a decrease in PPO activity lowers the H_2O_2 concentration, reducing lipid peroxidation and improving resistance against water stress.

6.3 Lettuce

Coelho et al. (2005) investigated the yield and bioactive amine content of American lettuce (*Lactuca sativa* cv. Lucy Brown) grown under greenhouse conditions and drip irrigation. Spermidine was the prevalent amine, followed by putrescine, cadaverine and agmatine. The contents of every amine, except agmatine, increased with water stress but not capable of negatively affecting the sensory quality of the lettuce. Changes in plant polyamine metabolism occur in response to a variety of abiotic stresses, however, the physiological significance of increased polyamine levels in abiotic stress responses is still unclear (Gill & Tuteja, 2010).

6.4 Cabbage

Supplying irrigation to achieve maximum cabbage (*Brassica oleracea* L. Capitata group) yield will also optimize sensory quality by minimizing the compounds responsible for pungency. However, glucosinolate concentration will be reduced (Radovich et al., 2005). Glucosinolates are amino acid-derived secondary metabolites that may exhibit antibiotic, anti-carcinogenic and organoleptic activity after hydrolysis. Sinigrin and progoitrin are the most important compounds with regard to flavour, since they are the primary determinants of pungency, bitterness and sulphurous aroma in cabbage (Buttery et al., 1976; Fahey et al., 2001; Talalay & Fahey, 2001).

6.5 Mustard

Similar to cabbage, mustard is a glucosinolate containing plant. This class of secondary metabolites are found almost exclusively in plants of the order *Brassicales*, including horticulturally important crop plants of the *Brassicaceae* family (Fahey et al., 2001). In order to evaluate the glucosinolate metabolism response to environmental stresses, Schreiner et al. (2009) submitted the Ethiopian mustard (*Brassica carinata*), lines Holeta-1 and 37-A, to water restriction treatments. Control plants were maintained at 80% of field capacity during the total growing period. Water restriction consisted of systematic decreases in soil water content (40, 23, 17, and 15% of field capacity in the 6-8, 11-12, 13-14, and 15-16 leaf stages, respectively). The most abundant glucosinolate found in both lines was 2-propenyl glucosinolate, followed by 3-indolyl methyl glucosinolate. The concentration of these two compounds remained constant in control plants from 6-8 to 15-16 leaf stage. In contrast, the stressing condition led plants to a distinct increase (80-120%) of 2-propenyl glucosinolate and 3-indolyl methyl glucosinolate in the leaves of both lines. The increases in leaf glucosinolates was inversely correlated to soil water content, in both lines, leading to severe yield losses.

6.6 Broccoli

We have demonstrated that low soil water content (0.40 MPa of soil water tension) during broccoli growth leads to leaf size reduction, without affecting weight or yield, and contributes to the maintenance of green colour, possibly due to induced cytokinin synthesis (Zaicovski et al., 2008). Cytokinins, which are considered senescence inhibitors acting as ethylene antagonists and protectors of membranes, mitochondria, and plastid metabolism are also known to be induced in response to stresses (Xu & Huang, 2009). However, severe stress (−0.6 MPa of soil water pressure) leads to negative effects on broccoli yield and morphology as observed by Wurr et al. (2002). In another study we cultivated broccoli under low (0.40 MPa) and normal (0.04 MPa, equivalent to field capacity) soil water content, stored it under low (1 °C) and room (23 °C) temperature, and assessed for changes in colour, bioactive compounds, and antioxidant activity. We concluded that low soil water content during plant growth and postharvest cold storage are the conditions that, combined, give the best preservation of colour, antioxidant activity, and l-ascorbic acid and 5-methyl-tetrahydrofolate contents (Cogo et al., 2011).

6.7 Eggplant

Kirnak et al. (2002) investigated the effects of deficit irrigation on fruit yield and quality in eggplant (*Solanum melongena* L.) cv. Pala. The experiment consisted of the following treatments: (1) well-watered, receiving 100% of plant evaporation on a daily basis (C); (2) water-stressed, receiving 90% of plant evaporation at 4-day intervals (WS1); (3) water-stressed, receiving 80% of plant evaporation at 8-day intervals (WS2); and (4) water-stressed, receiving 70% of plant evaporation at 12-day intervals (WS3). The highest yield, the largest and the heaviest fruit was observed in well-watered treatment (C). WS1 did not significantly affect fruit yield or fruit size but produced fruit slightly lighter, whilst nutrient and chlorophyll concentrations in leaves were the same as in C. WS1 presented higher soluble dry matter (SDM) than C. The WS2 and WS3 treatments caused reductions in most parameters, except SDM concentrations in the fruit, compared to C treatment. WS2 and WS3 reduced marketable yield by 12% and 28.6%, respectively, compared with C. The highest total water use efficiency and irrigation water use efficiency were in WS2, resulting in a 20.4% water saving compared with C.

7. Attributes commonly evaluated in fruit and vegetable under water stress

It became noticeable that in water stress related studies, vegetables are evaluated mainly in terms of health promoting compounds while fruit are evaluated mainly in terms of sensorial related attributes (Table 1)

Plant specie	Cultivar/Var.	Healthy related metabolites	Quality related properties	Yield	Reference
Kiwifruit	Hayward	-	TSS[a] (↑) Firmness (=)	(↓)	Miller et al. (1998)
	Hayward	-	Firmness (↑)	(=)	Reid et al. (1996)
Strawberry	Elsanta	-	Sweetness (↑)	(=)	Bordonaba & Terry (2010)
Apricot	Búlida	-	TSS (↑) TA[b] (↑) Colour (↑) PHDQ[c] (↑)	(=)	Perez-Pastor (2007)
Peach	-		TSS (↑) Colour (↑) TA (↑)	(=)	Gelly et al. (2003) Gelly et al. (2004)
Pear	Conference		TSS (↑) Firmness (↑) PHDQ[c] (↑)	(↓)	Lopez et al. (2011)
Plum	Black-Gold		TSS (↑	(=)	
Orange	Salustiano	-	TSS (↑) TA (↑)	(=)	García-Tejero (2010)
	Clementina de Nules	-	TSS (↑) TA (↑)	(=)	Velez et al. (2007)
Table grape	Rizamat	Ascorbic. acid (↑)	TSS (↑) TA (↓)	(=)	Du et al. (2008)
Grapevine	Moscatel	Anthocianins (↑) Phenols (↑)	Colour (↑)	(=)	Santos et al. (2007)
Tomato	Zarina	Flavonoids (↑)	-	-	Sánchez-Rodríguez et al. (2011)
Cabbage	Holeta-1 and 37-A	Glucosinolates (↑)	-	(↓)	Schreiner et al. (2009)
Broccoli	Green star	Ascorbic. Acid (↑) Folates (↑)	PHDQ (↑)	(=)	(Cogo et al., 2011)

[a] TSS: Total soluble solids.
[b] TA: Titratable Acidity.
[c] PHDQ: Post-harvest desirable qualities.

Table 1. Effect of limited water supply during the plant production cycle on postharvest quality of fruit and vegetables. The most relevant reference is cited.

8. Conclusion

Environmental sustainability is a current issue in global agenda. This concept comprises, among other concerns, the efficient use of water. Climate change is a reality and the pressure over water reserves will increase in next years, as well over high water consuming activities. Agricultural practices are among the biggest water consuming activities, considering that, alternatives to reduce water use in agricultural practices is of special interest in the present moment. In this review we exposed successful experiences with the intent to reduce water use with minimal losses in yield and quality. Irrigation practices often aims at total replacement of culture evapotranspiration in order to obtain maximum yield. In many occasions experiments demonstrate that is possible to reduce water use without significant losses in yield. In addition, the increases in health/quality related compounds and postharvest preservation are evident in response to environmental stress. The use of regulated stress (water stress, salinity, heat, cold, UV radiation) is a feasibly strategy to enhance accumulation of health promoting compounds in food. Another interesting perspective is the improvement of plant resistance against biotic stresses (e.g. pests and diseases) when submitted to controlled abiotic stresses, as was scientifically demonstrated in experiments presented in this review. However, more studies must be performed to determine the effects of different water stresses on different edible plants.

9. Acknowledgment

The authors acknowledge CAPES and CNPq from Brazilian government for the financial support.

10. References

Bejarano, L., Mignolet, E., Devaux, A., Espinola, N., Carrasco, E., Larondelle, Y. (2000). Glycoalkaloids in potato tubers: the effect of variety and drought stress on the α-solanine and α-chaconine contents of potatoes. *Journal of the Science of Food and Agriculture,* Vol. 80, No. 14, pp. 2096-2100, ISSN 1097-0010

Bordonaba, J.G. & Terry, L.A. (2010). Manipulating the taste-related composition of strawberry fruits (Fragaria x ananassa) from different cultivars using deficit irrigation. *Food Chemistry,* Vol. 122, No. 4, (Oct), pp. 1020-1026, ISSN 0308-8146

Buendía, B., Allende, A., Nicolás, E., Alarcón, J.J., Gil, M.I. (2008). Effect of Regulated Deficit Irrigation and Crop Load on the Antioxidant Compounds of Peaches. *Journal of Agricultural and Food Chemistry,* Vol. 56, No. 10, pp. 3601-3608, ISSN 0021-8561

Buttery, R.G., Guadagni, D.G., Ling, L.C., Seifert, R.M., Lipton, W. (1976). Additional volatile components of cabbage, broccoli, and cauliflower. *Journal of Agricultural and Food Chemistry,* Vol. 24, No. 4, (1976), pp. 829-832, ISSN 0021-8561

Chu, H.Y., Wegel, E., Osbourn, A. (2011). From hormones to secondary metabolism: the emergence of metabolic gene clusters in plants. *Plant Journal,* Vol. 66, No. 1, (Apr), pp. 66-79, ISSN 0960-7412

Coelho, A.F.S., Gomes, E.P., Sousa, A.D., Gloria, M.B. (2005). Effect of irrigation level on yield and bioactive amine content of American lettuce. *Journal of the Science of Food and Agriculture,* Vol. 85, No. 6, (Apr), pp. 1026-1032, ISSN 0022-5142

Cogo, S.L.P., Chaves, F.C., Schirmer, M.A., Zambiazi, R.C., Nora, L., Silva, J.A., Rombaldi, C.V. (2011). Low soil water content during growth contributes to preservation of green colour and bioactive compounds of cold-stored broccoli (Brassica oleraceae L.) florets. *Postharvest Biology and Technology*, Vol. 60, No. 2, pp. 158-163, ISSN 0925-5214

Costa, J.M., Ortuno, M.F., Chaves, M.M. (2007). Deficit irrigation as a strategy to save water: Physiology and potential application to horticulture. *Journal of Integrative Plant Biology*, Vol. 49, No. 10, (Oct), pp. 1421-1434, ISSN 1672-9072

D'Auria, J.C. & Gershenzon, J. (2005). The secondary metabolism of Arabidopsis thaliana: growing like a weed. *Current Opinion In Plant Biology*, Vol. 8, No. 3, (Jun), pp. 308-316

Dixon, R.A. & Paiva, N.L. (1995). Stress-Induced Phenylpropanoid Metabolism. *Plant Cell*, Vol. 7, No. 7, (Jul), pp. 1085-1097

Du, T., Kang, S., Zhang, J., Li, F., Yan, B. (2008). Water use efficiency and fruit quality of table grape under alternate partial root-zone drip irrigation. *Agricultural Water Management*, Vol. 95, No. 6, pp. 659-668, ISSN 0378-3774

Fahey, J.W., Zalcmann, A.T., Talalay, P. (2001). The chemical diversity and distribution of glucosinolates and isothiocyanates among plants. *Phytochemistry*, Vol. 56, No. 1, (Jan), pp. 5-51, ISSN 0031-9422

Friedman, M. (2006). Potato glycoalkaloids and metabolites: Roles in the plant and in the diet. *Journal of Agricultural and Food Chemistry*, Vol. 54, No. 23, (Nov 15), pp. 8655-8681

García-Tejero, I., Jimenez-Bocanegra, J.A., Martinez, G., Romero, R., Duran-Zuazo, V.H., Muriel-Fernandez, J. (2010). Positive impact of regulated deficit irrigation on yield and fruit quality in a commercial citrus orchard Citrus sinensis (L.) Osbeck, cv. salustiano. *Agricultural Water Management*, Vol. 97, No. 5, (May), pp. 614-622, ISSN 0378-3774

Gill, S.S. & Tuteja, N. (2010). Polyamines and abiotic stress tolerance in plants. *Plant Signaling & Behavior*, Vol. 5, No. 1, pp. 26-33

Godfray, H.C.J., Crute, I.R., Haddad, L., Lawrence, D., Muir, J.F., Nisbett, N., Pretty, J., Robinson, S., Toulmin, C., Whiteley, R. (2010). The future of the global food system. *Philosophical Transactions of the Royal Society B: Biological Sciences*, Vol. 365, No. 1554, (September 27, 2010), pp. 2769-2777

Hooper, L. & Cassidy, A. (2006). A review of the health care potential of bioactive compounds. *Journal of the Science of Food and Agriculture*, Vol. 86, No. 12, pp. 1805-1813, ISSN 1097-0010

Intrigliolo, D.S. & Castel, J.R. (2010). Response of plum trees to deficit irrigation under two crop levels: tree growth, yield and fruit quality. *Irrigation Science*, Vol. 28, No. 6, (Sep), pp. 525-534, ISSN 0342-7188

Kirnak, H., Tas, I., Kaya, C., Higgs, D. (2002). Effects of deficit irrigation on growth, yield, and fruit quality of eggplant under semi-arid conditions. *Australian Journal of Agricultural Research*, Vol. 53, No. 12, (2002), pp. 1367-1373, ISSN 0004-9409

Krouk, G., Ruffel, S., Gutierrez, R.A., Gojon, A., Crawford, N.M., Coruzzil, G.M., Lacombe, B. (2011). A framework integrating plant growth with hormones and nutrients. *Trends In Plant Science*, Vol. 16, No. 4, (Apr), pp. 178-182, ISSN 1360-1385

Lopez, G., Larrigaudiere, C., Girona, J., Behboudian, M.H., Marsal, J. (2011). Fruit thinning in 'Conference' pear grown under deficit irrigation: Implications for fruit quality at harvest and after cold storage. *Scientia Horticulturae*, Vol. 129, No. 1, (May), pp. 64-70, ISSN 0304-4238

Miller, S.A., Smith, G.S., Boldingh, H.L., Johansson, A. (1998). Effects of water stress on fruit quality attributes of kiwifruit. *Annals of Botany*, Vol. 81, No. 1, pp. 73-81, ISSN 0305-7364

Mpelasoka, B.S., Behboudian, M.H., Dixon, J., Neal, S.M., Caspari, H.W. (2000). Improvement of fruit quality and storage potential of 'Braeburn' apple through deficit irrigation.

Journal of Horticultural Science & Biotechnology, Vol. 75, No. 5, (Sep), pp. 615-621, ISSN 0022-1589

Mpelasoka, B.S., Behboudian, M.H., Ganesh, S. (2001a). Fruit quality attributes and their interrelationships of 'Braeburn' apple in response to deficit irrigation and to crop load. *Gartenbauwissenschaft*, Vol. 66, No. 5, (Sep-Oct), pp. 247-253, ISSN 0016-478X

Mpelasoka, B.S., Behboudian, M.H., Mills, T.M. (2001b). Effects of deficit irrigation on fruit maturity and quality of 'Braeburn' apple. *Scientia Horticulturae*, Vol. 90, No. 3-4, (Nov), pp. 279-290, ISSN 0304-4238

Mpelasoka, B.S., Behboudian, M.H., Mills, T.M. (2001c). Water relations, photosynthesis, growth, yield and fruit size of 'Braeburn' apple: Responses to deficit irrigation and to crop load. *Journal of Horticultural Science & Biotechnology*, Vol. 76, No. 2, (Mar), pp. 150-156, ISSN 0022-1589

Osbourn, A. (2010). Secondary metabolic gene clusters: evolutionary toolkits for chemical innovation. *Trends in Genetics*, Vol. 26, No. 10, (Oct), pp. 449-457, ISSN 0168-9525

Peleg, Z. & Blumwald, E. (2011). Hormone balance and abiotic stress tolerance in crop plants. *Current Opinion In Plant Biology*, Vol. 14, No. 3, pp. 290-295, ISSN 1369-5266

Perez-Pastor, A., Ruiz-Sanchez, M.C., Martinez, J.A., Nortes, P.A., Artes, F., Domingo, R. (2007). Effect of deficit irrigation on apricot fruit quality at harvest and during storage. *Journal of the Science of Food and Agriculture*, Vol. 87, No. 13, (Oct), pp. 2409-2415, ISSN 0022-5142

Perez-Sarmiento, F., Alcobendas, R., Mounzer, O., Alarcon, J., Nicolas, E. (2010). Effects of regulated deficit irrigation on physiology and fruit quality in apricot trees. *Spanish Journal of Agricultural Research*, Vol. 8, (2010), pp. S86-S94, ISSN 1695-971X

Pospíšilová, J. & Dodd, I.C. (2005). Role of plant growth regulators in stomatal limitation to photosynthesis during water stress, In *Handbook of Photosynthesis*, M. Pessarakli, (Ed.), 811-825, CRC Press, ISBN 0824758390, New York, USA

Proietti, S., Rouphael, Y., Colla, G., Cardarelli, M., De Agazio, M., Zacchini, M., Rea, E., Moscatellol, S., Battistelli, A. (2008). Fruit quality of mini-watermelon as affected by grafting and irrigation regimes. *Journal of the Science of Food and Agriculture*, Vol. 88, No. 6, (Apr 30), pp. 1107-1114, ISSN 0022-5142

Radovich, T.J.K., Kleinhenz, M.D., Streeter, J.G. (2005). Irrigation timing relative to head development influences yield components, sugar levels, and glucosinolate concentrations in cabbage. *Journal of the American Society for Horticultural Science*, Vol. 130, No. 6, (Nov), pp. 943-949, ISSN 0003-1062

Reid, J.B., Brash, D.W., Sorensen, I.B., Bycroft, B. (1996). Improvement in kiwifruit storage life caused by withholding early-season irrigation. *New Zealand Journal of Crop and Horticultural Science*, Vol. 24, No. 1, (Mar), pp. 21-28, ISSN 0114-0671

Sánchez-Rodríguez, E., Moreno, D.A., Ferreres, F., Rubio-Wilhelmi, M.d.M., Ruiz, J.M. (2011). Differential responses of five cherry tomato varieties to water stress: Changes on phenolic metabolites and related enzymes. *Phytochemistry*, Vol. 72, No. 8, pp. 723-729, ISSN 0031-9422

Santner, A. & Estelle, M. (2009). Recent advances and emerging trends in plant hormone signalling. *Nature*, Vol. 459, No. 7250, (June 2009), pp. 1071-1078, ISSN 0028-0836

Santos, T.P.d., Lopes, C.M., Rodrigues, M.L., Souza, C.R.d., Ricardo-da-Silva, J.M., Maroco, J.P., Pereira, J.S., Chaves, M.M. (2007). Effects of deficit irrigation strategies on cluster microclimate for improving fruit composition of Moscatel field-grown grapevines. *Scientia Horticulturae*, Vol. 112, No. 3, pp. 321-330, ISSN 0304-4238

Schreiner, M., Beyene, B., Krumbein, A., Stuetzel, H. (2009). Ontogenetic Changes of 2-Propenyl and 3-Indolylmethyl Glucosinolates in Brassica carinata Leaves as Affected by Water Supply. *Journal of Agricultural and Food Chemistry*, Vol. 57, No. 16, (Aug 26), pp. 7259-7263, ISSN 0021-8561

Stefanelli, D., Goodwin, I., Jones, R. (2010). Minimal nitrogen and water use in horticulture: Effects on quality and content of selected nutrients. *Food Research International*, Vol. 43, No. 7, (Aug), pp. 1833-1843, ISSN 0963-9969

Stoll, M., Loveys, B., Dry, P. (2000). Hormonal changes induced by partial rootzone drying of irrigated grapevine. *Journal Of Experimental Botany*, Vol. 51, No. 350, (Sep), pp. 1627-1634, ISSN 0022-0957

Talalay, P. & Fahey, J.W. (2001). Phytochemicals from cruciferous plants protect against cancer by modulating carcinogen metabolism. *Journal Of Nutrition*, Vol. 131, No. 11, (Nov), pp. 3027S-3033S, ISSN 0022-3166

Terry, L.A., Chope, G.A., Bordonaba, J.G. (2007). Effect of water deficit irrigation and inoculation with Botrytis cinerea on strawberry (Fragaria x ananassa) fruit quality. *Journal of Agricultural and Food Chemistry*, Vol. 55, No. 26, (Dec 26), pp. 10812-10819, ISSN 0021-8561

Thipyapong, P., Melkonian, J., Wolfe, D.W., Steffens, J.C. (2004). Suppression of polyphenol oxidases increases stress tolerance in tomato. *Plant Science*, Vol. 167, No. 4, (Oct), pp. 693-703, ISSN 0168-9452

Treeby, M.T., Henriod, R.E., Bevington, K.B., Milne, D.J., Storey, R. (2007). Irrigation management and rootstock effects on navel orange Citrus sinensis (L.) Osbeck fruit quality. *Agricultural Water Management*, Vol. 91, No. 1-3, (Jul 16), pp. 24-32, ISSN 0378-3774

Velez, J.E., Intrigliolo, D.S., Castel, J.R. (2007). Scheduling deficit irrigation of citrus trees with maximum daily trunk shrinkage. *Agricultural Water Management*, Vol. 90, No. 3, (Jun), pp. 197-204, ISSN 0378-3774

Wang, Y. & Frei, M. (2011). Stressed food - The impact of abiotic environmental stresses on crop quality. *Agriculture, Ecosystems & Environment*, Vol. 141, No. 3-4, pp. 271-286, ISSN 0167-8809

Wilkinson, S. & Davies, W.J. (2010). Drought, ozone, ABA and ethylene: new insights from cell to plant to community. *Plant Cell and Environment*, Vol. 33, No. 4, (April 2010), pp. 510-525, ISSN 0140-7791

Wink, M. (1999). Introduction: biochemistry, role and biotechnology of secondary metabolites, In *Biochemistry, role and biotechnology of secondary metabolites*, M. Wink, (Ed.), 1-16, Sheffield Academic Press and CRC Press, Annual Plant Reviews, ISBN 978-1-4051-8528-8, Sheffield, UK

Wurr, D.C.E., Hambidge, A.J., Fellows, J.R., Lynn, J.R., Pink, D.A.C. (2002). The influence of water stress during crop growth on the postharvest quality of broccoli. *Postharvest Biology and Technology*, Vol. 25, No. 2, (Jun), pp. 193-198

Xu, Y. & Huang, B.R. (2009). Effects of foliar-applied ethylene inhibitor and synthetic cytokinin on creeping bentgrass to enhance heat tolerance. *Crop Science*, Vol. 49, No. 5, (Sep-Oct), pp. 1876-1884, ISSN 0011-183X

Yamaguchi-Shinozaki, K. & Shinozaki, K. (2006). Transcriptional regulatory networks in cellular responses and tolerance to dehydration and cold stresses, In *Annual Review Of Plant Biology*, D.P. Delmer, (Ed.), 781-803, Annual Reviews, ISBN 1543-5008, New York, USA

Zaicovski, C.B., Zimmerman, T., Nora, L., Nora, F.R., Silva, J.A., Rombaldi, C.V. (2008). Water stress increases cytokinin biosynthesis and delays postharvest yellowing of broccoli florets. *Postharvest Biology and Technology*, Vol. 49, No. 3, (Sep), pp. 436-439, ISSN 0925-5214

4

Water Stress and Afforestation: A Contribution to Ameliorate Forest Seedling Performance During the Establishment

A. B. Guarnaschelli[1], A. M. Garau[1] and J. H. Lemcoff[2]
[1]*Dep. of Vegetal Production. Faculty of Agronomy, University of Buenos Aires, Buenos Aires*
[2]*Israel Gene Bank, The Volcani Center, Agricultural Research Organization, Bet Dagan*
[1]*Argentina*
[2]*Israel*

1. Introduction

For hundreds of years, forest ecosystems have been supplying human needs with timber and non-timber products such as oils, resins, tannins and other goods like wood or medicine. Beyond material goods, forests also provide a range of other relevant environmental benefits. The accelerating loss of forests represents one of the major environmental challenges. Intensive commercial logging focused mainly on timber products cause unfortunately degradation of extensive areas and, at the same time, the conversion of forest land for commercial agriculture, subsistence farming and logging for fuel wood are considered the main factors of deforestation. Both degradation and deforestation lead to a considerable reduction in the world forest resources. In the last decades worldwide concern about the necessity to protect native forests emerged and a shift in silviculture occurred, changing into a broader concern where environmental values and diversified interests are becoming more important (Food and Agriculture Organization of the United Nations [FAO], 2009, 2010).

Historically, an increase in economic growth and population has been the main force fuelling global wood consumption. The expected increase in world population in the next years and the rise in the standards of living will increase wood demand. As this additional wood demand cannot come from further increases in the harvest of natural forests, it must come from planted forests. Reforestation and afforestation seem to be effective alternatives to increase forest production land, thus covering those timber needs and at the same time contributing to reduce timber extraction in many natural forests (FAO, 2009, 2010; Paquette & Messier, 2010; United Nations Framework Convention on Climate Change [UNCCC].

Commercial plantations can provide large quantities of timber to keep up with the increasing demand of forest products (sawmills, pulp and particle board industries mainly), and can also be used to provide environmental benefits. It is important to consider that the potential supply of wood arising from forest plantations will depend on the capacity to maintain the rate of new planting and forest productivity. Forest plantations have greater structural diversity than agricultural crops (Lindenmayer et al., 2003; Viglizzo et al., 2011); they can reduce wind and soil erosion (Kort et al., 1998) and are particularly important for carbon sequestration, contributing to climate change mitigation (Laclau, 2005; Vitousek, 1991). Apart from these plantations benefits and according to environmental conditions of the site, some trade-offs need to be considered (Jackson et al., 2005; Jobbágy & Jackson, 2004).

Forest productivity is determined by genetic potential and is closely linked with environmental resources. Both genetic and environmental factors control plant physiological processes regulating the biomass production of forest stands. The growth depends on solar energy captured by the canopy and water and nutrients provided by the soil for converting CO_2 to photoassimilates (Kozlowski & Pallardy, 1997). Unfortunately, the environment imposes many resource limitations to growth; therefore, forest stands can rarely achieve their potential productivity for extended periods. Indeed, over a wide range of climatic and soil conditions there is a large variation in forest net primary productivity around the world (Gowers et al., 1992; Kozlowsky, 2002; Perry et al., 2008; Waring & Schlesinger, 1985).

Plantation performance is commonly hindered by both biotic and abiotic factors. The rapid growth rates of the species used in commercial plantations (like *Pinus, Eucalyptus, Populus*) are based on their high water demands (Braatne et al., 1992; Florence, 1996; Monclus et al., 2006; Whitehead & Beadle, 2004). But the availability of water, one of the main factors related to plant growth, varies both seasonally and from year to year, regulating forest productivity. During the rotation cycle trees are often subjected to periods of soil and atmospheric water deficit but the effects on plant performance are more serious during the establishment period because they may not only reduce seedlings growth but also compromise their survival. Besides climatic factors, soil water availability is determined by soil factors, stand density and levels of understory competition.

It is crucial to understand the physiological processes that determine growth and survival during the establishment and the response of tree seedling to environmental stimuli because they deeply influence future productivity. Nowadays, all of these issues face new challenges which are expected to come out in the actual context of climate change. In the future an increase in the frequency of drought events is predicted even outside semi-arid and arid areas (Intergovernmental Panel on Climate Change [IPCC], 2007). Under stressful and/or unexpected weather conditions a careful silvicultural planning is essential to ensure success. Silvicultural decisions including the selection of high quality plant material and practices applied during the establishment are tightly connected with plantation performance (Smith et al., 1997).

In this chapter, among the multiple factors that influence growth during the forest establishment, we chiefly discuss the importance of water supply on tree growth and how water stress compromises plant performance. We take into account the interactions between water deficiencies and other abiotic and biotic factors that in fact modulate survival and plant growth. We analyze the potential mechanisms that tree seedlings exhibit to overcome serious stressful conditions during the establishment phase and the effects of nursery techniques applied to produce good quality stock, particularly those imposed to acclimate seedlings to water stress before outplanting. We include results of several studies in species of great importance in afforestation programs.

2. Water stress on tree function and growth

Water is an essential resource for plant function and growth. Most of physiological processes are directly or indirectly regulated by tissue water content. Water is solvent of gases, salts and solutes within and among cells and from organ to organ; it is required to provide a substrate and a medium for biochemical reactions and for the transport of mineral ions, as well as to maintain cell turgor for cell enlargement, stomatal opening and the maintenance of the form of young leaves and slightly lignified structures. Only a small fraction of water that passes through the plant and is lost by transpiration during its whole life remains in plant tissue. When leaving the plant through transpiration, chiefly from the leaves, water has an additional cooling effect that allows a plant to maintain a temperature suitable for metabolic reactions and life (Kozloswki & Pallardy, 1997). Zhaner (1968) reviewed the importance of an adequate water supply for woody plants. There is a very high correlation between the amount of wood production and the available soil moisture. As a matter of fact, a large proportion of the variation in diameter growth can be attributed to variation in the availability of water. Although it is one of the most abundant substances in the environment, there is a great temporal and spatial variation in water supply for plants. Water deficit is mainly caused by droughts that refer to periods in which rainfall fails to keep up with potential evaporation, generating the exhaustion of soil water content. Insufficient water supply to plant leaves can also arise in response to other circumstances, either abiotic or biotic. For instance when, despite the existence of adequate water in the soil, there is large air humidity deficit (Hirasawa & Hsiao, 1999; Levin et al., 2009). Regardless of the amount of precipitation, water may not be available because it is frozen. Several soil properties, like soil texture, salinity and/or hydraulic conductivity reduce water supply. Plants can have difficulties in obtaining water from the soil if salts are present in the root zone (osmotic effect). Generally, the need of energy to allow water uptake is greater when the soil solution is saline than when it is not. Competition of neighboring vegetation is also another cause of water depletion to the tree crop (Nielsen & Orcutt, 1996; Passioura, 1996).

When soil water availability is scarce, the magnitude of plant transpiration losses is usually greater than root absorption and therefore plants are prone to experience water stress, which is mainly evidenced through a lowering in the plant relative water content and/or in the water potential, decreasing tissue turgor and compromising cellular expansion. Stomatal closure has been identified also as an early response to water deficit that limits the photosynthesis rate leading to a limitation in plant carbon fixation. It is still under debate whether stomatal closure is triggered by chemical signals synthesized in dehydrating roots or by changes in plant hydraulic conductivity (Brodribb, 2009; Brodribb & McAdam, 2011; Cochard et al., 2002; Davies et al., 1986). The depression in gas exchange also reduces water loss leading generally to higher water use efficiency, but the lowering in the photosynthesis rate as well as the decrease in plant leaf area due to leaf desiccation and shedding causes a reduction in plant growth (Chaves et al., 2003; Hsiao, 1973; Kozlowski, 1982; Kramer & Boyer, 1995). Excess of radiation constitutes another concomitant type of stress under water limited conditions. As water stress reduces the rate of photosynthesis, the absorption of light exceeds its utilization. If plants lack the ability to dissipate that excess of energy through external and internal mechanisms of protection, they undergo photoinhibition, i.e. the oxidative damage in the photosynthetic apparatus (Adir et al., 2003; Chaves et al., 2003; Powles, 1984).

As well as affecting plant growth, water scarcity modifies the pattern of carbon allocation (Cannell & Dewar, 1994; Turner, 1986a). More photoassimilates are transferred to belowground components, especially to the fine roots and higher root-shoot ratios can be found in plants experiencing water stress in comparison to plants under high water availability (Guarnaschelli, 2009; Guarnaschelli el al., 2006; Li et al., 2009). In addition, droughts may change diverse morphological and anatomical characteristics. For example, leaves are usually smaller, thicker, more cutinizied and their vessels diameter, whereas cell walls are thicker and more lignified. Plant chemical composition may be altered, which means that it can influence the way plants protect themselves against other stressful factors. Plant vigor and overall resistance to stress from insects and/or diseases are influenced by water status (Kozlowski, 1982; Kozlowski & Pallardy, 1997).
Tree species differ in their optimal water requirements and in their susceptibility to water stress. Certain stages of plant growth are more sensitive to water stress than others. While some species are well-adapted to short period of water deficit, they become vulnerable to prolonged water deficit decreasing stomatal conductance and leaf area, changing the pattern of biomass allocation and reducing stemwood production. Water stress compromises cambial growth in different ways. It slows or stops the production of xylem cells and influences the seasonal duration of xylem production and the time of initiation and duration of the latewood production (Kozloswki & Pallardy, 1997). Besides differences among species, the effects and the magnitude of loss caused by droughts vary among provenances or clones (Arend et al., 2011; Guarnaschelli, 2009; Guarnaschelli et al., 2010b) and depend on the season in which the water deficiency occurs (Guarnaschelli, 2009), as well as the intensity and duration of drought (Guarnaschelli et al., 2003c; Guo et al., 2010; Osório et al., 1998). Damage is also linked to the health and vigour of the plants previous to the drought. Plants with healthy root systems and adequate carbohydrate reserves will behave much better during and after a drought period (Kozlowski, 1992; Marshall, 2006).

3. Problems during the forest establishment

Planting has become the main method of reforestation in many parts of the world. It is more predictable and reliable than natural regeneration and allows a more effective control of stand density. When planting nursery grown seedlingsor cuttings, the most critical processes of natural regeneration, from seed production to the early seedling developmental phase, are skipped (Stoneman, 1994; Tappeiner & Helms, 1973). Forest plantation area has increased considerably in the last years and today includes different species selected due to the good availability of planting material, clear silvicultural management techniques and high productivity. In South America, for instance, forest plantations consist of different introduced species like *Eucalyptus* (*Eucalyptus grandis, Eucalyptus globulus, Eucalyptus camaldulesis, Eucalyptus urophylla* as well as interspecific hybrids), *Pinus* (*Pinus taeda, Pinus elliottii, Pinus caribaea, Pinus ponderosa*), *Populus* (*Populus deltoides, Populus* x*canadensis*), *Salix, Gmelina, Toona* among others. The development of forest nurseries, which can produce large quantities of tree seedlings annually, allowed the possibility of establishing large areas of forest plantations.
Planted material is generally exposed to, and has to overcome stressful conditions after planting. Several factors are found to affect the early establishment of forest plantations. However, among abiotic ones, water deficiency is the most common limitation for seedlings from outplanting and all over the establishment (Burdett, 1990; Close et al.,

2005; Margolis & Brand, 1900). Water limitations, derived from periods of soil and/or atmospheric water deficiencies, affect plantations both directly by reduced growth and increased mortality, and indirectly by increased susceptibility to, or damaged by, other abiotic and/or biotic stress factors.

Immediately after being planted, seedlings are subjected to transplant shock, a phenomenon characterized by a depression in the physiological status mainly associated with moisture stress. The alteration in plant water status, caused by limited contact between roots' seedlings and soil, impairment of seedling root function, high evaporative demand, low hydraulic conductance of suberized roots and/or root confinement, constitutes the main constraint for plant survival and growth after planting (Burdett, 1990; Grossnickle, 2005; Kozlowski & Davies, 1975; Rietveld, 1989; Sands, 1984). Low root-soil contact can occur even if soil water potential is near to zero. Therefore transplant shock is not necessarily associated with the soil water status, but low water availability or post-planting droughts intensifies that condition and aggravates it.

The first physiological symptom of transplanting shock is the lowering in predawn water potential. As water potential becomes more negative, other physiological symptoms appear like reduced stomatal conductance, photosynthesis, transpiration and growth (Grossnickle, 1988; Guehl et al., 1989; Jacobs et al., 2009; Sands, 1984). These symptoms were observed by Mena-Petite et al. (2005) after transplanting seedlings of *Pinus radiata* both under drought and under well-watered conditions. Reduced leaf area, leaf shedding, shoot growth and needle length area are the principal morphological symptoms (Haase & Rose, 1993; Struve & Joly, 1992).

The improvement in root-soil contact is mediated by the extension of new root growth. The ability to regenerate new roots after planting is a key process to reduce the effects of transplanting shock and assure plant survival after planting (Rietveld, 1989; Sands, 1987). It has been observed in some Conifer species that new root growth is largely dependent on current photosynthesis (Maillard, 2004; Phillipson, 1998; van den Driessche, 1987), also in some Hardwoods species (Sloan & Jacobs, 2008). Seedlings that develop their root systems after planting reestablish a proper water balance. With a favorable water status they can have a cycle of root growth supported by photosynthesis and photosynthesis supported by root growth. Therefore, high seedling water potential during the outplanting is a favorable condition to initiate new root growth. All plant attributes that ensure a better water balance will benefit the process of transplanting by granting more resistance to water stress (Burdett, 1990).

Bare root plantings are specially accompanied by specific transplanting stress. When plants are lifted for transplanting, a large percentage of the absorbing roots are severed and/or damaged, therefore, the newly transplanted tree suffers from water stress. Even trees that have been grown in containers and could retain their whole root system suffer from some degree of transplant shock. However, container seedlings have proved to have better performance than bare root plants, especially when conditions are more stressful (Barnett & McGilvray, 1993).

When transplanting shock has been overcome, additional drought events that may occur several times along the rest of the establishment period may cause depression in plant function and growth. Forest sites are frequently subjected to periodic droughts. On drought-prone sites a successful regeneration will depend mainly on the ability of plants to tolerate moisture stress.

As mentioned above, other abiotic factors can affect independently or interact with the water availability and cause serious damage in recently planted tree seedlings. Usually several stress act simultaneously with droughts such as heat and high irradiance intensifying moisture deficiencies. By contrast low temperatures may arise. Freezing temperatures induce cell dehydration, which is the most common damage. In specific tissue and organs damage can occur for other reasons, like cell death, separation of cell layers, creation of cavities and frost cracks, xylem embolism among other (Pearce, 2001). Drought as well as low temperature and water-logging reduce water uptake and hence the rate of photsynthesis. An increase in irradiance may cause photoinhibition. Nevertheless, photoinhibition may arise indirectly from the action of all stress factors that cause a reduction in photosynthesis and induce excess light absorption. The effects of competing vegetation, restricting resource availability and herbivory damaging leaf tissues on seedlings performances are discussed later (Burdett, 1990; Close et al., 2005).

4. Strategies to cope with water deficiencies during the establishment

As the establishment of forest plantations requires long-term investments, it is crucial to select the best management options in order to improve forest productivity in site-specific conditions and to make cost-effective decisions. Moreover, when stressful and/or unexpected weather conditions are predicted, careful silvicultural planning is essential to ensure success. Silvicultural decisions including the selection of high quality plant material and the application of adequate site preparation techniques, proper planting and plantation maintenance practices are the key to a successful establishment and are tightly connected with plantation performance (Blum, 2003; Smith et al., 1997; Tappeiner et al., 2007). In fact, they may be adjusted to withstand water stressful conditions since transplanting generates a severe physiological shock and moisture conditions exclude the establishment of trees. There are many textbooks on Silviculture and availability of information that discuss thoroughly the effects of alternative cultural practices before and after plantation. Our analysis will focus mainly on the strategies to improve seedling quality.

The use of high quality seedlings as well as the application of several cultural practices before and after planting can help to overcome stressful conditions and enhance plantation performance. Regarding particularly outplanting stress, although some level of transplant shock is unavoidable, stock with high performance will minimize the event (Burdett, 1990; Close et al., 2005; Grossnickle, 2005; Rietveld, 1989).

The quality of seedlings is the result of its genotype, growing conditions and cultural practices applied in nurseries. In the context of transplanting process, the utilization of provenances or clones that have been tested as drought tolerant and that have been subjected to appropriate nursery management will show a better performance.

4.1 Mechanisms associated with drought tolerance

Chaves et al. (2003) suggested that it is essential to have a holistic understanding of plant resistance to water stress to improve crop management and breeding techniques. As originally defined by Ludlow (1989), under drought conditions there are physiological processes and morphological attributes that could extend the period of active growth by controlling water uptake, water loss and hence cell turgor. Water loss can be effectively controlled by stomatal closure and, ultimately, by leaf shedding. To maintain to some extent cell turgor and stomatal conductance and a sizeable photosynthesis rate under drought conditions, plants have developed particular physiological processes. Species possessing these attributes are very sensitive to dehydration and avoid water deficits when soil moisture limitation occurs. Conversely, other species tolerate dehydration mainly through osmotic adjustment or changes in tissue elasticity.

Drought avoidance, a strategy held by many tree species, is characterized by relatively high lethal water potential or high relative water content with relatively little osmotic adjustment, while drought tolerance species have lower lethal water status values and relative water content and display much more active changes in osmotic potential when subjected to drought conditions. Not all plants fit closely in one category or another but the division in plant responses helps to understand consequences for the species like survival, potential for carbon fixation, maintenance of growth under drought, and metabolic costs of drought resistance mechanisms (Ludlow, 1989).

Schulte & Hinckley (1987) observed that the ability of stomatal closure varied among *Populus* species and their interspecific hybrids, and some of them required a large change in tissue water content for stomatal closure. For example, *P. deltoides* and its hybrids display a large number of strategies avoiding water deficiencies. Among them a reduction in stomatal conductance and transpiration (Marron et al., 2002; Silim et al., 2009), leaf abscission, and decreases in shoot-root ratio (Liu & Dickman, 1992).

For some species, greater biomass allocation to roots relative to shoots, that increase root growth and reduced leaf area, as well as stomatal regulation seem to be effective mechanisms to resist moisture stress (Jacobs et al., 2009). As it has been suggested, plants respond to shifts in resource supply by allocating carbon to the organ involved in capturing the limited resource. Root development is fundamental under water and nutrient deficiency, resulting in plants that have lower shoot:root ratios and greater capacity to absorb water and minerals relating to the shoots that must be supported. The possession of a deep and thick root is considered highly important because it allows access to water deep in the soil profile. The investment of carbon in a deep root system may have, however, a yield implication due to lost carbon allocation to the shoot.

Under water stress, plants of *Alnus rubra* may show leaf senescence and shedding. These processes lead to a favorable decrease in shoot:root ratio and high survival, whereas the reduction in the photosynthetic area resulted in a reduced shoot growth (Pezeshki & Hinkley, 1988).

Thus, mechanisms that allow plants to avoid water deficiencies present some disadvantages because they imply a reduction in leaf area and gas exchange, then the photosynthetic capacity is reduced and at the same time the change in the pattern of carbon allocation favoring root growth reduces shoot growth.

Among the drought tolerance mechanisms, osmotic adjustment implies the accumulation of organic and inorganic solutes in the cell that reduces water potential and allows plants to obtain water from a lower water potential medium and sustain the physiological processes. The capability of osmotic adjustment appears to be controlled by only one or two genes and is simply inherited (Morgan & Condon, 1986). This mechanism may be an advantageous strategy to maintain cell water status (Ashraf et al., 2011; Morgan, 1984). The maintenance of cellular turgor by lowering the osmotic potential in plants exposed to low water conditions is considered to be one of the most important mechanisms of plant adaptation to environmental stresses (Turner, 1986a, 1986b; Turner & Jones, 1980).

The drought induced lowering in osmotic potential has been observed in many coniferous tree species (Edwards & Dixon, 1995; Nguyen-Queyrens & Bouchet Lannat, 2003) as well as in hardwood species (Abrams, 1990; Arndt et al., 2000; Marron et al., 2002; Tschaplinski et al., 1998). In some cases the magnitude of adjustment was related to the maintenance of plant growth rate (Johnsen & Major, 1999; Meier et al., 1992; Pita & Pardos, 2001; Tan et al., 1992).

Considerable variation in the osmotic adjustment capacity has been observed. Among *Eucalyptus* species Lemcoff et al. (1994, 2009) detected different magnitude of osmotic adjustment among seedlings of *E. grandis*, *Eucalyptus tereticornis*, *Eucalyptus viminalis* and *E. camaldulensis*. Other studies that considered different species of *Eucalyptus* reached similar results confirming genetic variability for this mechanism (Merchant et al., 2006; Merchant et al., 2007; White et al., 2000). Variability is also present among provenances of a single species (Guarnaschelli et al., 2001; Guarnaschelli et al., 2006; Tuomela, 1997) or among clones (Pita & Pardos, 2001). Guarnaschelli (2009) compared the response to drought conditions in 13 provenances of *E. globulus* and found differences both among subspecies and also among provenances within some of the subspecies.

As originally stated by Turner (1986a), osmotic adjustment would only represent a useful strategy to tolerate drought if it also develops in the root system (Merchant et al., 2006; Nguyen & Lamant, 1989; Parker & Pallardy, 1988). Guarnaschelli & Lemcoff (2001) detected osmotic adjustment in both shoot and roots of *E. globulus* subsp. *maidenii* provenances and difference in the capacity to adjust cell-wall elasticity.

Gebré & Tchaplinski (2000) indicated that drought tolerance in *Populus* is not only related to the capacity of solute accumulation, but also to low values of osmotic potential itself. Consistent with this statement, Guarnaschelli et al. (2010b), evaluating the drought tolerance of several *P. deltoides* clones, recently showed that they have low osmotic adjustment capacity but in contrast there were consistent differences in osmotic potential among them.

Osmotic adjustment has been recognized as an important cellular drought-responsive trait and, despite past speculations, there is no definitive proof that the osmotic adjustment capacity entails a compromise to potential crop yield (Blum, 2005). Moreover, after stress relief it has been observed greater capacity of stem diameter and height growth in seedlings of *E. globulus* subesp. *globulus* that had shown higher osmotic adjustment capacity (Guarnaschelli, 2009), results that present similarities with those of Osorio et al. (1998).

Although the maintenance of leaf turgor is often linked to osmotic adjustment, changes in tissue elasticity may also contribute to turgor maintenance in plants with (Guarnaschelli, 2009; Lemcoff et al., 2002; Pita & Pardos, 2001) and without osmotic adjustment. The regulation of cell-wall elasticity, generally called elastic adjustment and measured as the change in maximun bulk modulus of elasticity, also aids some species to maintain tissue hydration (Pita & Pardos, 2001; Prior & Eamus, 1999; Stoneman et al., 1994)

Under moderate water stress an increase in tissue elasticity would allow the maintenance of tissue turgor and physiological functions and growth (Fan et al., 1994; White et al., 1996). But also a decrease could represent a useful strategy to overcome low water availability (Nielsen & Orcutt, 1996) while conditioning a decrease in plant water potential, increasing the water potential gradient between plant and soil allowing water uptake. An additional consequence of this kind of cell-wall adjustment, and the concomitant rapid lowering in the water potential, is the closure of stomata that prevents severe water deficit conditions (Dumbroff, 1999; Lemcoff et al., 2002). It has been observed that under moisture stress some species increase cell-wall elasticity while others decrease it (White et al., 1996; White et al., 2000); this behavior was also detected among provenances of a single species (Guarnaschelli, 2009). A decrease in cell-wall elasticity has been associated with tissue maturation (Bowman & Roberts, 1985; Parker et al., 1982), which commonly occurs during fall and winter. Water stress conditions of similar intensity and duration may exert different responses in cell-wall elasticity, like was observed in subspecies and provenances of *E. globulus* (Guarnaschelli et al., 2001; Guarnaschelli, 2009).

It is important to highlight that the expression of drought tolerance mechanisms may vary if multiple stress factors are simultaneously affecting plants, phenomenon which in fact occurs quite often in nature. Some studies under controlled conditions assessed the responses of several *Eucalyptus* provenances and *Salix* clones submitted to different water and light availability regimes, analyzing the genetic contribution as well as the level of restriction and the interaction effects. In saplings of *Salix* submitted to drought conditions, a decrease in osmotic potential at full turgor and at the turgor loss point occurred in plants growing under full sunlight but also in those that were growing under moderate shade conditions (Guarnaschelli et al., 2010b). These results contrast with those that argue that only dryness and high sunlight trigger a lowering in osmotic potential, stating that under high irradiance conditions there is a higher capacity to accumulate solutes and where osmotic adjustment can take place (Aranda et al., 2005; Uemura et al., 2000). Thus, osmotic adjustment contributed to turgor maintenance processes, allowing an increase in the drought tolerance under full sunlight and moderate shade conditions, without a compromise between the plants adjustments to cope with those two kinds of stress factors (Smith & Huston, 1989). These responses indicate that shade could alleviate the effects of drought (Guarnaschelli et al., 2007; Guarnaschelli et al., 2008), although this is not always the case. Recently lower water potential values were observed in droughted seedlings of *E. grandis* and *E. grandis* x *E. camaldulensis* clones that were growing under shade conditions in contrast to those that were under full sunlight. All shaded plants displayed high plasticity increasing their leaf area as an strategy to increase the uptake of light, but the decrease in shoot:root ratio observed in water stressed plants was not as effective as the change that occurred in water stress plants growing under full sunlight, which allowed better water acquisition under moisture deficit conditions (Guarnaschelli et al., unpublished data).

Breeding for drought tolerance

Tree selection can be a relevant component when dealing with moisture stress. In fact, species, provenances and/or clones genetically adapted to certain levels of water deficiencies can overcome soil and atmospheric water deficits. Plants originating from drier sites are more likely to survive drought than those from mesic sites because they have different mechanisms that help them to tolerate low water availability.

As discussed previously, trees possess a varied number of mechanisms to compensate for water limitation that allow the acquisition of water resources or limit water loss, and indeed represent useful tools in the context of the establishment. Genetic variability in the responses to water stress has been tested in different tree species, among populations or even clones of a single species. It is accepted that high levels of genetic variation within any species improve the potential to withstand abiotic stress, such as droughts and biotic stress as well. However, the ability of plants to sense stressful conditions and environmental change through plastic responses is part of this genetic variation. Trees may also exhibit certain level of physiological and/or morphological change, which is defined as the capacity of organisms to produce different phenotypes according to environmental changes (Schlichting, 1986; Valladares, 2006). Phenotypic plasticity has been recognized now as a heritable characteristic and can be genetically controlled (Bradshow, 2006; Lande, 2009). It has a great potential importance in plant evolution (Nicotra et al., 2010).

Mechanisms or traits that avoid or tolerate the dry conditions include a variety of morphological, physiological and biochemical attributes at biochemical, cellular, tissue and whole-organism levels, as discussed previously. Most of them are dependent on other mechanisms; in fact, a sequence of adjustments occurs simultaneously in plants.

Breeding for drought stress tolerance in forest trees should be given high research priority in plant biotechnology programs. Plant response to abiotic stress like water deficiency involves many genes and biochemical-molecular mechanisms. Plant modification to enhanced stress tolerance is based on manipulation of genes that protect and maintain the function and the structure of the cellular components. Due to the complex responses to stress it is more difficult to control and engineer in contrast to traits of engineered resistance to pests or herbicides. Although the improvement of stress tolerance by gene transformation has resulted in important achievements, the complexity of the mechanisms involved makes this task very difficult.

Despite the existence of multiple traits related to drought resistance, the selection of improved growth in water limited environments may not be generalized because it may lead to a trade-off between traits related to both drought resistance and growth. The selection of drought-resistant and productive material may not be simple to achieve, as we mentioned previously. Some species exhibit numerous drought resistance strategies that may impact on productivity differently, like lower leaf area, leaf abscission, enhanced root growth, stomatal closure among others as has been noted before.

But although drought tolerance may be considered as a penalty towards potential productivity, it is not necessarily the case (Blum, 2005). To maintain growth and productivity under water limited conditions, plants have to exercise specific tolerance mechanisms. Plant breeders have improved the performance of crops by breeding for improved yield or quality under conditions of water shortage; however the selection of this kind of plant material is nowadays becoming more important. The development of new tools for monitoring and understanding plant responses to water deficit, ranging from molecular, plant and ecosystem, will allow a better understanding of plant performance under stress, which will be very useful to breeding programs (Chaves et al., 2003).

Pita et al. (2005) discussed the use of particular physiological traits in breeding programs for improved yield under drought conditions focusing mainly in *E. globulus* experience in southern Spain. They explained that *E. globulus* must be considered a species with avoidance capacity because high survival and growth were observed through lower water use efficiency and embolism tolerance. They highlighted the relevance of several hydraulic characteristics, like maximum permeability, maximum leaf conductivity and vulnerability to cavitation and its relationship to stomatal conductance. Osmotic adjustment capacity, a drought tolerance strategy, has also been observed in this species (Guarnaschelli, 2009; Guarnaschelli et al., 2003; Guarnaschelli et al., 2006; Pita & Pardos, 2001; White et al., 1996).

4.2 Nursery conditions and culture

Nursery conditions and culture affect the structural and functional characteristics of tree seedlings. They can produce planting stock of different size and physiological state;

however, each ecological condition needs a different type of plant according to it. If field site conditions where seedlings are going to be transplanted differ from those at nurseries, they may be severely stressed because the process of acclimation in the field occurs over several days, or even weeks. Therefore, it is necessary to use plants suited to the environmental restrictions in which they will be planted (Burdett, 1990).

Nurseries adapt their environmental conditions and cultural practices to produce target seedlings that will assure survival and growth according to the species and site conditions. The alternative methods of seedling production are outlined in several manuals, which cover all phases of production from seed collection to seedling storage and provide detailed information about nursery practices, allowing to secure high quality material (Duryea & Landis, 1984; Landis et al., 1989, 1990, 1992).

Characteristics of target seedlings include height, stem diameter, root volume, root growth potential, plant water status, drought resistance and frost hardiness among others (Rose et al., 1990). Seedlings' height and stem diameter, the most common traits used to assess seedling quality are not always accurate predictors of performance after outplanting. As stated by Burdett (1990), it appears that the central processes in plantation establishment are root growth and photosynthesis in mutual dependence. All attributes that assure a better seedling water balance favor the maintenance of photosynthesis and root growth and benefit the process of transplanting shock in coping with water stress.

In the context of establishment commonly hindered by water stress, plants with root systems of high morphological and physiological standards enable them to establish rapidly and succeed upon outplanting. Large root volume, high root fibrosity and an increased number of first-order lateral roots have shown some correlation with improved field performance (Landis, 2010). Leaf area, shoot:root ratio, the capacity to regulate stomatal conductance as well as the development of osmotic and elastic adjustment will help seedlings performance. Water and nutrient regimes are of particular importance during seedlings production because they control both the rate and type of growth and hence have relevant impact on morphological and physiological attributes mentioned above (Duryea & Landis, 1984; Landis et al., 1989).

Drought acclimation

Plant drought acclimation is a process that results after the exposure to stress conditions that induce structural and functional adjustments, allowing the acquisition of tolerance to drought. Acclimation is a process that occurs spontaneously in nature, helping plants to survive and grow under stress. For example, many plants increase their freezing tolerance at the end of summer and fall upon exposure to low non-freezing temperatures and short days, a phenomenon known as cold acclimation that increases their cold tolerance. At the beginning of spring, when temperatures are rising, this process is reversed. Other plants experience drought acclimation when exposed to moisture stress (Kozlowski & Pallardy 2002; Yordanov et al., 2000).

Drought preconditioning or drought hardening is a common practice applied in nurseries, which entails exposure to sublethal stress and allows seedlings to trigger plastic responses associated with drought acclimation, finally resulting in protection against lethal stress. Submitting tree seedlings to nursery stress conditions may render them more protected from injuries or reduced growth when environmental stresses are abruptly imposed than plants not previously stressed. It is accepted that drought preconditioning helps seedlings to improve their performance, competitiveness and productivity under field conditions. The process of acclimation occurs over days or weeks. In contrast, poor drought acclimated seedlings will exacerbate transplanting shock.

Drought preconditioning consists in withholding irrigation or restricting the amount of water supplied for short periods. It is generally applied during the last weeks of seedlings production (Landis et al., 1989; Vilagrosa et al., 2006). An irrigation regime that restricts the water availability will induce adjustments in several morphological attributes such as aerial growth and the pattern of dry matter partitioning, reducing leaf area and shoot:root ratio (Lamhamedi et al., 2001; Stewart & Lieffers, 1993). Stock types with low leaf area and shoot:root ratio perform better under drought conditions, since a more favorable balance between water uptake and loss is reached (Cregg, 1994). In addition, physiological adjustments like stomatal regulation, osmotic adjustment and/or elastic adjustment may also contribute to better performance after planting. Several Conifer species deliberately exposed to water deficit displayed drought hardiness and were able to maintain more favorable water status and gas exchange (Edward & Dixon, 1995; van den Driessche, 1991; Zine El Abidine et al., 1994; Zwiazek & Blake, 1989), and greater survival after plantation compared to non-conditioned plants (van den Driessche, 1992).

Among hardwood species, in several *Salix* cultivars drought acclimation was manifested in decreased stomatal conductance, osmotic potential and leaf area to vessel internal cross-sectional area ratio, and increased shoot hydraulic conductance. An increase resistance to stem xylem cavitation was observed in only one clone (Wikberg & Ogren, 2007). Seedlings of three provenances of *E. globulus* that were submitted during one month to drought cycles of 6 and 9 days, considered as moderate and severe stress, displayed plastic changes: osmotic adjustment and reduced leaf area. The extent of osmotic adjustment was influenced by the degree of drought preconditioning; osmotic potential at full turgor in severe stressed plants was significantly lower than in moderate stressed plants. Thus, the magnitude of the adjustment increased with the intensity of water stress, suggesting an additive effect in this drought tolerance mechanism. When non conditioned plants (daily irrigated) and drought conditioned were all evaluated under a new drought cycle, the last showed higher water status and stomatal conductance (Guarnaschelli et al., 2003c). Similar results were observed with three representative Mediterranean species (Vilagrosa et al., 2003; Villar-Salvador et al., 2004).

Plants may display different phenotypic plasticity in their attributes according to the intensity of the drought preconditioning (Guarnaschelli et al., 2003c). Medium and moderate levels of water stress induced a higher level of hardiness in comparison with severe water stress conditioning (Villar Salvador et al., 2004). It is important to highlight that severe water stress preconditioning is likely to induce some level of tissue damage and a higher decrease in growth due to the high dehydration experienced or low growth rate after transplanting.

As for bareroot seedlings, the control of water irrigation may be more difficult to achieve, preconditioning includes root pruning. After being root-pruned, seedling of *Quercus rubra* and *Juglans nigra* experienced water stress, which triggered many changes in growth as well as morphological and physiological attributes. Undercutting reduced seedling growth and shoot:root ratio and increased the number of first order lateral roots, which increased field survival (Schultz & Thompson, 1997). These results show that drought preconditioning triggers morphological and physiological adjustments associated with an increase in drought tolerance.

Several studies have also tested the effects of preconditioning on growth and survival after planting. Some of them indicate that there is a positive effect on survival (Guarnaschelli et al., 2003a; Guarnaschelli et al., 2006; van den Driessche, 1992), whereas only a few show a positive effect on growth (Arnott et al., 1993).

Guarnaschelli et al. (2006) evaluated the effect of drought preconditioning after transplanting. Three provenances of *E. globulus* subsp. *bicostata* were subjected to moderate water deficit conditions. They evidenced drought acclimation capacity showing osmotic adjustment, a reduction in seedlings size, leaf area, shoot:root ratio and stomatal conductance. After being transplanted under moisture stress conditions, drought preconditioned plants showed better water status, gas exchange capacity and higher levels of survival than well irrigated plants. It was observed that midday relative water content (RWC) was closely correlated with survival as also found by Mena-Petite et al. (2005), while survival was inversely correlated with shoot:root biomass ratio. Both RWC and shoot:root ratio could be considered reliable indicators of potential initial survival and are closely associated with establishment success (Grossnickle & Folk, 1993).

Several processes involved in drought tolerance confer also an increase in cold tolerance (Bigras & Dumais, 2005; Moraga et al., 2006). It has been observed that drought resistant genotypes of *E. globulus* showed greater cold tolerance than drought susceptible ones (Costa e Silva et al., 2009). The application of drought hardening treatments may also increase cold tolerance. The accumulation of solutes that commonly occurs under water stress decreases the osmotic potential and may cryoprotect freezing labile cell structures. Coopman et al. (2010) applied two drought hardening treatments to different genotypes of *E. globulus* subsp. *globulus* under nursery conditions that resulted in an increase in the seedlings drought tolerance but also in their freezing tolerance. They observed that the freezing tolerance varied with the genotypes and the level of water stress preconditioning.

The interactive effects of water and other factors during seedlings production

When dealing with a process of drought acclimation, many other factors can influence plant responses to drought preconditioning, as is the case of nutrition, age and/or growing media conditions. In fact the nutrient regime as well as the growing media used to produce forest containerized seedlings can strongly affect plant during the drought preconditioning period and their performance after outplanting.

The growing media usually consists of a mix of organic materials, like sphagnum peat moss, composted pine bark, coco fiber or other local organic products with inorganic components as perlite, vermiculite or sand. The mix is selected in order to obtain a growing media of slightly acid pH, high cation exchange capacity, low inherent fertility, adequate porosity, and freedom of pests, all characteristics that allow the production of healthy seedlings. Both the water-holding and nutrient supplying properties are functions of the different growing medium components; therefore the irrigation and fertilization have to be adjusted according to their characteristics (Landis et al., 1990). Artificial mixes dry out much more rapidly than surrounding soil, so watering is very important. Verdaguer et al. (2011) observed that when seedlings of *Quercus coccifera* were grown in natural soil compared to standard nursery growing medium growth rates were higher. Results suggest that the former were acclimated to the soil and higher photosynthetic rate, transpiration and stomatal conductance were observed in those seedlings.

Besides irrigation regime, fertilization can have important effects on plant quality modifying their performance under drought during the establishment. Harvey & van den Driessche (1999) observed that increasing nitrogen (N) supply under dry conditions increase leaf loss and decrease water potential in *Populus trichocarpa,* making them more vulnerable to cavitation. The effects of N fertilization on drought and cold acclimation were assessed on several forest Mediterranean species. High N level decreased frost hardiness in *Pinus* species. In all species high N increased shoot:root ratio and in some of them stomatal conductance, which might impair seedlings water balance if soil water content is low after planting (Villar Salvador et al., 2005).

van Den Driessche (1992) analyzed the responses of *Pseudotsuga menziesii, Pinus contorta* and *Picea glauca* seedlings that were grown in containers in a greenhouse and submitted to two N treatments and three potassium (K) treatments with three drought treatments. A positive relation between shoot:root ratio and survival in *P. contorta* and *P. glauca* indicated that increase in N increased both shoot growth and drought resistance over the N range investigated, while *P. menziesii* showed an interaction between drought and N treatment and a small response in both survival and dry weight to potassium (K) application.

K plays important physiological functions in plants. It regulates cell membrane activity, it is an enzyme cofactor and one of the main ions that contribute to plants osmorregulation, improving the tissue water content and regulating stomatal closure among others (Morgan, 1984; Taiz & Zeiger, 2002). While high N fertilization stimulates rapid soft growth, adequate K promotes firmer tissue. K addition increases water uptake, decreases transpiration losses, leaf area ratio and dessication damage. It has been shown that K, with calcium and magnesium, increases leaf waxes, protecting plants during the hardening process. Thicker cuticles also protect to insect feedings and penetration fungi (Cakman, 2005; Landis, 2005). Nurseries have traditionally applied extra K as part of the hardening process.

Several studies showed that K fertilized seedlings had better performance under water stress conditions (Garau et al., 2004a; Garau et al. 2004b). The responses of *E. camaldulensis* and *E. globulus* seedlings submitted to different treatments of fertilization and water availability were analyzed at the nursery stage and after plantation. Nursery treatments involved two levels of K and two levels of water availability. Results indicated that drought conditioning and K fertilization in nursery improve seedlings growth after plantation (Garau et al., 2005).

Guarnaschelli et al. (2010b) observed a decrease in the osmotic potencial in water stressed and K fertilized plants of *E. globulus,* which would imply an increase in the water potencial gradient between the plant and the soil, facilitating water uptake. In addition, treated plants showed higher relative water content and stomatal conductance.

Recently, Oddo et al. (2011) analyzed the effects of short-term K fertilization on plants of *Laurus nobilis.* They observed an enhancement in hydraulic conductance following short-term K fertilization; phenomenon that can be quite advantageous for maintaining cell turgor, stomatal aperture and gas exchange rates under moderate drought stress.

Apart from previous results, Römheld & Kyrkby (2010) argued that it is still necessary to explore the influence of K on plants under stressful conditions. In fact, there is no clear evidence that K addition may increase cold tolerance as well.

5. Effects of vegetation competition and herbivory during the tree establishment

5.1 Weed competition: The main biotic cause of water and resource deficiencies

During plantation establishment conditions of high levels of light and, sometimes, increased availability of water and nutrients favor the development of opportunistic, fast-growing herbaceous and/or woody species that invade the disturbed site from wind-blow seeds, seeds stored in the seed soil-bank or by sprouting. Although the surrounding vegetation may play important functions in the forest system (reduce soil erosion, retain and recycle nutrients, add organic matter to the soil) and in certain environments can facilitate the growth of tree seedlings, it commonly interferes and captures resources at the expense of trees seedling performance (Cannell & Grace, 1993; Nambiar & Sands, 1993). Only shade tolerant tree species can become established beneath such vegetation. Herbaceous and/or woody vegetation which limits resource availability, hindering seedlings' expected growth and survival has been cited as one of the main causes of low plantation success (Smith et al., 1997; Close et al., 2005; Tappeiner et al., 2007).

Weed vegetation reduces the levels of water availability in the soil and affects negatively seedling water status (Dinger & Rose, 2009, 2010; Löf & Welander, 2004; Picon-Cochard et al., 2006), representing the main biotic cause of water stress (Lamhamedi et al., 1998; Picon-Cochard et al., 2001). Differences in weed density (Florentine & Fox, 2003; Garau et al., 2008b) and growth forms (Balandier et al., 2006; Coll et al., 2004; Provendier & Balandier, 2008) may cause variations in water restriction. Nevertheless, weeds will significantly interfere with forest seedlings only during "the critical period" (Adams et al., 2003).

Plant responses to competition are similar to those that allow coping with abiotic resource deficiencies and involve several physiological and/or morphological adjustments. Seedlings of shade intolerant trees, commonly used in many commercial plantations, are generally characterized by a great physiological plasticity that enables them to better utilize the higher light levels of open environments, and to withstand better the associated environmental stresses (Peltzer & Köchy, 2001; Picon-Cochard et al., 2006). Some responses are rapid and potentially lead to greater survival (i.e. stomatal closure), whereas others are delayed for hours or days, representing new capabilities and allocation patterns, providing some degree of resistance to the stress.

Numerous studies demonstrated that tree seedlings responded to weed imposed water deficit by anticipating stomatal closure and showing a reduction in leaf water potential (Coll et al., 2004; Dinger & Rose, 2009, 2010; Garau et al., 2008b; Picon-Cochard et al., 2001, 2006; Provendier & Balandier, 2008; Rey Benayas et al., 2003; Watt et al., 2003). Even with high soil water availability, the presence of weeds produced a significant decrease in leaf stomatal conductance (Garau et al., 2008b; Watt et al., 2003). Reductions between 30 and 80% in seedlings stomatal conductance reflect a useful response to limit water loss under soil water deficit (Garau et al., 2008b; Picon-Cochard et al., 2001).

Changes in tissue water relation parameters can help to withstand water stress induced by weed competition. A higher bulk modulus of cell wall elasticity and a lower osmotic potential or both were found in tree seedlings in response to neighboring competition (Rey Benayas et al., 2003). Garau et al. (2008b) observed that the restriction of soil water by weeds induced an osmotic adjustment of 0.38 MPa in seedlings of *E. globulus* subsp. *maidenii* and of 0.65 MPa when, additionally, a water restriction condition was imposed. Similarly, a cell wall stiffening of 8.0 MPa was induced independently of the level of water availability. Together, both strategies ensure water uptake while maintaining both cell turgor and volume. These features also allowed the maintenance of higher rates of gas exchange at low soil water content, being crucial for the establishment of tree seedlings (Lopez et al., 2009; Rodriguez-Calcerrada et al., 2010; Serrano & Peñuelas, 2005).

Some studies showed an increase in the intrinsic tree seedlings water-use efficiency (WUE) in response to soil dehydration caused by weed competition. Stomatal closure enabled seedlings to limit water stress, and since the decrease in stomatal conductance was faster than the decrease in CO_2 assimilation rate, an increase in the intrinsic WUE occurred (Garau et al., 2008b; Picon-Cochard et al., 2001; Picon-Cochard et al., 2006).

Other morphological adjustments that imply the postponement of dehydration by lessening the loss and increasing the uptake of water allow withstanding the low water availability. The presence of neighbouring vegetation may reduce seedlings above and below ground biomass to values 60- 80% lower than those obtained without competition (Garau et al., 2008b; Picon-Cochard et al., 2006; Watt et al., 2003). Under such conditions tree seedlings show lower leaf area and shoot:root ratio (Coll et al., 2004; Garau et al., 2008b; Picon-Cochard et al., 2006; Shipley & Meziane, 2002; Watt et al., 2003).

It is not possible to have water stress through competition without having some degree of nutrient stress, and although water stress has been implicated as the main regulator of seedlings performance, it is likely that competition for light also occurs. In fact, weeds are considered "multiple stressors" that compete not only for water but also for nutrients and light (Adams et al., 2003; Caldwell et al., 1995; Nambiar & Sands, 1993). Significant reductions in nutrients (mainly N) and light availability were observed in several studies attributed to competing vegetation (Cogliastro et al., 2006; Davis et al., 1999; Powell & Bork, 2004). Garau et al. (2008b) found a 50-75% reduction in photosynthetically active radiation when eucalypt seedlings were growing under a 100% of weed cover, condition that triggered the greatest leaf osmotic adjustment capacity. Significant increases in free amino acids that contribute to osmoregulation were observed in response to water stress and shade conditions (Showler, 2002; Valladares & Pearcy, 2002). Their synthesis may be associated with increasing weed competition.

Implication for plantation establishment

It is possible to mitigate the effects of competition effectively by selecting plant material and applying cultural practices during the establishment (Adams et al., 2003; Jacobs et al., 2004). There are species or genotypes that during the critical initial stage of plantation establishment can demonstrate a "strong tolerance ability" (Goldberg, 1996) because they can maintain a high water status, particularly in the presence of weed competition. Physiological and morphological differences within species can thus form a basis for matching plant material to site, although survival and growth will ultimately depend on the levels of water stress experienced.

As discussed above, differences in the response of water-relations parameters to water stress induced by weeds can potentially be used to match genetic material to specific sites. Net solute accumulation could hence favor drought-tolerance beyond the seedling stage and be used as preliminary prediction for genetic screening studies and of field performance studies (Lemcoff et al., 1994; López et al., 2009). As observed by Garau et al. (2008b) seedlings with greater capacity for osmotic adjustment and cell wall stiffening were able to maintain a higher water status under severe weed cover. Garau (2003) found that although two provenances *E. globulus* subsp. *maidenii* presented similar drought-tolerance mechanisms, they have different capacity of acclimation. Murrabrine seedlings showed a higher water-stress threshold (a better combination of processes for "reduced water loss" and "maintenance of water uptake") than Tantawanglo, which allowed Murrabrine seedlings to maintain higher rates of growth under weed competition.

In addition to genetic characteristics, nursery practices and silvicultural techniques can have an important effect on the seedling's growth and competition effects. Planting large seedlings allows withstanding competition from herbaceous or shrubby vegetation and minimizes the period of susceptibility to competition and also to animal damage (Noland et al., 2001; South et al., 2005). Cuesta et al. (2010) observed that large, nitrogen enriched seedlings of *Pinus halepensis* in the presence of weeds had higher water potential, gas exchange, and root growth and had finally a better transplanting performance.
Several measures are used also to regulate competition: weeding, plowing and application of herbicides (Smith et al., 1997; Tappeiner et al., 2007). Alternative methods of weed control are very effective because a greater number of seedlings can be established on treated sites. Different herbicides have been used to keep plantations weed-free. However, restrictions on the use of herbicides and worries about their environmental effects are generating some changes in silvicultural decisions. It has been agreed that it is necessary to adopt a new approach to weed control towards and "integrated vegetation management". For example, reducing the area of weed control according to the tree crop and weed growth habit characteristics seems to be an appropriate and effective option which causes less environmental impact and at the same time is less expensive (Garau et al., 2008a).

5.2 Herbivory and its interaction with water deficiencies

Trees are subjected to the negative impact of herbivory that removes biomass that might be allocated to grow and increases mortality. Different insect, rodent and mammalian species affect seedlings establishment both in natural forests and in commercial plantations (Becerra & Bustamante, 2008; Dulamsuren et al., 2008; Meiners et al., 2000). They are known to affect tree seedlings, especially during their first year of growth, but the intensity of damage is related to the population density of herbivores and to habitats' food offer. Plantation productivity is reduced due to the decrease in seedling growth, seedling survival and the development of multiple leaders. In Central and South America leaf-cutting ants are one of the main plagues that threaten seedlings growth and survival during the establishment (Della Lucia, 1993; Forti & Castelli Boaretto, 1997; Vasconcelos & Cherrett, 1997).
Plants have developed different kinds of defences, which allow their successful survival against herbivores. Some species have physical barriers (such as thorns and trichomes) or biomechanical properties in their leaves (such as strength and toughness) as anti-herbivore defences. Other plants produce chemical defences; mostly secondary metabolites like alkaloids, glucosinolates, terpens, phenols, hydroxamic acids, tiophenes, and cyanogenic glycosides, among others. Secondary metabolites represent adaptive characters that have been subjected to natural selection during evolution and are thus important for plant survival and reproductive fitness (Benett & Walsgrove, 1994; Thies & Lerdau, 2003; Wink, 2003). For instance, *Eucalyptus* species contain high concentrations of monoterpenic and sesquiterpenic compounds in their foliage, which have proved to be responsible for their ability towards different herbivores (Marsaro et al., 2004; Moore et al., 2004).
Genetic and environmental factors contribute to modulate the level at which different types of defence responses occur in plant species. Guarnaschelli et al. (2000) observed significant differences in leaf thickness and toughness among seedlings of provenances of *E. globulus* subsp. *maidenii*. Leaf-cutting ants caused higher damage in the provenance with lower toughness. Although nutritional quality of plant tissue can influence herbivore host choice, the structural and chemical defences are the major determinants of leaf and twigs

palatability. Most of those attributes are under genetic control (Raymond, 1995) and inter- and intra-provenance variation in resistance to insect damage has been reported in *Eucalyptus* sp. (Floyd et al., 1995).

Abiotic stresses such as drought can have important effects on plant-herbivore interactions, modifying herbivore population dynamics and anti-herbivore defences. Different hypotheses predict how plants under stress should respond to herbivory. However, it is difficult to generalize the effects of abiotic stress on host quality and herbivore damage (Wise & Abrahamson, 2007) since, although some herbivores are favored by unstressed plants, others are favored by moderately stressed plants or by severely stressed plants. Caffarini et al. (2006) observed in seven provenances of three subspecies of *E. globulus* that droughted plants or leaves (field and lab tests, respectively) were preferred by the leaf cutting ant *Acromyrmex lundi* to unstressed ones, overriding subspecies or provenance.

The decreased performance of herbivores on water stressed plants may be caused by several non-exclusive reasons: an increase in plant defense levels, a decrease of carbohydrate levels and/or a decrease in foliar water content (Scheirs & De Bruyn, 2006). But drought may increase or decrease secondary metabolism, increasing or decreasing host quality of trees for insect herbivores. For example, in several *Eucalyptus* species Muller da Silva et al. (2006) observed that water restriction decreased essential oil production while Stone & Bacon (1994) reported no modifications in total terpenoids yield.

The concentration of total phenolic glycosides increased in leaves of poplars that were growing under water stress conditions, however the growth of only one species of herbivore was negatively affected (Hale et al., 2005).

Sometimes the impact of herbivory is often greater if other original stress damage and lead to important tree losses. In young plantations of *E. camaldulensis* stressed by moisture deficit and in two young plantations of *E. dunnii* stressed by flooding and weed competition, Stone (2001) observed that the stress-inducing agents reduced canopy growth rates and architecture so that the proportion of leaf tissue damage by insects increased and the tree's ability to tolerate that damage decreases in all three cases.

Herbivory is thought to increase on water stress plants due to induced changes in plant physiology, specifically N availability. Water stress mediates N availability and then modifies the quality of the leaves and the population dynamics of insect herbivores (Huberty & Denno, 2004). Self-pruning of old leaves in combination with a reduction in leaf area of new leaves may allow the reallocation of foliar proteins to young leaves, explaining the enhanced foliar protein concentration in water stressed plants (Scheirs & De Bruyn, 2005). It was found that a greater availability of nutrients increased total essential oil and cineole levels in eucalypt leaves (Close et al., 2003; O´Reilly-Wapstra, 2005), however complex the effects of nutrients and secondary compounds on herbivores responses may be esteemed.

The susceptibility of pinyon pine (*Pinus edulis*) to the stem-and cone-borer increased in sites with intense water and nutrient deficiencies and the release from stress led to increased resistance to insect attack (Coob et al., 1997). *Eucalyptus* plants with high levels of fertilization showed higher levels of herbivory (Paine & Hanlon, 2010). Similarly, *Eucalyptus* seedlings, which had higher foliage N, lower tanin and higher essential oil levels were more browsed (Close et al., 2004).

Other abiotic stresses such as shade can affect the levels of damage by herbivores. A number of studies have reported lower tolerance of herbivory under shade conditions (Baraza et al.,

2004; McGraw et al., 1990; Norghaver et al., 2008; Salgado-Luarte & Gianoli, 2011). Differential herbivory in contrasting light environments may reflect light-induced differences in plant defensive traits such as leaf toughness and carbon-based secondary compounds. Plants' resistance may be greater in the sun probably due to their lower specific leaf area (SLA) (Salgado-Luarte & Gianoli, 2011). While comparing several characteristics in *Eucalyptus* seedlings growing under different levels of light, Guarnaschelli et al. (2003a, 2003b) found higher levels of herbivory damage in shaded plants, which had lower leaf thickness and toughness associated with higher SLA. Although Nichols-Orians (1991) detected much higher foliar concentrations of condensed tannins in plants growing in full sunlight compared to those growing in the understory, leaf-cutting ants found these leaves more acceptable because of the higher concentrations of foliar nutrients that override the benefits of increased concentrations of tannins. In other studies shade affected the chemical defences of *Acer, Quercus* and *Pinus* seedlings lowering the levels of total phenols and condensed tannins and increasing the levels of N (Baraza et al., 2004; Baraza et al., 2010).

Moreover, plant resistance to herbivory can be influenced not only by the independent effects of plant genotype and environmental variation, but also by interactions between the two. A significant genotype x fertilizer interaction was found in the defensive compounds of *E. globulus* and their resistance to mammalian herbivores (O´Reilly-Wapstra et al., 2005). Some differences were recently found in the levels of leaf damage caused by *Leptocybe invasa* in seedlings of several clones of *E. grandis* and *E. grandis* x *E. camaldulensis* that were growing under alternative water and light availability regimes. Attacks were concentrated only in plants of *E. grandis* x *E. camaldulensis* clones, particularly in plants under drought, shade and drought plus shade conditions (Guarnaschelli et al., unpublished data).

Management options

Interestingly enough, induction of intra-specific herbivore resistance is possible through manipulation of light, nutrients and water in the nursery environment or water and nutrient availability during the establishment. In fact, the potential of nursery preconditioning to enhance survival chances of future trees by reducing palatability or attracting beneficial insects as a result of changes in physical and/or chemical defences seems to be promising. Water availability manipulation during the nursery period of *E. camaldulensis* seedlings modified leaves essential oil composition, and drought triggered a significant increase in several oxygenated terpenes production (particularly linalool and 1,8-cineole) known to repel different defoliator species (Leicah et al., 2010).

Seedlings of *E. globulus* and *E. nitens* with high-fertilizer nursery treatment were browsed more than the low-fertilizer nursery treatment, the results being consistent with their differences in foliar N and tannins (Close et al., 2004). Similarly, the young foliage of *E. nitens* seedlings that receive medium and high levels of nutrient regime was intensively consumed. High levels of N in young leaves outweighed the presence of sideroxylonals and essential oils that generally act as deterrents to herbivores (Loney et al., 2006).

In contrast, low nutritional quality seems to be a plant defence against herbivores. Haukioja et al. (1991) demonstrated that the low nutritional quality is a potential active defence against herbivory in the mountain birch (*Betula pubescens* subsp. *tortuosa*), especially when it is correlated with repellent allelochemicals. In field trials Paine & Hanlon (2010) found that *Eucalyptus* trees treated with higher levels of irrigation and no fertilization demonstrated lower levels of damage by a psyllid.

Therefore, manipulation of seedlings or saplings chemistry through the application of particular fertilizer regimes as well as the regulation of water availability are useful tools for managers wishing to decrease damage by herbivory during the establishment. Alleviating tree stress through improved silvicultural practices or improved site selection techniques may indirectly reduce the impact of insect herbivory. In resource-limiting environments, an alternative approach may be planting species of slower growth that are predicted to have better defended foliage. Manipulation of these natural antiherbivore plant strategies is not exclusive of other management approaches, such as the genetic selection of natural insect resistance and selective chemical control techniques (Stone, 2001).

6. Conclusions

Water stress is the main cause of alteration of plant physiological processes and reduction of plant growth, which affects considerably tree seedlings performance immediately after plantation and during the whole establishment stage, causing serious losses and affecting tree regeneration and, consequently, future stand productivity. Water stress, itself and together with other abiotic and biotic factors, affects the dynamics of forest establishment, having an enormous impact on seedling performance. Multifactor stress conditions or sometimes the impact of secondary factors are often greater than the original drought effects, leading to greater tree losses. The enhancement of forest regeneration is a key process to ensure high forest productivity to supply timber needs in the present scenario of increasing demand of forest products. As discussed above, successful establishment depends on several management aspects. Breeding for drought tolerance should be given high priority in all research programs, considering genetic variation and phenotypic variation. In order to mitigate the transplanted shock, we also highlighted the importance of regulating water and nutrient regimes in nursery as useful tools not only to modify drought tolerance but also to reduce seedling's palatability, and hence herbivory damage in plantation forestry. Previous exposure to stress conditions, or a natural hardening period, can markedly influence future plant responses. Drought resistant seedling, nursery hardened, suited to the particular environmental condition, will have better field performance.

Some of the factors associated with successful forest establishment which have not been reviewed here, such as cultural practices applied before and over the establishment, have been widely studied; however, it is still necessary to foster new research in other areas. More information concerning seedling physiology is required, especially in relation of how seedlings adapt to particular plantation environments. The effects of water deficit and freezing temperatures, as well as the mechanisms that govern plant drought and frost tolerance, closely related to plant survival and growth during the establishment, have been analyzed in depth due to their great economic impact and relevance to stand productivity. Nowadays, climate change is altering environmental conditions. It is expected that drought conditions will become more frequent in extended areas. Changes in water availability and temperature will probably interact with other abiotic and biotic factors that may have a high effect during the forest establishment and the whole rotation. Under this environmental challenge it will be necessary for any species, provenances and clones, to understand the physiological basis of plant responses to water stress and together with interacting stressful factors that will be crucial also for predicting forest productivity. At the same time, nurseries' managers will be requested to adapt their cultural techniques in order to produce high quality stock seedlings according to good field performance. Irrigation and nutritional protocols at nurseries should also be adjusted not only taking into account their influence on abitioc stress tolerance but also considering their impact on plant nutritional status and the host quality of trees for insect herbivores. It is necessary to establish relationships between field performance and physiological and morphological properties of nursery stock for each species, which will allow the implementation of the most adequate practices to produce stock of high quality. Finally, foresters will be requested to adapt the silvicultural planning to the present environmental concerns applying low impact practices and implementing integrative vegetation and pests control programs according to the principles of sustainable management.

7. References

Abrams, M. D. (1990). Adaptations and responses to drought in *Quercus* species of North America. *Tree Physiology*, Vol.7, No.1-2-3-4, (December 1990), pp. 227-238, ISSN 0829-318X.

Adams, P. R.; Beadle, C. L. & Mendham, N. J. (2003). The impact of timing and duration of grass control on growth of a young *Eucalyptus globulus* Labill. plantation. *New Forests*, Vol.26, No.2, (September 2003), pp. 147-165, ISSN 0196-4286.

Adir, N.; Zer, H.; Shochat, S. & Chad, I. (2003). Photoinhibition - a historical perspective. *Photosynthesis Research*, Vol.76, No.1-3, (April 2003), pp. 343-370, ISSN 0166-8595.

Aranda, I.; Castro, L.; Pardos, M.; Gil, L. & Pardos, J. A. (2005). Effects of the interaction between drought and morphological traits in cork oak (*Quercus suber* L.) seedlings. *Forest Ecology and Management*, Vol.210, No.1-3, (May 2005), pp. 117-129, ISSN 0378-1127.

Arend, M.; Kuster, T.; Günthardt-Goerg, M. S.; Dobbertin, M. & Abrams, M. (2011). Provenance-specific growth responses to drought and air warming in three European oak species (*Quercus robur*, *Q. petraea* and *Q. pubescens*). *Tree Physiology*, Vol.31, No.3, (April 2011), pp. 287-297, ISSN 0829-318X.

Arndt, S. K.; Wanek, W.; Clifford, S. C. & Popp, M. (2000). Contrasting adaptations to drought stress in field-grown *Ziziphus mauritiana* and *Prunus persica* trees: water relations, osmotic adjustment and carbon isotope composition. *Functional Plant Biology*, Vol.27, No.11, (November 2000), pp. 985-996, ISSN 1445-4408.

Arnott, J. T.; Grossnickle, S. C.; Puttonen, P.; Mitchel, K. A. & Folk, R. S. (1993). Influence of nursery culture on growth, cold hardiness and drought resistance of yellow cypress. *Canadian Journal of Forest Research*, Vol.23, No.12, (December 1993), pp. 2537-2547, ISSN 0045-5067.

Ashraf, M.; Akram, N. A.; Al-Qurainy, F. & Foolad, M. R. (2011). Drought tolerance: roles of organic osmolytes, growth regulators, and mineral nutrients. *Advances in Agronomy*, Vol.111, (April 2011), pp. 249-296, ISBN 012-0007-95-9.

Balandier, P.; Collet, C.; Miller, J.; Reynolds, P. & Zedaker, S. (2006). Designing forest vegetation management strategies based on the mechanisms and dynamics of crop tree competition by neighbouring vegetation. *Forestry*, Vol.79, No.1, (January 2006), pp. 3-27, ISSN 0015-752X.

Baraza, E.; Gómez, J.; Hódar, J. & Zamora, R. (2004). Herbivory has a greater impact in shade than in sun: responses of *Quercus pyranaica* seedlings to multifactorial environmental variation. *Canadian Journal of Botany*, Vol.82, No.3, (March 2004), pp. 357-364, ISSN 1916-2790.

Baraza, E.; Zamora, R. & Hódar, J. (2010). Species-specific responses of tree saplings to herbivory in contrasting light environments: An experimental approach. *Ecoscience*, Vol.17, No.2, (June 2010), pp. 156-165, ISSN 1195-6860.

Barnett, J. P. & McGilvray, J. M. (1993). Performance of container and bareroot Loblolly pine seedlings on bottomlands in South Carolina. *Southern Journal of Applied Forestry*, Vol.17, No.2, (May 1993), pp. 80-83, ISSN 0148-4419.

Becerra, P. I. & Bustamante, R. O. (2008). The effect of herbivory on seedling survival of the invasive exotic species *Pinus radiata* and *Eucalyptus globulus* in a Mediterranean ecosystem of Central Chile. *Forest Ecology and Management*, Vol.256, No.9 (October 2008), pp. 1573-1578, ISSN 0378-1127.

Bennett, R. N. & Wallsgrove, R. M. (1994). Secondary metabolites in plant defence mechanisms. *New Phytologist*, Vol.127, No.4, (August 1994), pp. 617-633, ISSN 1469-8137.

Bigras, F. J. & Dumais, D. (2005). Root-freezing damage in the containerized nursery: impact on plantation sites - A review. *New Forests*, Vol.30, No.2-3, (September 2005), pp. 167-184, ISSN 0196-4286.

Blum, A. (2005). Drought resistance, water-use efficiency, and yield-potential-are they compatible, dissonant, or mutually elusive? *Australian Journal of Agricultural Research*, Vol.56, No.11, (November 2005), pp. 1159-1168, ISSN 0004-9409.

Braatne, J. H.; Hinckley, T. M. & Stettler, R. F. (1992). Influence of soil water on the physiological and morphological components of plant water balance in *Populus trichocarpa, Populus deltoides* and their F1 hybrids. *Tree Physiology*, Vol.11, No.4, (December 1992), pp. 325-339, ISSN 0829-318X.

Bradshow, A. D. (2006). Unraveling phenotypic plasticity - why should we bother? *New Phytologist*, Vol.170, No.4, (June 2006), pp. 644-648, ISSN 1469-8137.

Brodribb, T. J. (2009). Xylem hydraulic physiology: The functional backbone of terrestrial plant productivity. *Plant Science*, Vol.177, No.4, (October 2009), pp. 245-251, ISSN 0168-9452.

Brodribb, T. J. & Mc Adam, S. A. (2011). Passive origins of stomatal control in vascular plants. *Science*, Vol.331, No.6017, (February 2011), pp. 582-585, ISSN 0036-8075.

Burdett, A. N. (1990). Physiological processes in plantation establishment and the development of specification for forest planting stock. *Canadian Journal of Forest Research*, Vol.20, No.4, (April 1990), pp. 415-427, ISSN 0045-5067.

Caffarini, P.; Pelicano, A.; Carrizo, P. & Lemcoff, J. H. (2006). Impacto del estrés hídrico y la procedencia de *Eucalyptus globulus* Labill. sobre el comportamiento de herbivoría de *Acromyrmex lundi* Guérin. *Idesia*, Vol.24, No.1, (April 2006), pp. 7-11, ISSN 0073-4675.

Caldwell, J.; Sucoff, E. & Dixon, K. (1995). Grass interference limits resource availability and reduces growth of juvenile red pine in the field. *New Forests*, Vol.10, No.1, (July 1995) pp. 1-15, ISSN 0196-4286.

Cakman, I. (2005). The role of potassium in alleviating detrimental effects of abiotic stresses in plants. *Journal of Plant Nutrition and Soil Science*, Vol.168, No.4, (August 2005), pp. 521-530, ISSN 1522-2624.

Cannell, M. G. R. & Grace, J. (1993). Competition for light: detection, measurement, and quantification. *Canadian Journal of Forest Research*, Vol.23, No.10, (October 1993), pp. 1969-1979, ISSN 0045-5067.

Cannell, M. G. R. & Dewar, R. C. (1994). Carbon allocation in trees: a review of concepts for modeling, In: M. Begon & H. A. Fitter, (Eds.), pp.59-104. *Advances in Ecological Research* 25, Academic Press Ltd., ISBN 978-012-0139-25-5, London, UK.

Chaves, M. M.; Maroco, J. P. & Pereira, J. S. (2003). Understanding plant responses to drought: from genes to whole plant. *Functional Plant Biology*, Vol.30, No.3, (March 2003), pp. 239-264, ISSN 1445-4408.

Close, D. J.; McArthur, C.; Paterson, S.; Fitzgerald, H.; Walsh, A. & Kincade, T. (2003). Photoinhibition: a link between effects of the environment on eucalypt leaf chemistry and herbivory. *Ecology*, Vol.84, No.11, (November 2003), pp. 2952-2966, ISSN 0012-9658.

Close, D. J.; McArthur, C.; Pietrzykowski, E.; Fitzgerald, H. & Paterson, S. (2004). Evaluating effects of nursery and post-planting nutrient regimes on leaf chemistry and browsing of eucalypt seedlings in plantation. *Forest Ecology and Management*, Vol.200, No.1-3, (October 2004), pp. 101-112, ISSN 0378-1127.

Close, D.; Bail, I.; Hunter, S. & Beadle, C. L. (2005). Effects of exponential nutrient-loading on morphological and nitrogen characteristics and on after-planting performance of *Eucalyptus globulus* seedlings. *Forest Ecology and Management*, Vol.205, No.1-3, (February 2005), pp. 397-403, ISSN 0378-1127.

Close, D. J.; Beadle, C. L. & Brown, P. H. (2005). The physiological basis of containerized tree seedlings 'transplant shock': a review. *Australian Forestry*, Vol.68, No.2, (June 2005), pp. 112-120, ISSN 0004-9158.

Cochard, H.; Coll, L.; Le Roux, X. & Améglio, T. (2002). Unraveling the effects of plant hydraulics on stomatal closure during water stress in walnut. *Plant Physiology*, Vol.128, No.1, (January 2002), pp. 282-290, ISSN 0032-0889.

Cogliastro, A.; Benjamin, K. & Bouchard, A. (2006). Effects of full and partial clearing, with and without herbicide, on weed cover, light availability, and establishment success of white ash in shrub communities of abandoned pastureland in southwestern Québec, Canada. *New Forests*, Vol.32, No.2, (September 2006), pp. 197–210, ISSN 0196-4286.

Coll, L.; Balandier, P. & Picon-Cochard, C. (2004). Morphological and physiological responses of beech (*Fagus sylvatica*) seedlings to grass-induced belowground competition. *Tree Physiology*, Vol.24, No.1, (January 2004), pp. 45-54, ISSN 0829-318X.

Coob, N. S.; Mopper, S.; Gehring, C. A.; Caouette, M.; Christensen, K. M. & Whitman, T. G. (1997). Increased moth herbivory associated with environmental stress of pinyon pine at local and regional levels. *Oecologia*, Vol.109, No.3, (February 1997), pp. 389-397, ISSN 0029-8549.

Coopman, R. E.; Jara, J. C.; Escobar, R.; Corcuera, L. J. & Bravo, L. A. (2010). Genotypic variation in morphology and freezing resistance of *Eucalyptus globulus* seedlings subjected to drought hardening in nursery. *Electronic Journal of Biotechnology*, Vol.13, No.1, (January 2010), 9 pp., Available at:
http://www.ejbiotechnology.info/ index.php/ejbiotechnology/index.

Costa e Silva, F.; Shvaleva, A.; Broetto, F.; Ortuño, M. F.; Rodrigues, M. L.; Almeida, M. H.; Chaves, M. M. & Pereira, J. S. (2009). Acclimation to short-term low temperatures in two *Eucalyptus globulus* clones with contrasting drought resistance. *Tree Physiology*, Vol.29, No.1, (January 2009), pp. 77-86, ISSN 0829-318X.

Cregg, B. M. (1994). Carbon allocation, gas exchange, and needle morphology of *Pinus ponderosa* genotypes known to differ in growth and survival under imposed drought, *Tree Physiology*, Vol.14, No.7-8-9, (July 1994), pp. 883-898, ISSN 0829-318X.

Cuesta, B.; Villar-Salvador, P.; Puértolas, J.; Jacobs, D. J. & Rey Benayas, J. M. (2010). Why do large, nitrogen rich seedlings, better resist stressful transplanting conditions? A physiological analysis in two functionally contrasting Mediterranean forest species. *Forest Ecology and Management*, Vol.260, No.1, (June 2010) pp. 71-78, ISSN 0378-1127.

Davies, W. J.; Metcalfe, J.; Lodge, T. A. & da Costa, A. R. (1986). Plant growth substances and the regulation of growth under drought. *Australian Journal of Plant Physiology*, Vol.13, No.1, (January 1986), pp. 105-125, ISSN 1445-4408.

Davis, M.; Wrage, K.; Reich, P.; Tjoelker, M.; Schaeffer, T. & Muermann, C. (1999). Survival, growth, and photosynthesis of tree seedlings competing with herbaceous vegetation along a water-light-nitrogen gradient. *Plant Ecology*, Vol.145, No.2, (December 1999), pp. 341-350, ISSN 1385-0237.

Della Lucia, T. M. C. (Ed.). (1993). *As formigas cortadeiras*, De Folha da Viçosa, ISBN , Minas Gerais, Brazil.

Dinger, E. & Rose, R. (2009). Integration of soil moisture, xylem water potential, and fall-spring herbicide treatments to achieve the maximum growth response in newly planted Douglass-fir seedlings. *Canadian Journal of Forest Research*, Vol.39, No.7, (July 2009), pp. 1401-1414, ISSN 0045-5067.

Dinger, E. & Rose, R. (2010). Initial fall-spring vegetation management regimes improve moisture conditions and maximize third-year Douglass-fir seedling growth in a Pacific Northwest plantation. *New Zealand Journal of Forestry Science*, Vol.40, No.1, (July 2010), pp. 93-108, ISSN 0048-0134.

Dulamsuren, C.; Hauck, M. & Mühlenberg, M. (2008). Insect and small mammal herbivores limit tree establishment in northern Mongolian steppe. *Plant Ecology*, Vol.195, No.2, (April 2008), pp. 143-156, ISSN 1385-0237.

Duryea, M I. & Landis, T. D. (Eds.). (1984). *Forest Nursery manual: Production of bareroot seedlings*, Martinus Nijhoff/Dr M. Junk Publisher, ISBN 978-902-4729-13-5, The Hague, Netherlands.

Edwards, R. R & Dixon, M. D. (1995). Mechanisms of drought response in *Thuja occidentalis* L. I. Water stress conditioning and osmotic adjustment. *Tree Physiology*, Vol.15, No.2, (February 1995), pp. 121-127, ISSN 0829-318X.

Fan, S.; Blake, T. J. & Blumwald, E. (1994). The relative contribution of elastic and osmotic adjustment to turgor maintenance of woody species. *Physiologia Plantarum*, Vol. 90, No.2, (February 1994), pp. 414–419, ISSN 0031-9317.

Food and Agriculture Organization of the United Nations (FAO). (2009). *State of the world´s forests 2009*, FAO. ISBN 978-925-1060-57-5, Rome, Italy.

Food and Agriculture Organization of the United Nations (FAO). *Global Forest Resource Assessment 2010*, FAO. ISBN 978-925-3066-54-4, Rome, Italy.

Florence, R. G. (1996). *Ecology and Silviculture of eucalypt forest*. CSIRO Publishing, ISBN 978-064-3057-99-9, Collinwood, Victoria, Australia.

Florentine, S. & Fox, J. (2003). Competition between *Eucalyptus victrix* seedlings and grass species. *Ecological Research*, Vol.18, No.1, (January 2003), pp. 25-39, ISSN 0912-3814.

Floyd, R. B.; Farrow, R. A. & Neumann, F. G. (1995). Inter- and intra-provenance variation in resistance to insect feeding. *Australian Forestry*, Vol.57, No.1, (March 1995), pp. 45-48, ISSN 0004-9158.

Folgarait, P. J.; Dyer, L. E.; Marquis, T. J. & Braker, H. E. (1996). Leaf-cutting ant preferences for five native tropical plantation tree species growing under different light conditions. *Entomologia Experimentalis et Applicata*, Vol. 80, No.3, (September 1996), pp. 521-530, ISSN 0013-8703.

Forti, L. C. & Castellani Boaretto, M. A. C. (1997). Formigas cortadeiras. Biologia, ecologia, danhos e controle. Botucatu, San Pablo, Brazil.

Garau, A. M. (2003). Estrategias de tolerancia al estrés hídrico provocado por la competencia con malezas durante el período de implantación de eucalipto. M.Sc. Thesis, Facultad de Agronomía, Universidad de Buenos Aires, Argentina.

Garau, A. M.; Caccia, F. D. & Guarnaschelli, A. B. (2008a). Impact of standing vegetation on early establishment of willow cuttings in the flooded area of the Parana River Delta. *New Forests*, Vol.36, No.1, (July 2008), pp. 79-91, ISSN 0196-4286.

Garau, A. M.; Guarnaschelli, A. B.; Mema, V. & Lemcoff, J. H. (2004a). Tissue water relations in *Eucalyptus* seedlings: effects of species, K fertilization and drought. *Proceedings of International IUFRO Eucalyptus in a Changing World*, Aveiro, Portugal, October 11-15, 2004.

Garau, A. M.; Guarnaschelli, A. B.; Prystupa, P. & Lemcoff, J. H. (2004b). Relaciones hídricas en plantines de *Eucalyptus* sp. Sujetos a la interacción de fertilización nitrogenada y potásica con restricción hídrica. *Proceedings XXV Reunión Argentina de Fisiología Vegetal*. p.97, San Rosa, La Pampa, Argentina. September 22-24, 2004.

Garau, A. M.; Lemcoff, J. H.; Ghersa, C. M. & Beadle, C. L. (2008b). Water stress tolerance of *Eucalyptus globulus* Labill. subsp. *maidenii* (F.Muell.) saplings induced by water restriction imposed by weeds. *Forest Ecology and Management*, Vol.255, No.7, (April 2008), pp. 2811-2819, ISSN 0378-1127.

Garau, A. M.; Guarnaschelli, A. B.; Prystupa P.; Niveyro, I.; Landi, L.; Zorza, F. & Lemcoff, J. H. (2005). Fertilización y restricción hídrica en plántulas de *Eucalyptus camaldulensis*: modificaciones morfológicas y sus efectos sobre el estrés de plantación. *CD Proceedings Tercer Congreso forestal Argentino y Latinoamericano*, ISSN 1669-6786, Corrientes, Argentina, September 6-9, 2005.

Goldberg, D. (1996). Simplifying the study of competition at the individual plant level: consequences of distinguishing between competitive effect and response for forest vegetation management. *New Zealand Journal of Forestry Science*, Vol.26, No.1/2, pp. 19-38, ISSN 0048-0134.

Gower, S. T.; Vogt, K. A. & Grier, C. C. (1992). Carbon Dynamics of Rocky Mountain Douglas-Fir: Influence of Water and Nutrient Availability. *Ecological Monographs*, Vol.62, No.1, (March 1992), pp. 43–65, ISSN 0012-9615.

Grossnickle, S. C. (1988). Planting stress in newly planted jack pine and white spruce. 2. Changes in tissue water potential components. *Tree Physiology*, Vol.4, No.1, (March 1988), pp. 85-97, ISSN 0829-318X.

Grossnickle, S. C. (2005). Importance of root growth in overcoming planting stress. *New Forests* Vol.30, No.2-3, (September 2005), pp. 273-294, ISSN 0196-4286.

Grossnickle, S. C. & Folk, R. S. (1993). Stock quality assessment: Forecasting survival or performance on a reforestation site. *Tree Planter's Note*, Vol.44, No.3, (Summer 1993), pp. 113-121. Available from http://www.rngr.net/publications/tpn.

Guarnaschelli, A. B. (2009). Efecto del ajuste osmótico y la elasticidad de la pared celular sobre el crecimiento de *Eucalyptus globulus* durante el establecimiento. M.Sc. Thesis, Facultad de Agronomía, Universidad de Buenos Aires, Argentina.

Guarnaschelli, A. B. & Lemcoff, J. H. (2001). Shoot and root osmotic adjustment in seedlings of *E. globulus* subsp. *maidenii*. Proceedings *The 6th ISRR Symposium. Roots: The Dynamic Interface between plants and the earth*, pp. 480-481, ISBN 493-1358-07-1, Nagoya, Japan, November 11-15, 2001.

Guarnaschelli, A. B.; Garau, A. M. & Lemcoff, J. H. (2001). Tissue water relations in *Eucalyptus globulus* subsp. *maidenii*. *Proceedings IUFRO International Symposium Developing Eucalypts for the Future*, pp. 65, Valdivia, Chile, September 10-15, 2001.

Guarnaschelli, A. B.; Prystupa, P. & Lemcoff, J. H. (2006). Drought conditioning improves water status, stomatal conductance and survival of *Eucalyptus globulus* subsp. *bicostata*. *Annals of Forest Science*, Vol.63, No.8, (December 2006), pp. 941-950, ISSN 1286-4560.

Guarnaschelli, A. B.; Erice, F.; Battaglia, A. & Lemcoff, J. H. (2003a). Respuestas a la sequía y a la sombra en plantines de *Eucalyptus grandis*. *CD XVIII Jornadas Forestales de Entre Ríos*. ISSN 1667-9253, Concordia, Entre Ríos, Argentina, October 23-24, 2003.

Guarnaschelli, A. B.; Garau, A. M.; Caccia, F. D. & Cortizo, S. C. (2008). Physiological responses to shade and drought in young willow plants. *Proceedings 23rd Session of the International Poplar Commission. Poplars, willows and people's wellbeing*, Available from ftp://ftp.fao.org/docrep/fao/011/k3334e/k3334e.pdf, pp. 75 Beijing, China, October 26–30, 2008.

Guarnaschelli, A. B.; Garau, A. M.; Cortizo, S. C. & Lemcoff, J. H. (2010a). Differences in the ability of *Salix* clones to cope with water and light restriction. *IPS-V Books of Abstracts IUFRO Fifth International Poplar Symposium. Poplars and willows: research models to multipurpose trees for biobased society* (Available from http://ocs.entecra.it/index.php/IPS/index/pages/view/material), pp. 183, Orvieto, Italia, September 20-25, 2010.

Guarnaschelli, A. B.; Garau, A. M.; Lemcoff, J. H. & Pathauer, P. (2007). Impacto de la sombra y la sequía sobre especies fisiológicas y de crecimiento en diversos orígenes de *Eucalyptus globulus* subesp. *globulus*. *CIDEU Bulletin, Bulletin of Forest and Environmental Topic*, Vol.3, No.1, (March 2007), pp. 91-98, ISSN 1885-5237.

Guarnaschelli, A. B.; Gutiérrez G.; Garau, A. M. & Lemcoff, J. H. (2003b). Aclimatación a la sombra en plántulas de *E. globulus* subesp. *globulus*. *Proceedings Primer Simposio Iberoamericano* de *Eucalyptus globulus*, Montevideo, Uruguay, October 30-31, 2003.

Guarnaschelli, A. B.; Lemcoff, J. H.; Prystupa, P. & Basci, S. O. (2003c). Responses to drought preconditioning in *Eucalyptus globulus* Labill. provenances. *Trees*, Vol.17, No.6, (November 2003), pp. 501-509, ISSN 0931-1890.

Guarnaschelli, A. B.; Garau, A. M.; Mendoza, E. M.; Zivec, V. & Cortizo, S. C. (2010b). Influence of drought conditions on morphological and physiological attributes of *Populus deltoides* clones. *IPS-V Book of Abstracts IUFRO Fifth International Poplar Symposium. Poplars and willows: research models to multipurpose trees for biobased society*, Available from http://ocs.entecra.it/index.php/IPS/index/pages/view/material, pp. 155, Orvieto, Italia, September 20-25, 2010.

Guarnaschelli, A. B.; Mantese, A.; Barañao, J. J.; De Haro, A. M. & Lemcoff, J. H. (2000). Anatomical leaf characteristics related to herbivory in *Eucalyptus globulus* subesp. *maidenii* seedlings, In: *The tree, L'Abre 2000,* I. Quentin, (Ed.) pp. 59-63, IQ Collectif Institut de Recherche en Biologie Végétale, ISBN 292-2417-21-2, Montréal, Canada.

Guarnaschelli, A. B.; Ruiz Nuñez, J.; Chiavassa, J. A.; Fedotova, N. & Garau, A. M. (2010c). Aclimatación en vivero en plantas de *Eucalyptus* por restricción hídrica y fertilización potásica. *CD XXIV Jornadas Forestales de Entre Ríos,* ISSN 1667-9253 Concordia, Entre Ríos, Argentina, October 28-29, 2010.

Guehl, J. M.; Aussenac, G. & Kaushal, P. (1989). The effects of transplanting stress on photosynthesis, stomatal conductance and leaf water potential in *Cedrus atlantica* Manetti seedlings: the role of root regeneration. *Annals of Forest Science,* Vol.46, No.Suplemental, pp. 464-468, ISSN 1286-4560.

Guo, X. Y.; Zhang, X. S. & Huang, Z. Y. (2010). Drought tolerance in three hybrid poplar clones submitted to different watering regimes. *Journal of Plant Ecology,* Vol.3, No.2, (June 2010), pp. 79-87, ISSN 1752-9921.

Hale, B.; Herms, D.; Hansen, R.; Clausen, T. & Arnold, D. (2005). Effects of drought stress and nutrient availability on dry matter allocation, phenolic glycosides and rapid induced resistance of poplar to two Lymantriid defoliators. *Journal of Chemical Ecology,* Vol.31, No.11, (November 2005), pp. 2601-2620, ISSN 0098-0331.

Harvey, H. P. & van den Driessche, R. (1999). Nitrogen and potassium effects on xylem cavitation and water-use efficiency in poplars. *Tree Physiology,* Vol.19, No.14, (December 1999), pp. 943-950, ISSN 0829-318X.

Haukioja, E.; Ruohomaki, K.; Suomela, J. & Vuorisalo, T. (1991). Nutritional quality as a defense against herbivores. *Forest Ecology and Management,* Vol.39, (1991), pp. 237-245, ISSN: 0378-1127.

Hirasawa, T. & Hsiao, T. C. (1999). Some characteristics of reduced leaf photosynthesis at midday in maize growing in the field. *Field Crops Research,* Vol. 62, No.1, (June 1999), pp. 53-62, ISSN 0378-4290.

Hsiao, T. C. (1973). Plant responses to water stress. *Annual Review of Plant Physiology,* Vol.24, (June 1973), pp. 519-570, ISSN 0066-4294.

Huberty, A. & Denno, R. (2004). Plant water stress and its consequences for herbivores insects: a new synthesis. *Ecology,* Vol.85, No.5, (May 2004), pp. 1383-1398, ISSN 0012-9658.

Intergovernmental Panel on Climate Change (IPCC). (2007). *Climate Change 2007: Impacts, Adaptation, and Vulnerability. Contribution of Working Group II to the Fourth Assessment Report of the Intergovernmental Panel on Climate Change.* M. L. Parry; O. F. Canziani; J. P. Palutikof; P. J. van der Linden & C. E Hanson, C. E. (eds.). Cambridge University Press, ISBN 978-052-1705-97-7, Cambridge, UK.

Jacobs, D. F.; Ross-Davis, A. L. & Davis, A. S. (2004). Establishment success of conservation tree plantation in relation to silvicultural practices in Indiana, USA. *New Forests,* Vol.28, No.1, (July 2004), pp. 23-36, ISSN 0196-4286.

Jacobs, D. F.; Salifu, K. F. & Davis, A. S. (2009). Drought susceptibility and recovery of transplanted *Quercus rubra* seedlings in relation to root system morphology. *Annals of Forest Science,* Vol. 66, No.6, (November 2009), 504, 12 pp., ISSN 1286-4560.

Jackson, R. B.; Jobbágy, E. G.; Avissar, R.; Roy, S. B.; Barrett, D. J.; Cook, C. W.; Farley, K. A.; le Maitre, D. C., McCarl, B. A. & Murray, B. C. (2005). Trading Water for Carbon with Biological Carbon Sequestration. *Science*, Vol.310, No.5752, (November 2005), pp. 1944-1947, ISSN 0036-8075.

Jobbágy, E. G. & Jackson, R. B. (2004). Groundwater use and salinization with grassland afforestation. *Global Change Biology*, Vol.10, No.8, (August 2004), pp. 1299–1312, ISSN 1354-1013.

Johnsen, K. H. & Major, J. E. (1999). Shoot water relations of black spruce families displaying a genotype x environment interaction in growth rate. I. Family and site effects over three growing seasons. *Tree Physiology*, Vol.19, No.6, (May 1999), pp. 367-374, ISSN 0829-318X.

Kort, J.; Collins, M. & Ditsch, D. (1998). A review of soil erosion potential associated with biomass crops. *Biomass and Bioenergy*, Vol.14, No.4, (April 1998), pp. 351-359, ISSN 0961-9534.

Kozlowski, T. T. (1982). Water supply and tree growth. I Water deficits. *Forestry Abstracts*, Vol.43, pp. 57-95, ISSN 0015-7538.

Kozlowski, T. T. (1992). Carbohydrate sources and sinks in woody plants. *Botanical Review*, Vol.58, No.2, (April 1992) pp. 107-222, ISSN 0006-8101.

Kozlowski, T. T. (2002). Physiological ecology of natural regeneration of harvested and disturbed stands: implications for forest management. *Forest Ecology and Management*, Vol.158, No.1-3, (March 2002), pp. 195-221, ISSN 0378-1127.

Kozlowski, T. T. & Davies, D. (1975). Control of water balance in transplanted trees. *Journal of Arboriculture*, Vol.1, No.1, (January 1975), pp. 1-10, Available from http://joa.isa-arbor.com/browse.asp?Journals_ID=1.

Kozlowski, T. T. & Pallardy, S. G. (1997). *Physiology of woody plants*, Second edition, Academic Press, ISBN 012-4241-62-X, San Diego, California, USA.

Kozlowski, T. T. & Pallardy, S. G. (2002). Acclimation and adaptive responses of woody plants to environmental stresses. *Botanical Review*, Vol.68, No.2, (April 2002), pp. 270-334, ISSN 0006-8101.

Kramer, P. J. & Boyer, J. S. (1995). *Water relations of plants and soils*, Academic Press, ISBN 012-4250-60-2, San Diego, California, USA.

Laclau, P. (2003). Biomass and carbon sequestration of ponderosa pine plantations and native cypress forests in northwest Patagonia. *Forest Ecology and Management*, Vol.180, No.1-3, (July 2003), pp. 317-333, ISSN 0378-1127.

Lamhamedi, M. S.; Bernier, P.; Hebert, C. & Jobdon, R. (1998). Physiological and growth responses of three sizes of containerized *Picea Mariana* seedlings out planted with and without vegetation control. *Forest Ecology and Management*, Vol.110, No.1-3, (October 1998), pp. 13-23, ISSN 0378-1127.

Lamhamedi, M. S.; Lambany, G.; Margolis, H.; Renaud, M.; Veilleux, L. & Bernier, P. Y. (2001). Growth, physiology, root architecture and leaching of air-slit containerized *Picea glauca* seedlings (1+0) in response to time domain reflectrometry control irrigation regime. *Canadian Journal of Forrest Research*, Vol.31, No., pp. 1968-1980, ISSN 0045-5067.

Lande, R. (2009). Adaptation to an extraordinary environment by evolution of phenotypic plasticity and genetic assimilation. *Journal of Evolutionary Biology*, Vol.22, No.7, (July 2009), pp. 1435-1446, Online ISSN 1420-9101.

Landis, T. D. (2005). Macronutrients. Potassium. *Forest Nursery Notes*. Winter 2005. pp. 5-11, Available from http://www.rngr.net/publications/fnn.

Landis, T. D.; Tinus, R. W.; McDonald, S. E. & Barnett, J. P. (Eds.) (1989). *The Container Tree Nursery Manual*, Vol. 4, *Seedling nutrition and irrigation*, USDA, Forest Service, Agricultural Handbook 674, Washington D.C, USA. Available from http://www.rngr.net/publications/ctnm.

Landis, T. D.; Tinus, R. W.; McDonald, S. E. & Barnett, J. P. (Eds.), (1990). *The Container Tree Nursery Manual*, Vol. 2, *Containers and Growing media*, USDA, Forest Service, Agricultural Handbook 674, Washington D.C., USA. Available from http://www.rngr.net/publications/ctnm.

Landis, T. D.; Tinus, R. W.; McDonald, S. E. & Barnett, J. P. (Eds.). (1992). *The Container Tree Nursery Manual*, Vol. 3, *Atmospheric environment*, USDA, Forest Service, Agricultural Handbook 674, Washington D.C., USA. Available from http://www.rngr.net/publications/ctnm.

Landis, T. D.; Tinus, R. W. & Barnett, J. P. (Eds.). (2010). *The Container Tree Nursery Manual.* Vol. 7. *S Seedling Processing, Storage, and Outplanting*, USDA. Forest Service. Agricultural Handbook 674. Washington D.C. Available from http://www.rngr.net/publications/ctnm.

Leicach, S. R.; Garau, A. M.; Guarnaschelli, A. B.; Yaber Grass, M. A.; Sztarker, N. D. & Dato, A. (2010). Changes in *E. camaldulensis* essential oil as a response to drought. *Journal of Plant Interactions*, Vol.5, No.3, pp. 205-210 ISSN 1742-9145.

Lemcoff, J. H.; Garau, A. M. & Guarnaschelli, A. B. (2009). Afforestation under adverse biotic and abiotic environment. A contribution to ameliorate *Eucalyptus* performance during the establishment: The Argentinean Experience, *Proceedings The Dahlia Greidinger International Symposium on "Crop Production in the 21st Century: Global Climate Change, Environmental Risks and Water Scarcity"*, pp. 119-121, Available from http://dgsymp09.technion.ac.il/,Haifa, Israel, March 2-5, 2009.

Lemcoff, J. H.; Guarnaschelli, A. B.; Garau, A. M.; Bascialli, M. E. & Ghersa, C. M. (1994). Osmotic adjustment and its use as selection criterion in *Eucalyptus* seedlings. *Canadian Journal of Forest Research*, Vol.24, No.12, (December 1994), pp. 2404-2409, ISSN 0045-5067.

Lemcoff, J. H.; Guarnaschelli, A. B.; Garau, A. M. & Prystupa, P. (2002). Elastic and osmotic adjustment in rooted cuttings of *Eucalyptus camaldulensis*. *Flora*, Vol.197, No.2, pp. 134-142, ISSN 0367-2530.

Levin, M.; Resnick, N.; Rosianskey, Y.; Kolotilin, I.; Wininger, S.; Lemcoff, J. H.; Cohen, S.; Galili, G.; Koltai, H. & Kapulnik, Y. (2009). Transcriptional profiling of *Arabidopsis thaliana* plants' response to low relative humidity suggests a shoot-root communication. *Plant Science*, Vol.177, No.4, (November 2009), pp. 450-459, ISSN 0168-9452.

Li, F. L.; Bao, W. K. & Wu, N. (2009). Effects of water stress on growth, dry matter allocation and water-use efficiency of a leguminous species, *Sophora davidii*. *Agroforestry Systems*, Vol.77, No.3 (November 2009), pp. 193-201, ISSN 0167-4366.

Lindenmayer, D. B.; Hobbs, R. J. & Salt, D. (2003). Plantation forestry and tree biodiversity conservation. *Australian Forestry*, Vol.66, No.1, (March. 2003), pp. 62-66, ISSN 0004-9158.

Liu Z. & Dickmann, D. I. (1992). Responses of two hybrid poplar clones to flooding, drought and nitrogen availability. I. Morphology and growth. *Canadian Journal of Botany*, Vol.70, No.11, (November 1992), pp. 2265-2270, ISSN 1916-2790.

Loney, P. E.; McArthur, C.; Sanson, G. D.; Davies, N. W.; Close, D. C. & Jordan, G. J. (2006). How do soil nutrients affect within-plant patterns of herbivory in seedlings of *Eucalyptus nitens*? *Oecologia*, Vol.150, No.3, (December 2006), pp. 409-420, ISSN 0029-8549.

Löf, M. & Welander, N. (2004). Influence of herbaceous competitors on early growth in direct seeded *Fagus sylvatica* L. and *Quercus robur* L. *Annals of Forest Science*, Vol.61, No.8, (December 2004), pp. 781–788, ISSN 1286-4560.

Lopez, R.; Rodríguez-Calcerrada, J. & Gil, L. (2009). Physiological and morphological response to water deficit in seedlings of five provenances of *Pinus canariensis*: potential to select variation in drought-tolerance. *Trees*, Vol.23, No.3, (June 2009), pp. 509-519, ISSN 0931-1890.

Ludlow, M. M. (1989). Strategies in response to water stress. In: *Structural and functional responses under: water shortage*, K. H. Krebs, H. Richter & T. M. Hinckley, (Eds.), pp. 269-281, SPB Academic Press, ISBN 905-1030-27-4, The Hague, The Netherlands.

Maillard, P. D.; Garriou, E.; Deléens, P. G. & Guehl, J. M. (2004). The effect of lifting on mobilization and new assimilation of C and N during regrowth of transplanted Corsican pine seedlings. A dual ^{13}C and ^{15}N approach. *Annals of Forest Science*, Vol.61, No.8, (December 2004), pp. 795-805, ISSN 1286-4560.

Margolis, H. & Brand, D. (1990). An ecophysiological basis for understanding plantation establishment. *Canadian Journal of Forest Research*, Vol.20, No.4, (April 1990), pp. 375-390, ISSN 0045-5067.

Marron, N.; Delay, D.; Petit, J. M.; Dreyer, E.; Kalham, G.; Delmotte, F. M. & Brignolas, F. (2002). Physiological traits of two *Populus* x *eucamericana* clones, Louisa Avanzo and Dorskamp, during a water stress and a re-watering cycle. *Tree Physiology*, Vol.22, No.12, (August 2002), pp. 849-858, ISSN 0829-318X.

Marsaro, A.; Souza, R.; Della Lucia, T.; Fernandes, J.; Silva, M. & Vieira, P. (2004). Behavioural changes in workers of the leaf-cutting ant *Atta sexdens rubropilosa* induced by chemical components of *Eucalyptus* leaves. *Journal of Chemical Ecology*, Vol.30, No.9, (September 2009), pp. 1771-1780, ISSN 0098-0331.

Marshall, J. D. (1985). Carbohydrates status as a measure of seedling quality. In: *Evaluating seedling quality: Principles, procedures and predictive abilities of major tests, Proceedings of the Workshop Held October 16-18, 1984*, M. Duryea (Ed.), pp. 49-58, Forest Research Laboratory, Oregon State University, ISBN 9780874370003 Corvallis, Oregon, USA.

Marshall, J. G. & Dumbroff, E. B. (1999). Turgor regulation via cell wall adjustment in white spruce. *Plant Physiology*, Vol.119, No.1, (January 1999), pp. 313-319, ISSN 0032-0889.

McGraw, J.; Gottschalk, K.; Vavrek, M. & Chester, A. (1990). Interactive effects of resources availabilities and defoliation on photosynthesis, growth and mortality of red oak seedlings. *Tree Physiology*, Vol.7, No.1-2-3-4, (December 1990), pp. 247-254, ISSN 0829-318X.

Meiners, S. J.; Handel, S. N. & Pickett, S. T. A. (2000). Tree seedling establishment under insect herbivory: edge effect and inter-annual variation. *Plant Ecology*, Vol.151, No.2, (December 2000), pp. 161-170, ISSN 1385-0237.

Mena Petite, A.; Muñoz-Rueda, A. & Lacuesta, M. (2005). Effects of cold storage treatments and transplanting stress on gas exchange, chlorophyll fluorescence and survival under water limiting conditions of *Pinus radiata* stock-types. *European Journal of Forest Research*, Vol.124, No.2, (June 2005), pp. 73-82, ISSN 1612-4669.

Merchant, A.; Tausz M.; Arndt, S. K. & Adams, M. A. (2006). Cyclitols and carbohydrates in leaves and roots suggest contrasting physiological responses to water deficits. *Plant, Cell and Environment*, Vol.29, No.11, (November 2006), pp. 2017-2029, Online ISSN 1365-3040.

Merchant, A.; Callister, A.; Ardnt, S.; Tausz, M. & Adams, M. (2007). Contrasting physiological responses of six *Eucalyptus* species to water deficits. *Annals of Botany*, Vol.100, No.7, (December 2007), pp. 1507-1515, ISSN 0305-7369.

Meier, C. E.; Newton, R. J.; Puryear, J. D. & Sean, S. (1992). Physiological responses of Loblolly pine (*Pinus taeda* L.) seedlings to drought stress: osmotic adjustment and tissue elasticity. *Journal of Plant Physiology*, Vol. 140, No.6, pp. 754-760, ISSN 0176-1617.

Monclus, R.; Dreyer, E.; Villar, M.; Delmotte, F. M.; Delay, D.; Petit, J. M.; Barbaraux, C.; Le Thiec, D.; Bréchet, C. & Brignolas, F. (2006). Impact of drought on productivity and water use efficiency in 29 clones of *Populus deltoides* x *Populus nigra*. *New Phytologist*, Vol.169, No.4, (February 2006), pp. 765-777, ISSN 1469-8137.

Moore, B.; Wallis, I.; Pala-Paul, J.; Brophy, J.; Willis, R. & Foley, W. (2004). Antiherbivore chemistry of *Eucalyptus*- Cues and deterrents for marsupial folivores. *Journal of Chemical Ecology*, Vol. 30, No.9, (September 2009), pp. 1743-1769, ISSN 0098-0331.

Moraga. P.; Escobar, R. & Valenzuela, S. (2006). Resistance to freezing in three *Eucalyptus globulus* Labill. subspecies. *Electronic Journal of Biotechnology*, Vol.9, No.3, (June 2006) pp. 310-314, Available from http://www.ejbiotechnology.info/index.php/ejbiotechnology/index.

Morgan, J. M. (1984). Osmoregulation and water stress in higher plants. *Annual Review of Plant Physiology*, Vol.35, (June 1984), pp. 299-319, ISSN 0066-4294.

Morgan, J. M. & Cordon, A. G. (1986). Water use, grain yield and osmoregulation in wheat. *Australian Journal of Plant Physiology*, Vol.13, No.4, (August 1986), pp. 523-532, ISSN 1445-4408.

Muller da Silva, P.; Brito, J. & da Silva, F. (2006). Potential of eleven *Eucalyptus* species for the production of essential oils. *Scientia Agricola* Vol. 63, No.1, (Jan./Feb.2006), pp. 85-89, ISSN 0103-9016.

Nambiar, E. K. S. & Sands, R. (1993). Competition for water and nutrients in forests. *Canadian Journal of Forest Research*, Vol.23, No.10, (October 1993), pp. 1955-1968, ISSN 0045-5067.

Nguyen, A. & Lamant, A. (1989). Variation in growth and osmotic regulation of roots of water-stressed maritime pine (*Pinus pinaster* Ait.) provenances. *Tree Physiology*, Vol.5, No.1, (March 1989), pp. 123-133, ISSN 0829-318X.

Nguyen-Queyrens, A. & Bouchet Lannat, F. (2003). Osmotic adjustment in three-year-old seedlings of five provenances of maritime pine (*Pinus pinaster*) in response to drought. *Tree Physiology*, Vol.23, No.6, (April 2003), pp. 397-404, ISSN 0829-318X.

Nicotra, A. B.; Atkin, O. K.; Bonser, S. P.; Davidson, A. M.; Finnegan, E. J.; Marthesius, U.; Poot, P.; Purugganan, M. D.; Richards, C. L.; Valladares, F. & van Kleunen, M.

(2010). Plant phenotypic plasticity in a changing climate. *Trends in Plant Science*, Vol.15, No.12, (December 2010), pp. 684-692, ISSN 1360-1385.

Nielsen, E. T. & Orcutt, D. M. (1996). *The physiology of plants under stress. Abiotic factors.* J. Wiley & Sons, ISBN 047-1035-12-6, New York, USA.

Noland, T. L.; Mohammed, G. H. & Walter, R. G. (2001). Morphological characteristics associated with tolerance to competition from herbaceous vegetation for seedlings of jack pine, black spruce and white pine. *New Forests*, Vol.21, No.2, (May 2001), pp. 199-215, ISSN 0196-4286.

Norghaver, J.; Malcom, J. & Zimmerman, B. (2008). Canopy cover mediates interactions between a specialist caterpillar and seedlings of a neotropical tree. *Journal of Ecology*, Vol.96, No.1, (January 2008), pp. 103-113, ISSN 1365-2745.

Oddo, E.; Inzerillo, S.; La Bella, F.; Grisafi, F.; Salleo, S.; Nardini, A. & Goldstein, G. (2011). Short-term effects of potassium fertilization on the hydraulic conductance of *Laurus nobilis* L. *Tree Physiology*, Vol.31, No.2, (February 2011), pp. 131-138, ISSN 0829-318X.

O´Reilly-Wapstra, J.; Potts, B.; McArthur, C. & Davies, N. (2005). Effects of nutrient variability on the genetic-based resistance of *Eucalyptus globulus* to a mammalian herbivore and on plant defensive chemistry. *Oecologia*, Vol.142, No.4, (May 2005), pp. 597-605, ISSN 0029-8549.

Osório, J.; Osório, M. L.; Chaves, M. M. & Pereira, J. S. (1998). Water deficits are more important in delaying growth than in changing patterns of carbon allocation in *Eucalyptus globulus*. *Tree Physiology*, Vol.18, No.6, (June 1998), pp. 363-373, ISSN 0829-318X.

Paine, T. & Hanlon, C. (2010). Integration of tactics for management of *Eucalyptus* herbivores: influence of moisture and nitrogen fertilization on red gum lerp psyllid colonization. *Entomologia Experimentalis et Applicata*, Vol.137, No.3, (December 2010), pp. 290-295, Online ISSN 1570-7458.

Paquette, A. & Messier, C. (2010). The role of plantations in managing the world´s forests in the Anthropocene. *Frontiers in the Ecology and Environment*, Vol.8, No.1, (February 2010), pp. 27-34, ISSN 1540-9295.

Parker, W. C. & Pallardy, S. G. (1988). Leaf and root osmotic adjustment in drought-stressed *Quercus alba*, *Q. macrocarpa*, and *Q. stellata* seedlings. *Canadian Journal of Forest Research*, Vol.18, No.1, (January 1988), pp. 1-5, ISSN 0045-5067.

Passioura, J. B. (1996). Drought and drought tolerance. *Plant Growth Regulation*, Vol.20, No.2, (December 1996), 79-83, ISSN 0169-6903.

Peltzer, D. & Köchy, M. (2001). Competitive effects of grasses and woody plants in mixed-grass prairie. *Journal of Ecology*, Vol.89, No.4, (August 2001), pp. 519-527, ISSN 1365-2745.

Perry, D. A.; Oren, R. & Hart, S. C. (2008). *Forest Ecosystems. Second Edition.* The John Hopkins University Press. ISBN 978-080-1888-40-3, Maryland. USA.

Pezeshki, R. & Hinckley, T. M. (1988). Water relation characteristics of *Alnus rubra* and *Populus trichocarpa*: responses to field drought. *Canadian Journal of Forest Research*, Vol.18, No.9, (September 1988), pp. 1159-1166, ISSN 0045-5067.

Phillipson, J. (1988). Root growth in Sitka Spruce and Douglas-fir transplants: dependence on the shoot and stored carbohydrates. *Tree Physiology*, Vol.4, No.2, (June 1988), pp. 101-108, ISSN 0829-318X.

Picon-Cochard, C.; Nsourou-Obame, A.; Collet, C.; Guehl, J. & Ferhi, A. (2001). Competition for water between walnut seedlings (*Juglans regia*) and rye grass (*Lolium perenne*) assessed by carbon isotope discrimination and $\delta^{18}O$ enrichment. *Tree Physiology*, Vol.21, No.2-3, (June 2001), pp. 183–191, ISSN 0829-318X.

Picon-Cochard, C.; Coll, L. & Balandier, P. (2006). The role of below-ground competition during early stages of secondary succession: the case of 3-year-old Scots pine (*Pinus sylvestris* L.) seedlings in an abandoned grassland. *Oecologia*, Vol.148, No., pp. 373-383, ISSN 0029-8549.

Pita, P. & Pardos, J. A. (2001). Growth, morphology, water use and tissue water relations of *Eucalyptus globulus* clones in response to water deficit. *Tree Physiology*, Vol.21, No.9, (June 2001), pp. 599-607, ISSN 0829-318X.

Pita, P.; Cañas, I.; Soria, F.; Ruiz, F. & Toval, G. (2005). Use of physiological traits in tree breeding for improved yield in drought-prone environments. The case of *Eucalyptus globulus*. *Investigaciones Agrarias. Sistemas y Recursos Forestales*, Vol.14, No.3, pp. 383-393, ISSN 1131-7965.

Powell, G. & Bork, E. (2004). Above- and below-ground effects from alfalfa and marsh reedgrass on aspen seedlings. *Forest Ecology and Management*, Vol.199, No.2-3, (October 2004), pp. 411-422, ISSN 0378-1127.

Powles, S. B. (1984). Photoinhibition of photosynthesis induced by visible light. *Annual Review of Plant Physiology*, Vol.35, No., pp. 15-44, ISSN 0066-4294.

Provendier, D. & Balandier, P. (2008). Compared effects of competition by grasses (Graminoids) and broom (*Cytisus scoparius*) on growth and functional traits of beech saplings (*Fagus sylvatica*). *Annals of Forest Science*, Vol.65, No., pp. 510-508, ISSN 1286-4560.

Prior, L. D. & Eamus, D. (1999). Seasonal changes in leaf water characteristics *of Eucalyptus tetrodonta* and *Terminalia ferdinandiana* saplings. *Australian Journal of Botany*, Vol.47, No.4, pp. 587-599, ISSN 0067-1924.

Raymond, C. A. (1995). Genetic variation in *Eucalyptus regnans* and *Eucalyptus nitens* for level of observed defoliation caused by *Eucalyptus* leaf beetle, *Chysophtharta bimaculata* Olivier, in Tasmania. *Forest Ecology and Management*, Vol.72, No.1, (March 1995), pp. 21-29, ISSN 0378-1127.

Rey Benayas, J.; Espigares, T. & Castro-Diez, P. (2003). Simulated effects of herb competition on planted *Quercus faginea* seedlings in Mediterranean abandoned cropland. *Applied Vegetation Science*, Vol.6, No.2, (December 2003), pp. 213-222, ISSN 1402-2001.

Rietveld, R. J. (1989). Transplanting stress in bareroot conifer seedlings: its development and progression to establishment. *Northern Journal of Applied Forestry*, Vol.6, No.3, (September 1989), pp. 99-107, ISSN 0742-6348.

Rodríguez-Calcerrada, J.; Pardos, J. & Aranda, I. (2010). Contrasting responses facing peak drought in seedlings of two co-ocurring oak species. *Forestry*, Vol.83, No.4, (October 2010), pp. 369-378, ISSN 0015-752X.

Römheld V. & Kirkby, E. A. (2010). Research on potassium in agriculture: needs and prospects. *Plant and Soil*, Vol.335, No.1-2, (October 2010), pp. 155-180, ISSN 0032-079X.

Rose, R.; Campbell, S. & Landis, T. D. (Eds.). (1990). *The Target seedling Symposium: Proceedings, Combined Meeting of the Western Forest Nurseries Associations, August 13-*

17, 1990, Roseburg, Oregon, General Technical Report GTR-RM-200. Rocky Mountain Forest and Range Experiment Station, Forest Service, USDA, Available from http://www.rngr.net/publications/proceedings/1990/WFCNA, Fort Collins, Colorado, USA.

Salgado-Luarte, C. & Gianoli, E. (2011). Herbivory may modify functional responses to shade in seedlings of a light-demanding tree species. *Functional Ecology,* Vol.25, No.3, (June 2011), pp. 492-494, Online ISSN 1365-2745.

Sands, R. (1984). Transplanting stress in radiate pine. *Australian Forest Research,* Vol.14, pp. 67-72, ISSN 0004-914X.

Scheirs, J. & De Bruyn, L. (2005). Plant-mediated effects of drought stress on host preference and performance of a grass miner. *Oikos,* Vol.108, No.2, (August 2005), pp. 371-385, Online ISSN 1600-0706.

Schlichting, C. D. (1986). The evolution of phenotypic plasticity in plants. *Annual Review of Ecology and Systematics,* Vol.17, pp. 677-693, ISSN 0066-4162.

Schulte, P. J. & Hinckley, T. M. (1987). The relationships between guard cell water potential and the aperture of stomata of *Populus. Plan Cell and Environment,* Vol.10, No.4, (June 1987), pp. 313-318, ISSN 1364-1344.

Schultz, R. C. & Thompson, J. R. (1997). Effect of density control and undercutting on root morphology on 1+0 bareroot harwood seedlings: five year field performance or root-graded stock in the central USA. *New Forests,* Vol.13, No.1-3, (May 1997), pp. 301-314, ISSN 0196-4286.

Scheirs, J. & De Bruyn, L. (2005). Plant-mediated effects of drought stress on host preference and performance of a grass miner. *Oikos,* Vol.108, No.2, (February 2005), pp. 371-385, Online ISSN 1600-0706.

Serrano, L. & Peñuelas, J. (2005). Contribution of physiological and morphological adjustments to drought resistance in two Mediterranean tree species. *Biologia Plantarum,* Vol.49, No.4, (December 2005), pp. 551-559, ISSN 1573-8264.

Shipley, B. & Meziane, D. (2002). The balanced-growth hypothesis and the allometry of leaf and root biomass allocation. *Functional Ecology,* Vol.16, No.3, (June 2003), pp. 326-331 Online ISSN 1365-2745.

Showler, A. (2002). Effects of water deficit stress, shade, weed competition, and kaolin particle film on selected foliar free amino acid accumulations in cotton, *Gossypium hirsutum* (L.). *Journal of Chemical Ecology,* Vol.28, No.3, (March 2002), pp. 631-651, ISSN 0098-0331.

Silim, S.; Nash, R.; Reynard, D.; White, B. & Shroeder, W. (2009). Leaf gas exchange and water potential responses to drought in nine poplar (*Populus* spp.) clones with contrasting drought tolerance. *Trees,* Vol.23, No.5, (October 2009), pp. 959-969, ISSN 0931-1890.

Sloan, J. L. & Jacobs, D. F. (2008). Carbon translocation patterns associated with new root growth proliferation during episodic growth of transplanted *Quercus rubra* seedlings. *Tree Physiology,* Vol.28, No.7, (July 2008), pp. 1121-1126, ISSN 0829-318X.

Smith, D. M.; Larson, B. C.; Kelty, M. J. & Ashton, P. M. S. (1997). *The practice of Silviculture: Applied Forest Ecology.,* 9th Edition, John Wiley & Sons Inc, ISBN 047-1109-41-X, New York, USA.

Smith T. & Huston, M. (1989). A theory of the spatial and temporal dynamics of plant communities. *Vegetatio,* Vol.83, No.1-2, (October 1989), pp. 49-69, ISSN 0042-3106.

South, D. B.; Harris, S. W.; Barnett, J. P.; Hainds, M. J. & Gjerstad, D. H. (2005). Effect of container type and seedling size on survival and early height growth of *Pinus palustris* seedlings in Alabama, U.S.A. *Forest Ecology and Management*, Vol.204, No.2-3, (January 2005), pp. 385-398, ISSN 0378-1127.

Stone, C. (2001). Reducing the impact of insect herbivory in eucalypt plantations through management of extrinsic influences on tree vigor. *Austral Ecology*, Vol.26, No.5, (October 2001), pp. 482-488, ISSN 1442-9985.

Stone C. & Bacon, P. E. (1994). Relationships Among Moisture Stress, Insect Herbivory, Foliar 1,8-cineole Content and the Growth of River Red Gum *Eucalyptus camaldulensis*. *Journal of Applied Ecology*, Vol.31, No.4, (November 1994), pp. 604-612, ISSN 1365-2664.

Stoneman, G. L. (1994). Ecology and physiology of establishment of eucalypt seedlings from seed: A review. *Australian Forestry*, Vol.57, No.1, (March 1994), pp. 11-30, ISSN 0004-9158.

Stoneman, G. L.; Turner, N. C. & Dell, B. (1994). Leaf growth, photosynthesis and tissue water relations of greenhouse-grown *Eucalyptus marginata* seedlings in response to water deficits. *Tree Physiology*, Vol.14, No.6, (June 1994), pp. 633-646, ISSN 0829-318X.

Stewart, J. D. & Lieffers, V. J. (1993). Preconditioning effects of nitrogen addition rate and drought stress on container-grown lodgepole seedlings, *Canadian Journal of Forest Research*, Vol.23, No.8, (August 1993), pp. 1663-1671, ISSN 0045-5067.

Taiz, L. & Zeiger, E. (2002). *Plant Physiology. Third edition.* Sinauer Associates Inc. Publishers, ISBN 087-8938-23-0 Sunderland, Massachussetts, USA.

Tan, W.; Blake, T. J. & Boyle, T. B. (1992). Drought tolerance in faster- and slower-growing black spruce (*Picea mariana*) progenies: II. Osmotic adjustment and changes of soluble carbohydrates and amino acids under osmotic stress. *Physiologia Plantarum*, Vol. 85 No.4, (August 1992), pp. 645-651, ISSN 0031-9317.

Tappeiner, J. C. & Helms, J. A. (1971). Natural regeneration of Douglas fir and white fir in exposed sites in the Sierra Nevada of California. *American Midland Naturalist*, Vol. 86, No.2, (October 1971), pp. 358-370, ISSN 1938-4238.

Tappenier, J. C.; Maguire, D. A. & Harrington, T. B. (2007). *Silviculture and Ecology of Western U. S. Forests.* Oregon State University Press, ISBN 978-087-0711-87-9, Corvallis, Oregon, USA.

Theis, N. & Lerdau, M. (2003). The evolution of function in plant secondary metabolites. *International Journal of Plant Sciences*, Vol.164, No.S3, (May 2003), pp. S93-S102, ISSN 1058-5893.

Tschaplinski, T. J.; Tuskan, G. A.; Gebré, G. M. & Todd, D. E. (1998). Drought resistance of two hybrid *Populus* clones grown in a large-scale plantation. *Tree Physiology*, Vol.18, No.10, (October 1998), pp. 653-658, ISSN 0829-318X.

Tuomela, K. (1997). Leaf water relations in six provenances of *Eucalyptus microtheca:* a greenhouse experiment. *Forest Ecology and Management*, Vol.92, No.1-3, (May 1995), pp. 1-10, ISSN 0378-1127.

Turner, N. C. (1986a). Adaptation to water deficit: A changing perspective. *Australian Journal of Plant Physiology*, Vol.13, No.1, pp. 175-190, ISSN 1445-4408.

Turner, N. C. (1986b). Crop water deficits: a decade of progress. *Advances in Agronomy*, Vol.39, pp. 1–51, ISBN 012-0007-95-9.

Turner, N. C. & Jones, M. M. (1980). *Turgor maintenance by osmotic adjustment: a review and evaluation*. In: *Adaptation of Plants to Water and High Temperature Stress*, N. C. Turner & P. J. Kramer, (Eds.), pp. 78–103, John Wiley & Sons, ISBN 978-047-1053-72-9, New York, USA.

Uemura, A.; Ishida, A.; Nakano, T.; Terashima, I.; Tanabe, H. & Matsumoto, Y. (2000). Acclimation of leaf characteristics of *Fagus* species to previous-year and current-year solar irradiance. *Tree Physiology*, Vol.20, No.14, (Agosto 2000), pp. 945-951, ISSN 0829-318X.

United Nations Framework Convention on Climate Change (UFCCC). (2007). Investment and financial flows to address climate change. 273 pp.

Valladares, F.; Vilagrosa, A.; Peñuelas, J.; Ogaya, R.; Camarero, J. J.; Corcuera, L.; Siso, S. & Gil-Pelegrin, E. (2004). *Estrés hídrico: Fisiología y escalas de la sequía*, In: *Ecología del bosque mediterráneo en un mundo cambiante*, F. Valladares, (Ed.), pp. 163-190, Ministerio de Medio Ambiente, EGRAF. S.A., ISBN 84-8014-552-8, Madrid, España.

Valladares, F.; Sánchez-Gómez, D. & Zavala, M. A. (2006). Quantitative estimation of phenotypic plasticity: bringing the gap between the evolutionary concept and its ecological applications. *Journal of Ecology*, Vol.94, No.6, (November 2006), pp. 1103-1116, ISSN 1365-2745.

van den Driessche, R. (1987). Importance in current photosynthate to new root growth in planted conifer seedlings. *Canadian Journal of Forest Research*, Vol.17, No.8, (August 1989), pp. 776-782, ISSN 0045-5067.

van den Driessche, R. (1991). Influence of container nursery regimes on drought resistance of seedlings following planting. II. Stomatal conductance, specific leaf area and root growth capacity. *Canadian Journal of Forest Research*, Vol.21, No.5, (May 1995), pp. 566-572, ISSN 0045-5067.

van den Driessche, R. (1992). Changes in drought resistance and root growth capacity of container seedlings in response to nursery drought, nitrogen and potassium treatments. *Canadian Journal of Forest Research*, Vol.22, No.5, (May 1992), pp. 740-749, ISSN 0045-5067.

Vasconcelos, H. L. & Cherrett, J. M. (1997). Leaf-cutting ants and early forest regeneration in central Amazonia: effects of herbivory on tree seedling establishment. *Journal of Tropical Ecology*, Vol.13, No.3, (May 1997), pp. 357-370, ISSN 0266-4764.

Verdaguer, D.; Vilagran, J.; Lloansi, S. & Fleck, I. (2011). Morphological and physiological acclimation of *Quercus coccifera* L. Seedlings to water availability and growing medium. *New Forests*, Vol.42, No.3, pp. 363-381, (November 2011), ISSN 0196-4286.

Viglizzo, E. F.; Frank, F. C.; Carreño, L. V.; Jobbágy, E. G.; Pereyra, H.; Clatt, J.; Pincén, D. & Ricard, M. F. (2011). Ecological and environmental footprint of 50 years of agricultural expansion in Argentina. *Global Change Biology*, Vol.17, No.2, (February 2011), pp. 959-973, ISSN 1354-1013.

Vilagrosa, A.; Cortin, J.; Gil-Pelegrin, E. & Bellot, J. (2003). Suitability of drought-preconditioning techniques in Mediterranean climate. *Restoration Ecology*, Vol.11, No.2, (June 2003), pp. 208-216, Online ISSN 1526-100X.

Vilagrosa, A.; Villar-Salvador, P. & Puértolas, J. (2006). *El endurecimiento en vivero de especies forestales mediterráneas*. In: *Calidad de planta forestal para la restauración en ambientes mediterráneos. Estado actual de conocimientos*, J. Cortina, J. L. Peñuelas, J. Puértolas, R.

Savé & A. Vilagrosa, (Eds.), pp. 119-140, DGB Ministerio de Medio Ambiente, Serie Forestal, ISBN 848-0146-70-2, Madrid, España.

Villar-Salvador, P.; Planelles, R.; Oliet, J.; Peñuelas, J. L.; Jacobs, D. F. & M. González. (2004). Drought tolerance and transplanting performance of holm oak (*Quercus ilex*) seedlings after drought hardening in the nursery. *Tree Physiology*, Vol.24, No.10, (October 2004), pp. 1147-1155, ISSN 0829-318X.

Villar-Salvador, P.; Puértolas, J.; Peñuelas, J. L. & Planelles, R. (2005). Effects of nitrogen fertilization in the nursery on the drought and frost resistance o Mediterranean forest species. *Investigación Agraria. Sistemas y Recursos Forestales*, Vol.14, No., pp. 408-418, ISSN 1131-7965.

Vitousek, P. M. (1991). Can planted forests counteract increasing atmospheric carbon dioxide? *Journal of Environmental Quality*, Vol.20, No.2, (April-June1991), pp. 348-354, ISSN 0047-2425.

Waring, R. H. & Schlesinger, W. H. (1985). *Forest Ecosystems. Concepts and management.* Academic Press Inc., ISBN 012-7354-40-9, New York, USA.

Watt, M.; Whitehead, D.; Mason, E.; Richardson, B. & Kimberley, M. (2003). The influence of weed competition for light and water on growth and dry matter partitioning of young *Pinus radiata* at a dryland site. *Forest Ecology and Management*, Vol.183, No.1-3, (September 2003), pp. 363-376, ISSN 0378-1127.

White, T. C. (1969). An index to measure weather-induced stress by trees associated with outbreak of psylids in Australia. *Ecology*, Vol.50, No.5, (September 1969), pp. 905-909, ISSN 0012-9658.

White, D. A.; Beadle, C. L. & Worledge, D. (1996). Leaf water relations of *Eucalyptus globulus* and *E. nitens*: seasonal, drought and species effects. *Tree Physiology*, Vol.16, No.5, (May 1996) pp. 469-476, ISSN 0829-318X.

White, D. A.; Turner, N. C. & Galbraith, J. H. (2000). Leaf water relations and stomatal behavior of four allopatric *Eucalyptus* species planted in Mediterranean southwestern Australia. *Tree Physiology*, Vol.20, No.17, (November 2000), pp. 1157-1165, ISSN 0829-318X.

Whitehead, D. & Beadle, C. L. (2004). Physiological regulation of productivity and water use in Eucalyptus: a review. *Forest Ecology and Management*, Vol.193, No.1-2, (May 2004), pp. 113-140, ISSN 0378-1127.

Whiteman, A. & Brown, C. (1999). The potential role of forest plantations in meeting future demands for industrial wood products. *International Forestry Review*, Vol.1, No.3, (September 1999), pp. 143-152, ISSN 1465-5489.

Wikber, J. & Ogren, R. (2007). Variation in drought resistance, drought acclimation and water conservation in four willow cultivars used for biomass production. *Tree Physiology*, Vol.27, No.9, (September), pp. 1339-1346, ISSN 0829-318X.

Wink, M. (2003). Evolution of secondary metabolites from an ecological and molecular phylogenetic perspective. *Phytochemistry*, Vol.64, No.1, (September 2003), pp. 3-19, ISSN 0031-9422.

Wise, M. J. & Abrahamson, W. G. (2007). Effects of resource availability on tolerance of herbivory: a review and assessment of three opposite models. *American Naturalist*, Vol.169, No.4, (April 2007), pp. 443-454, ISSN 00030147.

Yordanov, I.; Velikova, V. & Tsonev, T. (2000). Plants responses to drought, acclimation and stress tolerance. *Photosynthetica*, Vol.38, No.2, (June 2002), pp. 171-186, ISSN 0300-3604.

Zine El Abidine, A.; Bernier, P. Y.; Stewart, J. D. & Plamondon, A. P. (1994). Water stress preconditioning of black spruce seedlings from lowland and upland sites, *Canadian Journal of Botany*, Vol.72, No.10, (October 1994), pp. 1511-1518, ISSN 1916-2790.

Zwiazek, J. J. & Blake, T. J. (1989). Effects of preconditioning on subsequent water relations, stomatal sensitivity, and photosynthesis in osmotically stressed black spruce. *Canadian Journal of Botany*, Vol.67, No.8, (August 1989), pp. 2240-2244, ISSN 1916-2790.

5

Systemic Signaling Under Water Deficit Condition and Its Exploitation in Water Saving Agriculture

Bingbing Li and Wensuo Jia
College of Agronomy and Biotechnology,
China Agricultural University, Beijing
China

1. Introduction

To live and thrive in an ever changing environment, a living organism must be capable of sensing the environmental stimuli and making corresponding adaptive responses. This "stimulus-response" process is mediated by a series of signaling events. It appears that the more advanced the organism is, the more positive and quicker the responses display. This can be best demonstrated in animals where the nerve-mediated signaling enable animal to react swiftly and purposefully, thus preserving them from possible harms. The nerve-mediated signaling is characterized by a pattern of long-distance signaling, i.e., when one part is stimulated another distinct part of the body may take corresponding actions. Systematic signaling can be interpreted as a pattern of "stimulus-response" that when local part is stimulated the whole body may take corresponding actions. Clearly, long-distance signaling is the basis of the systemic signaling and embodies a kind of advanced responding behavior to an ever changing environment. While the systemic signaling has been well known for its crucial roles of in animals, it has been largely overlooked in plants owing to the inability of plants to move. In recent years, with a rapid progress in molecular biology, the molecular mechanisms for stress-resistance in plants have attracted considerable attentions. Studies on the molecular basis of stress tolerance have revealed that in response to environmental stresses expressions of numerous genes associated with stress tolerances can be regulated, indicating that plant cells are indeed able to sense and respond to environmental stresses. Like animals, the growth and development of advanced plants depends a cooperative regulation of different tissues and organs, therefore, plants should have evolved the capability to sense and respond to various stress stimuli through the systemic signaling. Nevertheless, while particular interests and attentions have been paid on the cellular signaling transduction, knowledge about the roles of the systemic signaling in plant adaptation responses is relatively scarce.

2. The root to shoot signaling plays critical roles in plant adaptive responses to drought stress

It has been traditionally thought that soil drying will limit water uptake by roots and this will inevitably result in the declination of leaf water status. However, much work suggests that in many cases the leaf water status is not closely correlated with the water availability of the root system. This has raised a question of whether the leaf water status may be regulated by an unknown mechanism other than by the hydraulic control, and it has been increasingly suggested that this unknown mechanism is the root to shoot chemical signaling.

Indirect evidence supporting the root to shoot signaling came from the observations on the disconnected relationship between the leaf and soil water status and that between leaf physiology and leaf water potential. It has been generally accepted that leaf physiology, such as leaf growth and stomatal movement, is predominantly regulated by the leaf water potential, which is again directly regulated by the soil water status. However, for many years much data suggest that leaf growth and stomatal movement may be regulated even when no perceptible changes in the leaf water potential occur, and also, the changes in leaf water potential may not be closely linked to to that in the soil water status. In contrast, it is often observed that variation in leaf conductance and growth may be linked more closely to changes in soil water status than to changes in leaf water status. This appears to suggest that leaf stomatal movement or growth may be regulated by a signal sourced from roots.

Direct evidences came from studies where the experiments were subtly designed so that the coupling relationship between the leaf and soil water status was broken. A classic experiment to do this is the split-root experiments. A well demonstrated example is split-root experiment with clonal apple trees (Gowing, 1990), where the root systems of individual plants were split into approximately two half parts, with each part contained in a individual container. With-holding water from half of the root system reduced the rate of leaf growth while no significant reduction in the leaf water potential could be observed. Further more, rewatering restored the rates of leaf growth to that of the well watered control plants. Interestingly, excising the roots in dry soil also restored the rates of leaf growth, which strongly suggested that the reduced rate of leaf growth was a result of inhibition by the roots in drying soil. Numerous studies on split-root have provided evidences supporting the existence of root to shoot signaling in different plant species. Besides the spit-root experiment, strong evidences also came from the soil pressure chamber experiment, where the coupling relationship between the reduction of leaf water status and soil drying can be destructed by supplying a pressure on the roots. In this way, a reduction in the leaf water potential of the unwatered plants can be avoided while the soil was allowed to dry, but the roots of these plants were still in contact with drying soil. In an investigation on wheat and sunflower plants, Passioura and colleagues (Passioura and Munns, 1984) found that soil drying could cause a significant reduction in leaf conductance of pressurized plants while it had no effect leaf water potential, suggesting the existence of root to shoot signaling. Collectively, in the past may years with different approaches and plant species it has been well established that root to shoot signaling plays critical roles in the regulation of shoot responses to soil drying.

3. The nature of signals mediating root to shoot signaling

3.1 ABA signaling

3.1.1 Evidences for ABA signaling

It is well known that soil drying will cause a declination in the root water status and this will eventually cause a declination in the leaf water status, hence, the leaf water status has been commonly used as an indicator of the soil water status. Clearly, the nexus between leaf and soil water status is coupled directly through the variations of the water potential along vascular system. Therefore, a variation in the hydraulic nexus between leaf and root can be the best signal (hydraulic signal) to mediate the responses of the leaf physiology to the water deficit of roots. However, for the root to shoot signaling concerned here particularly refers to the case that is before the hydraulic signaling occurring, or the case in which the leaf growth or stomatal movement is not coupled with the variation in the leaf water potential. The only logical hypothesis is that such a root to shoot signaling is mediated by some chemical substances. It is well known that the transportation of materials from root to shoot is achieved though transpiration stream, therefore, it is reasonable to think that the chemical signals mediating the root to shoot signaling should be contained in the transpiration stream. A wide analysis of the effect of soil drying on the xylem sap compositions indicated that concentrations of most components decline as the soil around the roots is allowed to dry except for the plant hormone abscisic acid (ABA). Mild soil drying will cause a significant increase in xylem sap ABA concentration (Neales, 1989). It has long been known that ABA can be synthesized in nearly all plant species and tissues. Given the potential roles in root to shoot signaling, the regulation of ABA synthesis in roots has attracted particular attentions (Cornish and Zeevaart,1985; Hubic et al., 1986; Lachno and Baker, 1986; Robertson, et al., 1985; Walton et al., 1976). Dehydration of detached roots is able to induce ABA accumulation (Neales, 1989) and comprehensive analysis of ABA contents of roots in different parts of the soil profile shows a close relationship between the ABA content and water status in the root profile (Atkinson, et al., 1989; Davies 1987). These evidences suggest that ABA may be a candidate signal mediating the root to shoot signaling.

More strong evidences supporting ABA to be a root to shoot signal is the close relationship between a variation in xylem ABA concentration and the leaf physiology. For example, it was reported that the ABA concentration in xylem sap from well-watered maize and sunflower plants is only around 10 nM, and this concentration can rise by one or two orders of magnitude (Schurr and Gollan, 1990; Zhang and Davies, 1989; 1990) as a function of mild soil drying. Besides maize and sunflower, mild soil drying can induce a substantial increase in xylem ABA concentration in many other plant species (Hartung and Radin, 1989; Munns and King,1988). Many studies demonstrate that xylem ABA concentration may be a more closely related to stomatal movement than either root ABA or leaf ABA is. Conclusive evinces for the ABA signaling came from bioassay experiments. No matter what ways for ABA supplying, the ABA treatment is able to induce stomatal closure (Zhang and Davies, 1989; 1990). More importantly, feeding xylem sap collected from unwatered maize plants caused an inhibition of transpiration, and removing the ABA from xylem sap by passing through an immunoaffinity column composed of ABA antibodies was able to relieve the antitranspirant activity (Ogunkanmi et al., 1973), strongly indicating that ABA can be a root to shoot signal regulating stomatal movement.

3.1.2 Arguments about ABA signaling

It is well known that leaf cells contain a lot of ABA. If root-sourced ABA can be a root to shoot signal mediating the regulation of stomatal movement or leaf growth, how is the ABA synthesized in leaf cells distinguished from the root-sourced ABA? This is actually not a problem because ABA in plant tissues or cells is normally sequestered within some specific spaces as a function of the effect of anion trap (Cowan et al, 1982; Hartung, 1990), e.g. in leaf cells ABA is mainly sequestered within chloroplast. Owing to this sequestration, ABA is not able to reach its action sites and therefore not able to play corresponding functions. Carrying by xylem stream, root-sourced ABA is believed to pass through the apoplastic spaces in leaves, where ABA may be able to directly interact with its receptor therefore effectively regulating stomatal movement or leaf growth.

Although it has been well established that root-sourced ABA is capable of regulating stomatal movement, it is not clear how stomatal movement respond to this ABA signal. Xylem ABA concentration, ABA flux entering leaves and the accumulated amount of ABA (mass of ABA) within the leaves are three parameters closely related to each other. All these parameters can be changed as a function of soil drying, and so, all of them may become a factor to regulate stomatal movement. It is estimated that the accumulated mass of ABA within an individual leaves per day may be over than the amount of leaf ABA even under the well watered condition. Supposing stomatal movement is capable of responding to the accumulated mass of ABA, such a large amount of ABA accumulated per day would expect to completely inhibit stomatal opening even when plants are well watered. As for the ABA flux, supposing it can be a stomatal responding factor, the root-sourced ABA would also be expected to affect stomatal movement under the well watered condition because a change in ABA flux normally occurs owing to a timely variation in the plant transpiration. There is evidence that stomatal movement may be able to respond to ABA flux and the amount of ABA, but many studies suggest that stomatal movement can only respond to xylem ABA concentration other than the ABA flux or mass. Whatever, substantial studies have suggested that xylem ABA concentration is a predominant factor regulating stomatal movement or other leaf behaviors. With xylem ABA concentration being a major responding factor of stomatal movement, the root sourced ABA can be well as a signal mediating the root to shoot signaling under the water deficit condition, whereas it will not perturb the normal stomatal movement under the non-stressed condition.

3.2 pH signaling

As mentioned above, ABA is not evenly distributed but sequestered in plant cells. The sequestration of ABA is a result of an uneven distribution of pH within plant cells, i.e. the distribution of ABA in plant cells was pH-dependent (Kaiser and Hartung 1981; Hartung et al. 1988; Hartung and Radin 1989). Based on the determination of ABA uptake in response to pH in isolated cells or tissues, Hartung's group generated a mathematic model, in which it was predicted that an increase as detected *in vivo* in response to drought would be enough to induce stomatal closure (Slovik and Hartung 1992a, 1992b). Because of this, it can be hypothesized that pH may be able to act as a signal if the cellular pH can be affected by soil drying. A few early works found that drought might indeed be able to trigger a pH change in different plant species. For example, Hartung and Radin (1989) observed that in response to drought stress the pH of xylem sap from *Phaseolus coccineus* roots increased from 6.3 to 7.2; and Gollan et al. (1992) found that in sunflower plants the pH of xylem sap from shoot increased from a range between 5.8 and 6.6 in well watered plants to 7.0 in the drought-affected plants. In the *Commelina communis* plant, it was reported that in response to soil drying, the pH of xylem sap raised from 6.1 to 6.7, and when artificial xylem sap buffered to different pH was fed to detached leaves of *Commelina communis*, and it was found that an increase in pH from 6.0 to 7.0 caused a reduction of transpiration rate by about 50% in the presence of low concentrations of ABA. This indicated that a change in xylem pH may be able to act as a signal mediating the root shoot signaling (Wilkinson and Davies, 1997). The stomatal movement in relation to the pH functioning is complex. pH may be able to act as a signal to mediate the root to shoot signaling and it may also function but not as a signal to modulate the ABA signaling. The mechanisms for theses two processes are different, and more detailed description will be given in the corresponding parts below.

3.3 Other signals

While many studies have suggested that an increase in xylem ABA concentration may be able to mediate root to shoot signaling (Zhang and Davies 1991; Tardieu et al. 1992, Tardieu and Davies 1993; Zhu and Zhang 1997), there are also evidences demonstrating that the root to shoot signaling may be not mediated by the ABA based signaling (Munns et al. 1993; Munns and Cramer 1996; Holbrook et al. 2002). For example, using the graft technique, Holbrook et al. (2002) constructed different kinds of plants grated with the ABA deficient (*sitiens*) and wild type (*flacca*) tomato. No matter what method of drying, differences were observed in the changing pattern in the stomatal conductance among the different kinds of grafting combinations, and with this Holbrook et al. (2002) clearly demonstrated that the stomatal closure in response to drought stress does not require ABA production by the root. Besides ABA and pH, there are possibly some other signals that may be involved in the root to shoot signaling. Ethylene, cytokinin and some biologically active substance may be candidate signals in some cases, but the information about this is much less. Given that pH can be affected by many factors, it is likely that some pH-regulatory factors may be possibly involved in the root to shoot signaling. For example, it was reported that strong ion differences-SID may be able to alkalize the apoplastic fluid hence serving as a signal to mediate the root to shoot signaling (Hartung and Radin, 1989; Schurr and Gollan, 1990). More detailed information about this will be given in the related sections below.

4. Modulation of the chemical signaling

While chemical signaling has been widely accepted to mediate stomatal responses to soil dying, it should be noted that the root to shoot signaling is not just a simple matter of the ABA transmission from root to shoot. A major issue about it is the great variation in the apparent sensitivity of leaf conductance to the root signals. For examples, in some plants chemical signaling may account for a decrease in stomatal conductance by more than 50% (Zhang and Davies 1989, 1990) or even more than 70% (Khalil and Grace, 1993), but in some other plants it can only account by less than 20% (Gollan *et al.*, 1986; Auge *et al.*, 1995; Yao *et al.*, 2001). Theoretically, this can be explained by two major reasons. One is that soil drying is not able to induce significant accumulation of ABA in the roots and another is that stomatal movement is able to sensitively respond to the root-sourced ABA. It appears that soil drying is capable of giving rise to a significant accumulation of ABA in nearly all plants, so it is more likely that the great variation in the regulation of stomatal movement by root signals is a result of the variation of the stomatal sensitivity to the root-sourced ABA.

It is well known that the functioning of ABA depends on a direct interaction of ABA molecule with its receptor. This means that the capability of the root ABA signal to regulate stomatal movement should be determined by both the guard cell itself (i.e. the activity of ABA receptor) and the intensity of the ABA signal in the guard cell. Theoretically, the root to shoot ABA signaling can be divided into three major phases: one is the accumulation and loading of ABA into xylem vessels in the root system, the second is the long-distance transmission of the ABA carried by transpiration stream from root to shoot and the third is the unloading and loading of the ABA from xylem vessels and to guard cells. Intensity modification of the ABA signal may occur in each phase above. It has been increasingly suggested that intensity of the ABA signal can be modified by many factors, and these modifications may have important effects on the stomatal sensitivity to the root-sourced ABA signal (Jiang and Hartung, 2008).

4.1 Modulation of ABA signaling

4.1.1 Effect of radial transport

The production of ABA in root cells and its loading into xylem vessels is first step in the long-distance transport of ABA. As mentioned above, xylem ABA concentration is a predominant factor being responded by stomatal movement, and hence, modulation of the xylem ABA concentration is a major concern for the long-distance transport of ABA. It was hypothesized that the stele and cortex possess an equal capacity to synthesize ABA, and for a long-distance transport ABA needs to be radially transported from the cortex to the xylem vessel. Xylem ABA concentration should be determined by the radial transport rate of both ABA and water, and hence the transpiration-caused changes in the rate of the lateral water flow may affect xylem ABA concentration (Else et al. 1994, 1995; Hartung et al. 2002). One concern is the role of Casparian in the radial transport of ABA in root. It is proposed that Casparian may retard the loading of ABA from cortex into xylem vessel and this may cause a high apoplastic ABA concentration built up in the cortex, and the high concentration of ABA in the cortex may facilitate an ABA efflux from the roots (Hartung et al. 2002). However, it has been increasingly suggested that ABA may be primarily produced in vascular bundle within the stele other than the cortex (Hartung et al. 2002; HB Ren, KF Wei and WS Jia, unpubl. data, 2007), such that the presence of exodermis and endodermis (site of Casparian formation) will largely retard the efflux of ABA into soil. In addition, the production of ABA in the vascular bundle will make it easier to for ABA to be loaded into the vascular vessel.

In regard to the long-distance transport of ABA from the root to the shoot, it seems that much attention should be paid to the lateral inter-transport of ABA between the xylem and stem parenchyma. There is evidence that ABA can laterally transport between xylem vessel and parenchyma cells in the stem (Sauter and Hartung 2002). The rate of ABA lateral transport is dependent on the ABA concentration gradient between xylem and stem parenchyma, thus under soil drying conditions, the ABA transport from xylem to stem parenchyma will be greatly promoted because the xylem ABA concentration is elevated, and this may give rise to losses of ABA from the xylem and decrease the xylem ABA concentration. Logically, this reduction of the xylem ABA concentration would be directly correlated with the stem length. There is plenty of evidence to suggest that herbaceous plant species compared with woody species are less sensitive to the root ABA signal (Saliendra et al. 1995; Fuchs and Livingston 1996; Comstock and Mencuccini 1998; Yao et al. 2001). Schulze (1991) suggested that large woody species may lack a chemical root signal. It is not known whether the stomatal insensitivity in woody plants may be a result of the modulation of xylem ABA concentration by the radial transport of ABA.

4.1.2 Effect of catabolism

It is well known that ABA content in plant cells is determined by a dynamic equilibrium between biosynthesis and catabolism (Zeevaart 1980, 1983; Zeevaart and Creelman 1988). It has been well established that a 9_*cis*-epoxycarotenoid dioxygenase (NCED) catalyzed reaction is a rate-limiting step in the ABA biosynthesis pathway due to the presence of a large pool of ABA precursors upstream from the NCED, and that the activation of NCED is responsible for the dehydration-induced ABA accumulation (Schwartz et al. 1997; Qin amd Zeevaart 1999; Seo and Koshiba 2002). Because dehydration-induced ABA accumulation is triggered by the activation of NCED, it has been commonly thought that the amount of ABA accumulated in the dehydrated cells is determined by ABA biosynthesis.

However, considering of the dynamic equilibrium between biosynthesis and catabolism, the important role of ABA catabolism in the regulation of ABA accumulation should not be over looked. Logically, supposing that ABA catabolism is kept unchanged or becomes lower in the dehydrated than non-dehydrated cells, the accumulated level of ABA would depend on the sustained time of the dehydration. A sustained soil drying would expect to result in an unlimited accumulation of ABA. Clearly, this will not happen in reality. The key point lies in the absolute rate of ABA catabolism and its variation in response to dehydration stress. Using an 18O-labelling method, Creelman et al. (1987) determined the rate of ABA catabolism, and a half-life of 15.5 h for the ABA catabolism was calculated in Xanthium leaves. Dehydration-triggered ABA accumulation is very fast, i.e. within several hours the ABA accumulation may reach its maximum (Gowing et al. 1993; Jia et al. 1996; Jia and Zhang 1997; Zhang and Jia 1997; Zhang et al. 1997). Compared with this, the half-life of 15.5 h for the ABA catabolism implies that the rate of ABA catabolism is very low. In contrast to the observation by Creelman et al. (1987), several lines of evidence strongly suggest that the ABA catabolism is actually very fast (i.e. only about 1 h in many plant species) (Gowing et al. 1993; Jia et al. 1996; Jia and Zhang 1997; Zhang et al. 1997). A study by Ren et al. (2007) suggested that the dynamic process of ABA catabolism exhibited a pattern of exponential decay and dehydration stress had no effect on the catabolic half-life in maize. This means that the absolute rate of ABA catabolism (i.e. amount of ABA catabolized per unit time) is proportional to the amount of ABA, i.e the higher the accumulated level of ABA is, the faster the absolute rate of the ABA catabolism is, and this further means that the ABA accumulation would, sooner or later, make the rate of ABA catabolism be rate of ABA biosynthesis, which will make the ABA accumulation reach its maximum thus preventing an unlimited accumulation of ABA.

Besides playing important roles in the regulation of stress-induced ABA accumulation in roots, ABA catabolism may also play crucial roles in the regulation of ABA transport along the stem. As mentioned above, xylem ABA concentration may be modulated along the stem as a function of its radial transport. The rate of radial transport of ABA is predominantly determined by the ABA concentration gradient between xylem vessel and its surrounding cells (stem parenchyma), and hence, the key point for the modulation of xylem ABA concentration lies in the regulation of this ABA gradient. Supposing that ABA can not be effectively removed in the stem parenchyma surrounding xylem vessel, an equilibrium of the ABA concentration between the xylem vessel and parenchyma will eventually occur, such that the lateral transport of ABA will no longer posses effect on the xylem ABA concentration. In contrast, if the parenchyma ABA can be removed at a rate large enough to support the presence of the radial gradient of ABA concentration, an efflux of ABA from the xylem vessel will always occur during the long-distance transport of the root-sourced ABA, and logically, the longer the root-sourced ABA is transported, the lower the xylem ABA concentration would become. Given the fast rate of ABA catabolism, it is likely that removal of ABA in the stem parenchyma is predominantly by the ABA catabolism. However, whether the rate of this removal is large enough to cause a radial gradient of ABA and whether this radial gradient of ABA is indeed large enough to cause significant decrease in the xylem ABA concentration is not known. The issue about the stomatal sensitivity in relation to the modulation of xylem ABA concentration during its long-distance transport needs to be investigated.

It is well known that for a molecular substance to be a signal, besides having a triggering mechanism it must have a deactivation mechanism. Likely, once the root-sourced ABA enters into leaves it should be removed sooner or later otherwise an unlimited accumulation of ABA would be caused, and this would lead to a permanent inhibition of the stomatal movement even when plants are not stressed. As mentioned above, the absolute rate of ABA catabolism is not only quite fast but also increases proportionally to the accumulated amount of ABA. So, the ABA catabolism should play critical roles in the regulation of ABA accumulation in leaves thus modulating the stomatal sensitivity to the root sourced ABA signal. The effect of ABA catabolism on the regulation of stomatal movement is complicated. On one hand, the ABA catabolism may play a central role in the removal of ABA thus preventing stomatal movement from permanent inhibition, and on the other hand, the removal of ABA as a function of the ABA catabolism may reduce the level of ABA accumulated in its action sites thus reducing the stomatal sensitivity to the root-sourced ABA signal. The removal of ABA to prevent a permanent inhibition of stomatal movement is clearly not a problem to be concerned, and the major concern is to what an extent the ABA catabolism may be able to modulate the stomatal sensitivity. Regulation of the root to shoot signaling associated with ABA catabolism is an importance issue, which should be paid particular attentions.

4.2 Modulation of pH signaling

4.2.1 Effect of chemical compositions in xylem stream

It is well known that the pH in solution is determined by a balance among ions and anions, such that a change in the composition of ions or anions may possibly result in a change in pH. More importantly, many transport processes of chemical composition across the cell membrane are associated with H^+ bumps (H^+-ATPase), and this must cause a change of pH in different compartments of the cells. Many ions, such as malate together with other organic acid, nitrate and ammonium are known to be powerful regulators of pH. L-Malate is a prominent organic acid in many plant tissues especially the dicotyledonous plants (Vickery 1963;Buttz and Long 1979), and in monocotyledonous plants *trans*aconitate is the major organic acid with L-malate present in smaller quantities (Clark 1969). Clearly, any changes in the uptake, transport and assimilation of these components may be able to cause a pH shift.

It has been increasingly suggested that N-containing compounds play crucial roles in the pH regulation. It was reported that under nitrate nutrition in castor oil plants, the organic acids were present in xylem sap only with trace amounts, and nitrate deprivation considerably increased the amounts of organic acids, thus increasing the pH from 5.6 to 7.3 (Kirkby and Armstrong 1980). In pepper plants, Dodd et al. (2003) also found that N-deprivation caused an alkalization of xylem sap. It was proposed that when nitrate is plentiful, nitrate will mainly be reduced in leaf cells, but when soil nitrate availability is low the nitrate reduction will be switched from shoot to root (Lips 1997), and this will produce hydroxyl ions, which are then converted to malate, thus leading to an alkalization of xylem sap (Wilkinson and Davies 2002). These observations seem to suggest that the soil drying-induced pH increase in xylem sap is closely related with possible changes in the rate of nitrate reduction in root. Consistent with this hypothesis, there is plenty of evidence demonstrating that soil drying can strongly affect the activity of the nitrate reductase in root, but in contrast, most of the studies demonstrate that soil drying results in a decrease, not an increase in nitrate reductase (Solomonson and Barber 1990). Whether the soil drying-induced pH increase in the xylem sap is directly governed by nitrate reduction remains to be further investigated.

As mentioned above, many studies demonstrate that soil drying can usually reduce the uptake of nitrate. Theoretically, the reduction in nitrate uptake should give rise to a decrease in the leaf apoplstic pH, and this seems unreasonable, because the reduction of apoplastic pH would decrease the stomatal sensitivity to ABA. To explain this question, Wilkinson and Davies (2002) proposed that the N availability may influence the species of the N-containing molecule, and under drought conditions, low nitrate availability would contribute to increases in the content of malate or other organic acid in xylem sap, thus giving rise to a higher apoplastic pH. It should be noted that the soil drying-induced xylem pH increase and the changes of xylem chemical composition or content may be two different concepts. This is because NO3 −, NH4 + and some other ions have actually no significant effect on xylem pH, but once transported to the leaf they may strongly modify the appolastic pH, and such a viewpoint has been proved by the recent study of Jia and Davies (2007). In regards to pH signaling in response to soil drying, we should not only pay attention to whether soil drying may directly affect xylem pH, but also pay close attention to whether it may affect the chemical composition or content, and also whether a type of soil may directly affect the chemical composition or content.

4.2.2 Effect of H^+ -exchange between xylem vessel and the surrounding cells

It is known that for the root to shoot signaling of ABA it is the significant variation in xylem ABA concentration that is predominantly responsible for mediation of stomatal responses to the soil drying stimulus. Likely, for pH signaling a shift of xylem pH has been normally thought to be responsible for the mediation of stomatal responses to the soil drying stimulus (Wilkinson 1999). This proposition is actually based on a hypothesis that xylem sap pH should be approximately the same with that in the apoplastic sap, i.e. the pH at the action sites of ABA signal. However, it has been increasingly suggested that such a hypothesis is not necessarily correct. Several studies reported that the leaf apoplastic pH is actually much different from that in xylem sap (Hoffmann and Kosegarten 1995; M¨ uhling and Lauchli 2000). Wilkinson and Davies proposed that the climatically induced changes in sap pH are a result of changes within the leaf apoplast rather than from the incoming xylem sap itself (Wilkinson and Davies 2002). There are also some reports that some factors that affect transpiration (such as high vapour pressure deficit [VPD], photosynthetic photon flux density [PPFD] and temperature) can give rise to a changed pH in sap expressed from shoot in *F. intermedia* and *H. macrophylla* (Wilkinson and Davies 2002). It is not known how the aerial factor can cause the changes in sap pH. It is hypothesized that high VPD may affect leaf cell H^+- ATPase activity by causing slight changes in localised water relation, and high PPFD may increase the removal of CO2 from apopalst, thus causing an apoplastic alkalization (Hartung and Radin 1989; Wilkinson and Davies 2002). Jia and Davies (2007) found that a great pH gradient exists between the stem base xylem and the leaf vascular system in many plant species. Also, it was found that a change in transpiration rate would be able to cause a change in the leaf xylem sap or apoplastic pH. They provided evidences that the great pH gradient along the stem is a result of pH modulation by the H^+-pump in the vascular parenchyma cells, i.e. the proton in xylem stream can be significantly pumped out thus causing more alkalization with the stem height. Understandably, the removal rate of proton as a function of H^+-pump should be affected by sap flow rate, and this may be why transpiration may be able to affect leaf apoplastic pH. Collectively, these data indicate that xylem sap pH can be strongly modified when passing through plant vascular systems, and moreover, the leaf cells may have a mechanism for the leaf apoplastic pH to be independently regulated and not necessarily controlled by the level of xylem sap pH.

5. Coordination of root signals in the regulation of stomatal movement

ABA has been thought to be a major signal mediating the stomatal responses to soil drying, but this does not mean that the decreased leaf conductance as a function of the root to shoot signaling is mainly caused by the ABA signals. It has been increasingly suggested that the stomatal regulation upon soil drying is actually controlled by multiple signals, among which ABA is a necessary signal to trigger stomamal closing, but not definitely a signal that is mainly responsible for the decrease of stomatal closure. The key points lie in the synergistic action or coordination among different signals. Detailed discussions above this are given below.

5.1 Coordination of ABA and pH signaling

Although it has been established that pH can be a signal mediating the root to shoot signaling, there is evidence that the pH signal may not be able to independently regulate stomatal movement. A well demonstrated example for this is the research by Wilkinson et al. (1998). Using tomato mutant as the research material, Wilkinson et al. found that soil drying resulted in an increase in xylem sap pH from 5.0 to 8.0, but when artificial xylem sap buffered to different pH was fed to the detached leaves of either wild type or ABA-deficient mutant *flacca,* it was found that the wild type but not *flacca* leaves exhibited reduced transpiration rates with the increase of pH. On the other hand, a well watered concentration of ABA added in the sap, just like the wild type *flacca* was able to exhibit a transpirational reduction with pH increased from 6.25 to 7.75. These data demonstrated that the stomatal regulation by xylem pH absolutely requires the presence of ABA.

It has been proposed that the coordination between the pH and ABA signaling is based on the effect of pH-regulation of ABA accumulation at its action sites, i.e. soil drying-induced pH increase in xylem would make apoplast of the leaf more alkaline, which would contribute to a sequestration of more ABA in the apoplast of guard cells, thus promoting the stomatal closure in the presence of ABA (Wilkinson 1999; Wilkinson and Davies 2002). However, owing to the modulation of xylem pH along the stem as described above, the soil drying-induced shift in the xylem pH may not necessarily lead to a variation in the leaf apoplastic pH that is big enough to pose an effect on the ABA accumulation. In addition, it should be noted that the pattern of this modulation may vary greatly among different plant species or under different culture conditions. For example, while soil drying is able to increase the pH of xylem sap in tomato, *Commelina communis* and barley, it has no effect on the pH of xylem sap in *Hydrangea macrophylla* cv Bluewave and *Cotinus coggyria* cv Royal Purple, and in *Forsythia intermedia* cv Lynwood it can even lead to a decrease in the pH (Wilkinson and Davies 2002). As mentioned above, stomatal sensitivity to root ABA signal often varies greatly with plant species or plant culture conditions. It is not known whether this variation may be correlated to the big difference in patterns of pH variation among different plant species or culture conditions.

Besides functioning to modulate ABA accumulation within leaves, pH may also play a role in the modulation of the ABA signal intensity. A well demonstrated example for this is a recent research carried out by Li et al (2010). Using grapevine (*Vitis riparia* × *Vitis labrusca*) as the research plant, they found that the intensity of ABA signal (i.e. xylem ABA concentration) increased rather than decreased as commonly thought during the long-distance transport of the root-sourced ABA, and in the mean while, they also found that there existed a basipetal pH gradient along the stem. The basipetal pH gradient along the stem may be related to the different activity of H^+-ATPase in the parenchyma of the vascular bundle. No matter mechanism for the presence of the basipetal pH gradient, this study has provided strong evidences that the enforced ABA signal along stem is a result of pH modulation, i.e. the increased pH towards the apex contributed to reduce the rate of ABA efflux from the xylem vessel. Collectively, the result of this study indicated that the intensity of the root-sourced ABA signal can be modulated by pH along its long-distance transmission in the stem.

The coordination of the ABA and pH signaling has complicated the shoot to shoot signaling. The reason for this is that pH itself can be modulated by many factors. Logically, any pH regulator may have a possibility to pose an effect on the regulation of the root to shoot signaling. A well demonstrated example for this is the effect of nitrate nutrition on the modulation of pH signaling as well as the stomatal sensitivity to the root ABA signal. Kosegarten et al. (1999) found that nitrate nutrition could cause a high apoplastic pH in immature sunflower leaves, but they were not able to observe such a phenomenon in mature leaves or in immature leaves with sole NH^+ or NH^{4+}/NO^{3-}as nutrition. M¨ uhling and Lauchli (2001) also found that nitrate nutrition caused more alkaline leaf apoplastic sap than ammonium nutrition, in both *Phaseolus vulgaris* and sunflower, but not in *Vicia faba* or *Zea mays*. More recently, using a pH ratio imaging technique, Jia and Davies (2007) clearly showed that feeding NO^{3-} induced a significant alkalization of the leaf apoplast, while the feeding of NH^{4+} had a contrary effect (i.e. it gave rise to a significant reduction of the apoplastic pH in *Commelina communis*). Importantly, feeding NH^{4+}or NO^{3-} significantly decreased (for NH^{4+}) or increased (for NO^{3-}) the stomatal responses to ABA, suggesting that the pH-regulator is indeed able to modulate stomatal sensitivity.

5.2 Hydraulic signal

A major issue about the chemical signaling is its argument with a hydraulic signaling. While numerous studies have strongly demonstrated the important roles of the chemical signaling in the root to shoot communication, are also plenty of evidences supporting the potential roles of the hydraulic signaling in the root to shoot signaling (Petersen et al. 1991; Saliendra et al. 1995; Fuchs and Livingston 1996; Yao et al. 2001; Comstock 2002; Sperry et al. 2002). In studies emphasizing the hydraulic regulation of stomatal behavior, it is even thought that a lack of response to shoot water potential would be potentially fatal to plants (Comstock 2002). For example, in Douglas fir Fuchs and Livingston (1996) found that the reduction in leaf conductance as a result of soil drying could be progressively reversed by the pressurization of the root system, and furthermore, once the pressurization was released the leaf conductance would return to its prepressurization levels within minutes, strongly indicating that the hydraulic signal was a predominant regulator of the stomatal behavior. Similar findings were made by Saliendra et al. (1995) in woody plants *Betula occidentalis* and Yao et al. (2001) in bell pepper (*Capsicum annuum L. vau. Maor*). Recently, using ABA-deficient mutant of Arabidopsis and a noninvasive imaging system for ABA action in combination with grafting technique, Christmann et al. (2007) have provided strong evidences that the stomatal behavior is controlled by a hydraulic signal rather than root-sourced ABA signal.

Although it is difficult to clearly address the debate between the chemical and hydraulic signaling, it seems that the three points below may contribute to understanding some of the key points about this debate. The first is the big difference in the stomatal behaviors among different species, i.e. while a change in xylem ABA concentration can greatly regulate stomatal movement in some plant species, it may have no effects on the stomatal movement in other plant species. The second is that even for a specific plant species different experimental condition and design may possibly give rise to different results. For example, the developmental stage, the size, the root to shoot ratio and so on all contribute to affect the stomatal behavior and the water relationship of whole plant. Additionally, the nature of soil may also likely pose an effect on the pH signaling as discussed above, thus modulating the stomatal sensitivity. Thirdly, the lack of a unitive and identical description for the stomatal behavior may possibly affect the assessment on the nature of signals. For example, to describe the stomatal responses to soil drying, the phrase 'stomatal closure' was commonly used in published reports. Actually, a mild drought can only, to some degree, cause a decrease in the stomatal conductance but not a stomatal closure. Given that the decreased degree of the stomatal conductance is correlated to the nature of the signals, a different evaluation of the stomatal closure would be likely to affect the determination of the root to shoot signals. No matter what reason for the different conclusions drawn in different researches, it seems that these two theories don't actually conflict with each other, i.e. different theories may be applicable to different cases depending on plant species, developmental stages or even their culture conditions.

According to the basic theory on the water transport in plants, the decreased water availability of roots as a result of soil drying would inevitably lead to a decrease in the leaf water potential, and this would again inevitably cause a decrease in the stomatal conductance. Therefore, irrespective of evidences for the chemical signaling, the hydraulic signal must play crucial roles in regulation of stomatal movement in response to soil drying. The key point lies in that to what a degree the chemical signaling may be able to control the stomatal behavior and at what sage of the soil drying the hydraulic signaling starts to play a role in the regulation of the stomatal movement. Additionally, much attention should be paid on the practical significance of the two different kinds of signaling in the water-saving agriculture not just on their relative importance in the stomatal regulation. The practical significance of the chemical signaling is embodied in its potential roles in reducing the transpiration without negative impacts on the food production. Although a mild decrease in the leaf water potential may affect, to some extent, plant biomass, it may not significantly affect the food production as a function of the changed assimilate transport and allocation. However, a mild decrease in the leaf water potential may be able to strongly induce stomatal closing, thus may also significantly improve WUE. In this context, it is the cooperation of the two different kinds of signaling that may play a maximum positive role in the water saving agriculture. Interestingly, there is evidence that a mild decrease in leaf water potential could increase the stomatal sensitivity to ABA. For example, in a research by Tardieu et al., epidermal pieces of *Commelina communis* were incubated in media with different water potential adjusted by polyethylene glycol. In the media without ABA, the water potentials between −0.3 and −1.5 megapascals had no significant effect on stomatal aperture, however, when ABA was added to the media with a decrease of the water potential from −0.3 to −1.5 megapascals, the stomatal aperture significantly decreased, which strongly suggested that the stomatal sensitivity to ABA could be modified by water potential (Tardieu et al., 1992). The same observation was obtained by feeding ABA into the field grown plants. There is evidence that the two kinds of signaling may be able to interact with each other and jointly control the stomatal behavior.lants over different ranges of leaf water potential (Tardieu and Davies 1993). These studies suggest that the hydraulic signaling may function to strengthen the chemcial signaling, thereby jointly regulating the stomatal responses to soil drying. For a given plant species, It is of high importance to quantitatively determine the critical point of the decrease in leaf water potential, at which a maximum increase in WUE can be achieved.

6. Manipulation of the systematic signaling in water saving agriculture

6.1 Exploitation of the systematic signaling to increase water use efficiency

Food production is at the cost of water resources. With the fast reduction of the limited water resources, the only way for the development of a sustainable agriculture is to increase the water use efficiency (WUE). The practical significance of the systemic chemical signaling lies in its effective regulation of the transpiration without a strongly negative impact on the food production, and hence, exploitation of the systemic chemical signaling is a good strategy to increase WUE. As described above, systemic signaling is embodied in that, in responses to a stimulus in one part of the plant, the corresponding responses may take place in another part of the plant. Therefore, for an exploitation of the root to shoot signaling to increase the WUE, we need to manipulate water deficit appropriately. The spilt-root experiment is known to be a good demonstration of the root to shoot signaling. Such an idea can be adopted in practical agriculture. Based on the theory of the systemic signaling, two water-saving irrigation techniques have been developed and widely used in many sectors of the agriculture in many counties, one commonly used technique is the regulated deficit irrigation (RDI) and another is the partial rootzone drying (PRD) (Loveys, 1991; Loveys et al., 2004; Dodd et al., 2006; Fereres and Soriano, 2007; Sadras, 2009).

RDI works on the premise that a mild water deficit would greatly reduce plant transpiration while it does not significantly affect the food production as a result of improved harvest index, such that an improvement of WUE would be achieved. However, in practice a sublimate control of the water deficit is difficult to be realized, and it is more likely that RDI would give rise to a great decrease in biomass, thus significantly reducing crop yield. RDI may be successfully used in the production of some horticulture plants, such as tomato and some fruit trees, where a vigorous vegetative growth may not be needed and even adverse to fruit production (Davies et al., 2001). In comparison with RDI, PRD is easier to be appropriately manipulated in practice and more widely used in water-saving agriculture. PRD works on the principle that when part of the root system is allowed to dry and the remaining roots are kept well-watered, the drying roots would trigger chemical signaling by which transpiration can be reduced, and meanwhile, the food production may be not affected as a function of the maintaining of a relatively high leaf water status by the wet roots (Dodd et al., 2006; Dry et. al., 2001; Wilkinson and Hartung 2009). PRD techniques can be adapted to the use of either drip or furrow irrigation and a commonly adopted method is alternative irrigation in crop production (Costa et al., 2007; Kang et al., 2000).

A major issue about the chemical signaling is its capability to regulate stomatal behavior, and additionally this capability often varies greatly among different plant species and different circumstances. In particular, it appears that chemical signaling plays little roles in most of woody plants (Comstock 2002, Ren et al., 2007). Therefore, trying to increase the stomatal sensitivity to the root chemical signals is a major concern for a better use of PRD. The stomatal sensitivity may be modulated by many factors as noted above. However, for a given plant species, a practical strategy may be pH modulation. Modulation of pH may be achieved by an appropriate manipulation of fertilizing, but until now, no much information is available about how PRD can be improved by an appropriate manipulation of fertilizing. More importantly, it should be noted that the water-saving effect of PRD may be not accounted for solely by the chemical signaling because the possibility of the involvement of hydraulic signaling can not be excluded in this process. PRD technique is based on a hypothesis that wet part of the root system is able to keep leaf water potential relatively constant as the remaining part of the root system dries. Under the field conditions, however, this may be not allays the case as expected. Whether a decrease in leaf water potential occurs should depend on the balance between water loss and income of the lamia. This balance can be affected by many factors not just by stotmtal movement, such as air temperature, humidity, shoot/root ratio, water conductance of roots and so on. A change in any of these factors may contribute to breaking the balance, and thus, posing an effect on the leaf water potential, .e.g. dry and hot whether may possibly give rise to a significant decrease in the leaf eater potential especially for those plants with a low root/shoot ratio even if the plants are kept well-watered. Given the potential roles of hydraulic signaling in the stomatal regulation, more attentions should be paid on the ratio of the dry/wet roots in RPD, so that degree of the water deficit should be finely controlled to the critical point at which a maximum decrease in the stomatal conductance, and in the mean time, a minimum negative impact on the food production can be achieved. In this context, it is the coordination of the chemical and hydraulic signaling that functions to effectively save water during the implement of PRD technique.

6.2 Exploitation of the systematic signaling to increase fruit quality.

As mentioned above, the systemic chemical signaling seems to be a good theory that can be exploited to save water in practical agriculture production. However, the water-saving capability of chemical signaling may be very limited for many plant species or under many developmental stages especially for woody plants. To effectively save water, PRD should be appropriately manipulated so that a coordination of the chemical and hydraulic signaling functions to induce a relatively maximum reduction in the stomatal conductance. Effectively saving water would inevitably pose negative impact on plant growth and development, and this would greatly limit the use of PRD technique for many crop plants. It seems that saving water is always contradictory to food production. Excitingly, this can be well resolved in the case of fruit production in many sectors of agriculture. In china, for example, no matter in cultivated areas or yield, the apple production is the first in the world, however its apple market only shares less than 5 % in the world. The major reason for this is its poor quality. In many sectors, it is the fruit quality but not yield that has become a factor limiting the development of fruit industry.

Application of RDI and PRD inevitably reduces vegetative growth and this will again pose negative impacts on crop yield. Unlike the crop yield, fruit quality is less negatively affected by vegetative growth, and by contrast, vigorous vegetative growth may be adverse to fruit quality even to the yield as a result of assimilate competition. Additionally, light intensity is known to be a predominant factor in the regulation of fruit quality and a reduction in vegetative growth may contribute to reducing the canopy density hence resulting in better light penetration to the bunch zone (Davies et al., 2001; Dry *et al.*, 1996). More importantly, it has been well established that mild water deficit would be able to increase sugar content of fruits, which is a major index of the fruit quality. Further more, it is worth noting that most arid agricultural land is in developing countries, where the economic income is a major concern for the local farmers. Increasing fruit quality will greatly increase the farmers' economic income, and this will surely save land and water to grow more crop plants for an expected economic income to be achieved. In this context, application of RDI and RPD in fruit production is of more significance than in crop plants as regards to their water-saving effects. In Australia and many other counties, RDI and PRD techniques have been successfully used for many years to save water and improve fruit quality in grape production (Dry *et al.*, 1996; Davies et al., 2002). Collectively, while there are many arguments and concerns about the application of RDI and PRD techniques with an aim to effectively improve WUE of many crop plants, the application of the two techniques seems to be a perfect strategy for the production of fruit trees.

7. Acknowledgment

This work is supported by grants from National Crops Transgenic Special Grant (2009ZX08009-0738 and 2009ZX08003-009B) and the National Natural Science Foundation (31171921)

8. References

Auge RM, Stodola AJW, Ebel RC, Duan X (1995) Leaf elongation and water relations of mycorrhizal sorghum in response to partial soil drying: two *Glomus* species at varying phosphorus fertilization. *J. Exp. Bot.* 46, 297–307.

Blackman PG, Davies WJ (1985) Root to shoot communication in maize plants of the effects of soil drying. *J. Exp. Bot.* 36, 39–48.

Buttz RG, Long RC (1979) L-Malate as an essential component of the xylem fluid of corn seedling roots. *Plant Physiol.* 64, 684–689.

Clark RB (1969) Organic acids from leaves of several crop plants by gas chromatography. *Crop Sci.* 9, 341–343.

Comstock JP (2002) Hydraulic and chemical signaling in the control of stomatal conductance and transpiration. *J. Exp. Bot.* 53, 195- 200.

Comstock J, Mencuccini M (1998) Control of stomatal conductance by leaf water potential in *Hymenoclea salsola* (T. & G.), a desert subshrub. *Plant Cell Environ.* 21, 1029–1038.

Creelman RA, Gage DA, Stults JT, Zeevaart JAD (1987) Abscisic acid biosynthesis in leaves and roots of Xanthium strumarium. *Plant Physiol.* 85, 726–732.

Davies WJ, Zhang J (1991) Root signals and the regulation of growth and development of plants in drying soil. *Annu. Rev. Plant Physiol. Plant Mol. Biol.* 42, 55–76.

Davies WJ, Tardieu F, Trejo CL (1994) How do chemical Signal work in plants that grow in drying soil? *Plant Physiol.* 104, 309–314.

Dodd IC, Tan LP, He J (2003) Do increases in xylem sap pH and/or ABA concentration mediate stomatal closure following nitrate deprivation? *J. Exp. Bot.* 54, 1281–1288.

Else MA, Davies WJ, Whitford PN, Hall KC, Jackson MB (1994) Concentrations of abscisic acid and other solutes in xylem sap from root systems of tomato and castor-oil plants are affected by the method of sap collection. *J. Exp. Bot.* 45, 317–323.

Else MA, Hall KC, Arnold GM, Davies WJ, Jackson MB (1995) Export of abscisic acid, 1-aminocyclopropane-1-carboxylic acid, phosphate, and nitrate from roots to shoots of flooded tomato plants. *Plant Physiol.* 107, 377–384.

Fawzi A, Razem, Luo M, Liu JH, Abrams SR, Hill RD (2004) Purification and characterization of a Barley Aleurone abscisic acidbinding protein. *J. Biol. Chem.* 279, 9922–9929.

Fuchs EE, Livingston NJ (1996) Hydraulic control of stomatal conductance in Doughlas fir and alder seedlings. *Plant Cell Environ.* 19,1091–1098.

Gollan T, Passioura JB, Munns R (1986) Soil water status affects the stomatal conductance of fully turgid wheat and sunflower plants.*Aust. J. Plant Physiol.* 13, 459–464.

Gollan T, Schurr U, Schulze ED (1992) Stomatal response to drying soil in relation to changes in the xylem sap composition of *Helianthus annuus*. I. The concentration of cations, anions, amino acids in, and pH of, the xylem sap. *Plant Cell Environ.* 15, 551–559.

Gowing DJ, Davies WJ, Jones HG (1990) A positive root sourced signal as an indicator of soil drying in apple, *Malus x domestica* Borkh. *J. Exp. Bot.* 41, 1535–1540.

Gowing DJG, Jones HG, Davies WJ (1993) Xylem-delivered abscisic acid: the relative importance of its mass and its concentration in the control of stomatal aperture. *Plant Cell Environ.* 16, 453–459.

Hartung W, Radin JW (1989) Abscisic acid in the mesophyll apoplast and in the root xylem sap of water-stressed plants: the significance of pH gradients. *Curr. Topics Plant Biochem. Physiol.* 8, 110– 124.

Hartung W, Radin JW, Hendrix DL (1988) Abscisic acid movement into the apoplastic solution of water-stressed cotton leaves: Role of apoplastic pH. *Plant Physiol.* 86, 908–913.

Hartung W, Sauter A, Hose E (2002) Abscisic acid in the xylem: where does it come from, where does it go to? *J. Exp. Bot.* 53, 27–32.

Hoffmann B, Kosegarten H (1995) FITC-dextran for measuring apoplast pH and apoplastic pH gradients between various cell types in sunflower leaves. *Physiol. Plant.* 95, 327–335.

Hornberg C, Weiler EW (1984) High affinity binding sites for abscisic acid on the plasmalemma of *Vicia faba* guard cells. *Nature* 310, 321–324.

Holbrook NM, Shashidhar VR, James RA, Munns R (2002) Stomatal control in tomato with ABA-deficient response of grafted plants to soil drying. *J. Exp. Bot.* 53, 1503–1514.

Jackson MB (1993) Are plant hormones involved in root to shoot communication? *Adv. Bot. Res.* 19, 103–187.

Jia WS, Zhang J (1997) Comparison of exportation and metabolism of xylem-delivered ABA in maize leaves at different water status and xylem sap pH. *Plant Growth Regul.* 21, 43–49.

Jia WS, Zhang JH (1999) Stomatal closure is induced rather by prevailing xylem abscisic acid than by accumulated amount of xylemderived abscisic acid. *Physiol. Plant.* 106, 268–275.

Jia WS, Davies WJ (2007) Modification of leaf apoplastic pH in relation to stomatal sensitivity to root-sourced abscisic acid signals. *Plant Physiol.* 143, 68–77.

Jia WS, Zhang J, Zhang DP (1996) Metabolism of xylem-delivered ABA in relation to ABA flux and concentration in leaves of maize and *Commelina communis L. J. Exp. Bot.* 47, 1085–1091.

Jones RJ, Mansfield TA (1970) Suppression of stomatal opening in leaves treated with abscisic acid. *J. Exp. Bot.* 21, 714–719.

Kaiser WM, Hartung W (1981) Uptake and release of abscisic acid by isolated photoautotrophic mesophyll cells, depending on pH gradients. *Plant Physiol.* 68, 202–206.

Khalil AAM, Grace J (1993) Does xylem ABA control the stomatal behaviour of water-stressed sycamore (*Acer pseudoplatanus* L.) seedlings? *J. Exp. Bot.* 44, 1127-1134.

Kirkby EA, Armstrong MJ (1980) Nitrate uptake by roots as regulated by nitrate assimilation in the shoot of castor oil plants. *Plant Physiol.* 65, 286–290.

Kosegarten HU, Hoffmann B, Mengel K (1999) Apoplastic pH and Fe_{3+} reduction in intact sunflower leaves. *Plant Physiol.* 121, 1069–1079.

Li Y, Walton DC (1987) Xanthophylls and abscisic acid biosynthesis in water-stressed bean leaves. *Plant Physiol.* 85, 910–915.

Lips SH (1997) The role of inorganic nitrogen ions in plant adaptation processes. *Russ. J. Plant Physiol.* 44, 421–431.

Loveys BR(1991) What use is a knowledge of ABA physiology for crop improvement? In: Davies WJ, Jones HG, eds. Abscisic acid, Oxford: Bios Scientific Publishers, 245-259.

Loewenstein NJ, Pallardy SG (1998) Drought tolerance, xylem sap abscisic acid and stomatal conductance during soil drying: a comparison of canopy trees of three temperate deciduous angiosperms. *Tree Physiol.* 18, 431–439.

Munns R, Cramer GR (1996) Is coordination of leaf and root growth mediated by abscisic acid? *Opin. Plant Soil* 185, 33–49.

Munns R, Passioura JB, Milborrow BV, James RA, Close TJ (1993) Stored xylem sap from wheat and barley contains a transpiration inhibitor with a large molecular size. *Plant Cell Environ.* 16, 867–872.

M¨ uhling KH, Lauchli A (2000) Light-induced pH and K_+ changes inthe apoplast of intact leaves. *Planta* 212, 9–15.

M¨ uhling KH, Lauchli A (2001) Influence of chemical form and concentration of nitrogen on apoplastic pH of leaves. *J. Plant Nutr.* 24,399–411.

Norman SM, Maier VP, Pon DL (1990) Abscisic acid accumulationand carotenoid and chlorophyll content in relation to water stress and leaf age of different types of citrus. *J. Agric. Food Chem.* 38, 1326–1334.

Parry AD, Horgan R (1991) Carotenoids and abscisic acid biosynthesis in higher plants. *Physiol. Plant.* 82, 320–326.

Parry AD, Babiano MJ, Horgan R (1990) The role of ciscarotenoids in abscisic acid biosynthesis. *Planta* 182, 118–128.

Passioura JB (1987) The use of the pressure chamber for continuously monitoring and controlling the pressure in the xylem sap of the shoot of intact transpiring plants. *Zn Proceedings of the International Conference on Measurement of Soil and Plant Water Status*. University Press, Logan. pp. 31–34.

Passioura JB (1980) The transport of water from soil to shoot in wheat seedlings. *J. Exp. Bot.* 31, 333–345.

Passioura JB (1988) Water transport in and to roots. *Annu. Rev. Plant Physiol. Plant Mol. Biol.* 39, 245–265.

Passioura JB, Tanner CB (1985) Oscillations in apparent hydraulic conductance of cotton plants. *Aust. J. Plant Physiol.* 12, 455–461.

Petersen KL, Moreshet S, FuchsM(1991) Stomatal responses of field grown cotton to radiation and soil moisture. *Agron. J.* 83, 1059–1065.

Qin XQ, Zeevaart JAD (1999) The 9-cis-epoxycarotenoid cleavage reaction is the key regulatory step of abscisic acid biosynthesis in water-stressed bean. *Proc. Natl Acad. Sci.* 96, 15 354- 15 361.

Ren HB, Gao ZH, Chen L, Wei KF, Liu J,Fan YJ, William J Davies, WS Jia and JH Zhang (2007) Dynamic analysis of ABA accumulation in relation to therate of ABA catabolism in maize tissues under water deficit. *J. Exp. Bot.* 58, 211–219.

Saliendra NZ, Sperry JS, Comstock JP (1995) Influence of leaf water status on stomatal response to humidity, hydraulic conductance, and soil drought in *Betula occidentalis*. *Planta* 196, 357- 366.

Sauter A, Hartung W (2002) The contribution of internode and mesocotyl tissues to root-to-shoot signaling of abscisic acid. *J. Exp. Bot.*53, 297–302.

Sauter A, Davies WJ, Hartung W (2001) The long-distance abscisic acid signal in the droughted plant: the fate of the hormone on its way from root to shoot. *J. Exp. Bot.* Special Issue, 52, 1991–1997.

Schulze ED (1991) Water and nutrient interactions with plant water stress. In: Mooney HA, Winner WE, Pell EJ, eds. *Response of Plants to Multiple Stresses* Academic Press, New York, pp. 89- 103.

Schumaker KS, Sze H (1987) Decrease of pH gradients in tonoplast vesicles by NO_3^- and Cl^-: Evidence for H^+-coupled anion. *Transp.Plant Physiol.* 83, 490–496.

Schwartz SH, Tan BC, Gage DA, Zeevaart JAD, McCarty DR (1997) Specific oxidative cleavage of carotenoids by Vp14 of maize. *Science* 276, 1872–1874.

Seo M, Koshiba T (2002) Complex regulation of ABA biosynthesis in plants. *Trends Plant Sci.* 7, 41–48.

Slovik S, Hartung W (1992a) Compartmental distribution and redistribution of abscisic acid in intact leaves. II. Model analysis. *Planta* 187, 26–36.

Slovik S, Hartung W (1992b) Compartmental distribution and redistribution of abscisic acid in intact leaves. III. Analysis of the stress signal chain. *Planta* 187, 37–47.

Solomonson LP, Barber MJ (1990) Assimilatory nitrate reductase: Functional properties and regulation. *Annu. Rev. Plant Physiol. Plant Mol. Biol.* 4, 225–253.

Sperry JS, Hacke UG, Oren R, Comstock JP (2002) Water deficits and hydraulic limits to leaf water supply. *Plant Cell Environ.* 25, 251- 263.

Tardieu F, Davies WJ (1993) Integration of hydraulic and chemical signaling in the control of stomatal conductance and water status of droughted plants. *Plant Cell Environ.* 16, 341–349.

Tardieu F, Zhang J, Davies WJ (1992) What information s conveyed by an ABA signal from maize roots in drying field soil? *Plant Cell Environ.* 15, 185–191.

Trejo C, Clephan AL, Davies WJ (1995) How do stomata read abscisic acid signals? *Plant Physiol.* 109, 803–811.

van Beusichem ML, Baas R, Kirkby EA, Nelemans JA (1985) Intracellular pH regulation during NO_{3-} assimilation in shoot and roots of *Ricinus communis*. *Plant Physiol.* 78, 768–773.

Vickery HB (1963) Metabolism of the organic acids of tobacco leaves. XIX. Effect of culture of excised leaves in solutions of potassium glutamate. *J. Biol. Chem.* 238, 2453–2459.

Wright STC (1977) The relationship between leaf water potential and levels of abscisic acid and ethylene in excised wheat leaves. *Planta* 134, 183–189.

Wilkinson S (1999) pH as a stress signal. *Plant Growth Regul.* 29, 87–99

Wilkinson S, Davies WJ (1997) Xylem sap pH increase: a drought signal received at the apoplastic face of the guard cell that involves the suppression of saturable abscisic acid uptake by the epidermal symplast. *Plant Physiol.* 113, 559–573.

Wilkinson S, Davies WJ (2002) ABA-based chemical signaling: the co-ordination of responses to stress in plants. *Plant Cell Environ.* 25, 195–210.

Wilkinson S, Corlett JE, Oger L, Davies WJ (1998) Effects of xylem pH on transpiration from wild-type and *flacca* tomato leaves: a vital role for abscisic acid in preventing excessive water loss even from well-watered plants. *Plant Physiol.* 117, 703–709.

Yao C, Moreshet S, Aloni B (2001) Water relations and hydraulic control of stomatal behavior in bell pepper plant in partial soil drying. *Plant Cell Envrion.* 24, 227–235.

Zeevaart JAD (1980) Changes in the levels of abscisic acid and its metabolites in excised leaf blades of Xanthium strumarium during and after water stress. *Plant Physiol.* 66, 672–678.

Zeevaart JAD (1983) Metabolism of abscisic acid and its regulation in Xanthium leaves during and after water stress. *Plant Physiol.* 71, 477–481.

Zeevaart JAD, Creelman RA (1988) Metabolism and physiology of abscisic acid. *Annu. Rev. Plant Physiol. Plant Mol. Biol.* 39, 439–473.

Zhang J, Davies WJ (1989) Abscisic acid produced in dehydrating roots may enable the plant to measure the water status of the soil. *Plant Cell Environ.* 12, 73–81.

Zhang J, Davies WJ (1990) Changes in the concentration of ABA in xylem sap as a function of changing soil water status com account for changes in leaf conductance and growth. *Plant Cell Environ.* 13, 271–285.

Zhang J, Davies WJ (1991) Antitranspirant activity in the xylem sap of maize plants. *J. Exp. Bot.* 42, 317–321.

Zhang J, Jia W (1997) Effects of water status and xylem pH on metabolism of xylem transported abscisic acid. *Plant Growth Regul.*21, 51–58.

Zhang J, Jia W, Zhang DP (1997) Re-export and metabolism of xylemdelivered ABA in attached maize leaves under different transpirational fluxes and xylem ABA concentration. *J. Exp. Bot.* 48, 1557– 1564.

Zhu X, Zhang J (1997) Anti-transpiration and anti-growth activities in the xylem sap from plants under different types of soil stress. *New Phytol.* 137, 657–664.

6

The "Pot-in-Pot" System Enhances the Water Stress Tolerance Compared with Above-Ground Pot

J. Miralles[1], J.J. Martínez-Sánchez[1,2] and S. Bañón[1,2]
[1]*Universidad Politécnica de Cartagena*
[2]*Unidad Asociada al CEBAS-CSIC*
Spain

1. Introduction

In the nursery, plant cultivation in pots or containers is a very common practice. On the one hand, this is to satisfy market demands for herbal plants and shrubs, and on the other because of its many advantages compared with ground cultivation. These advantages include lower stress for plants during transport and manipulation, space reduction in the nursery, the increased possibility of mechanization, a longer supply period, and greater transplantation success (Davidson et al., 1988). However, container cultivation is more affected by environmental conditions in the substrate-root complex, where extreme temperatures can negatively influence root development. The climatic season (winter or summer) will determine thermal stress through cold or heat. Container characteristics (material, color, form, drainage holes, etc.) also influence temperature in the root system, and many studies have looked at the use of different container types (Franco et al., 2006).

In contrast to traditional cultivation in above-ground pots (AGP), pot-in-pot (PIP) production, introduced around 1990 in the USA (Parkerson, 1990), is a nursery production method that combines some of the benefits of both field and container production. In a PIP system, a holder or socket pot is permanently placed in the ground with the top rim remaining above. The container-grown plant is then placed within the holder pot for the production cycle (Ruter, 1998a). Previous research into PIP compared with above ground potting (AGP) determined that PIP improves biomass production (Ruter, 1998b), reduces root zone temperature stress (Young and Bachman, 1996) and enhances efficient water use by decreasing container evapotranspiration (Martin et al., 1999). An additional advantage over AGP is the elimination of extensive staking and blowover and a more easy mechanization. Recent studies by Neal (2010) reported that crabapple and lilac root or shoot mass were greater in PIP compared with another four production systems: field-grown, plastic container, bag-in-pot and above ground system.

The disadvantages of the PIP system include high initial cost of pots and installation, potential drainage problems of socket pots in poorly drained soils, and the possibility of root elongation into the socket pot and surrounding soil. In PIP production, containers may stick together; the bottom of the insert pot may sag, causing an uneven base; and there is little or no spacing flexibility once the socket pots are established (Adrian et al., 1998). So, careful planning of the layout should be undertaken because of the large initial outlay for production and the cost of changing the system (Tilt et al., 1994). The economic analysis reported by Adrian et al. (1998) concluded that the PIP system had the highest total capital outlay and fixed cost compared with AGP, which related primarily to higher costs associated with purchasing and installing socket pots. The PIP system, however, was least costly on a per harvested plant basis due to less intensive, labor-saving cultural practices and the ability to grow larger plants quickly.

The avoidance of extreme temperatures in the root sphere was stressed as one of the most advantages of PIP system (Zinati, 2005). The importance of avoiding extreme temperatures in the substrate is well documented, especially because of their effect on root development. Ruter and Ingram (1992) observed that the normal growth of holly trees stopped when temperature exceeded 35°C, whereas in other species this value was lower, at 32°C (Levitt, 1980). Other authors, such as Kuroyanagi and Paulsen (1988), described how wheat plants that receive high temperatures in the root sphere suffer chlorophyll loss and show lower protein production in shoots. Mathers (2003) observed that roots on the west sides of pots are usually injured or destroyed by high temperatures. Zhu et al. (2004) suggested that the moderating of root temperature by PIP prevents the death of *Acer rubrum* roots in contact with the pot wall, whether in winter or summer.

Moderation of the substrate temperature influences the crop water balance, reducing evapotranspiration in the substrate-plant system (Ruter, 1998b) and, as a consequence, maintaining higher substrate moisture content (Fain et al., 1998; Ruter, 1997). Less extreme temperatures at root level as well as good water availability in the substrate can accelerate and improve plant growth and development. Indeed, one of the most interesting and practical aspects of PIP for nursery growers is plant growth magnification (Ruter, 1997; Martin et al., 1999). Ruter (1995) related the moderation of temperature with increased root biomass, although the extent of the effect varied according to the species (Ruter, 1993). Zinati (2005) observed root biomass increases of 50% in PIP grown plants compared with 20% in plants grown in AGP.

It is well known that one of the most important aspects for improving water irrigation efficiency is the correct choice of cultivation technique. Techniques that improve root function and minimize water consumption are considered key factors for improving pot cultivation (Mathers, 2003), and are especially important in landscape and gardening projects to ensure success in transplanting and establishment. In this example, a PIP system may help improve water management efficiency in nursery production, which is especially important in drought conditions and where the water is of low quality, such as in arid and semiarid areas of the planet.

The objective of this research was to determine, during eight months, differences in substrate temperature and water content in native *Rhamnus alaternus* cultivated in two systems: PIP and AGP. We also studied how each system affected growth and development, and water status and gas exchange parameters and we compared the results with a previous study with *Myrtus communis* (Miralles et al., 2009).

2. Material and methods

2.1 Plant material and growing conditions

Seedlings of two year old *Rhamnus alaternus* L. grown in 45 multi-pot forest trays (Plasnor S.A., Gipuzkoa, Spain) were used. The seeds were from natural populations growing in southeast Spain; they were collected in November 2005 and stored dry at 5 °C. The pots were arranged in a 9 × 5 configuration, had an inverted pyramid form, and measured 60 × 30 × 17 cm (240 cm^3 volume). The plants were transplanted to black PVC pots (cultivation pot) of 2.5 L volume, 16 cm upper external diameter, and 15 cm height. The pots were filled with a mixture of white peat (40%), clay loam soil (30%), and sand (30%). After transplantation, all the plants where cut back to approximately 20 cm height.

The experiment was performed in an open-air plot of 70 m^2 at the Tomás Ferro Experimental Agro-Food Station of the Polytechnic University of Cartagena (UPCT) (37° 35' N, 0° 59' W). Transplantation of seedlings to cultivation pots was performed on 15 March 2009, and the experiment took placed from 1 April 2009 to 4 December 2009. Weather conditions were taken from a meteorological station sited 100 m from the experimental plot. The mean hourly values of temperature, relative humidity, and solar radiation were registered (Fig. 1).

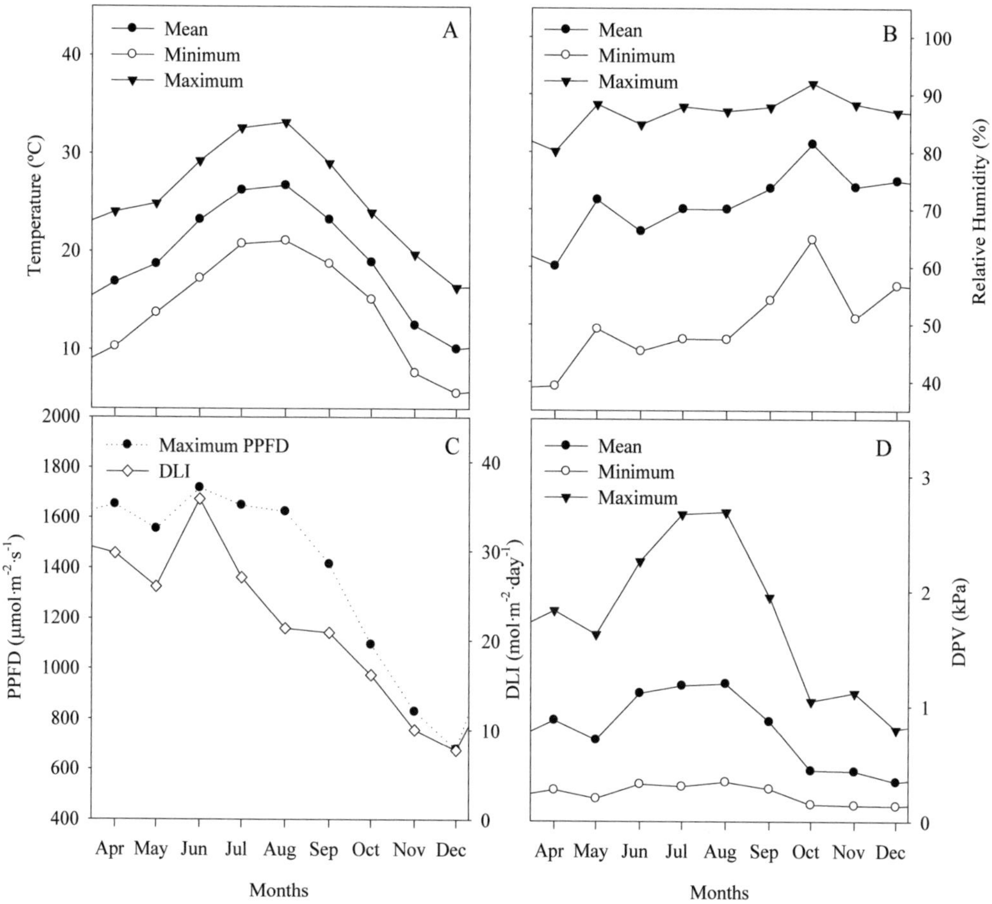

Fig. 1. Mean, minimum and maximum monthly environmental temperature (A), relative humidity (B) and vapor pressure deficit (DPV) (D), and maximum photosynthetic photon flux density (PPFD) and daily light integral (DLI) (C).

A drip irrigation system was installed, with one dripper per plant (2 L·h^{-1}) connected to two spaghetti tubes (one each side of every pot). Local irrigation water (pH 7.2; electric conductivity 1.7 dS m^{-1}) was used, containing Ca^{2+} (95 mg L^{-1}), Mg^{2+} (69 mg L^{-1}), Na^{+} (145 mg L^{-1}), Cl^{-} (232 mg L^{-1}), and HCO_3^{-} (110 mg L^{-1}). Both treatments were irrigated between 12:00 and 14:00 h with the same frequency and volume of water. Irrigation frequency was set so that soil matric potential (SMP) reached values of -60 and -80 kPa in AGP. To meet this criterion, irrigation frequency varied according to the season: two irrigations per week in spring and autumn, and three irrigations per week in summer. Irrigation amounts were programmed to obtain leaching of 15% to 20% in AGP, which produced irrigation water volumes between 400 and 700 mL per pot. Greater volumes of water were applied in summer and when the time between irrigations was greater (e.g., after the weekend). The leachate in PIP was not collected.

2.2 Experimental design and statistical analysis

The PIP system consisted of placing cultivation pots in pots already buried in the ground. The buried pots were made of black PVC and contained many small drainage holes to ensure drainage (5.5 L volume, 17 cm upper exterior diameter, and 30 cm height). An air chamber of 15 cm separated the bases of both pots. Once the pots were buried in the ground, the plot was covered with a plastic permeable mulch (Horsol 140 g m^{-2}; Projar S.A., Valencia, Spain), which was covered with a 4 cm layer of gravel (~2 cm dia.) (Fig. 2).

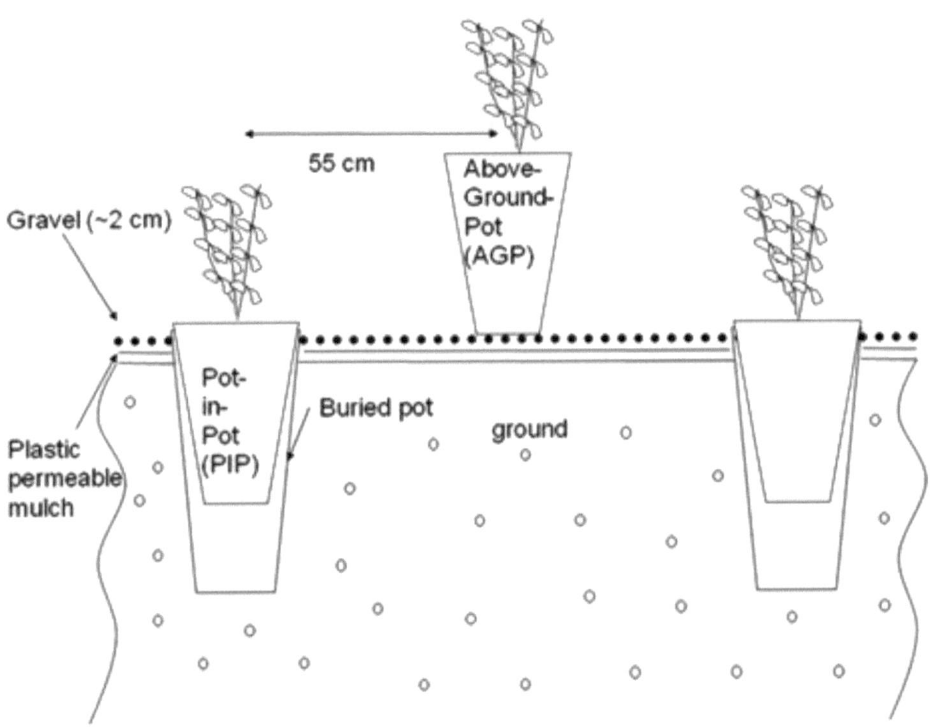

Fig. 2. Pot-in-Pot general design.

A total of 220 cultivation pots were placed in 10 rows, 60 cm apart, so that each row had 22 cultivation pots (Picture 1). These were placed 55 cm apart, buried pots (PIP) alternating with above-ground pots (AGP). A CR1000 datalogger and an AM16/32 multiplexer (Campbell Scientific, Logan, UT) were installed in the center of the plot connected to eight temperature probes (Termistor 107, Campbell Scientific S.L., Barcelona, Spain) and 16 watermark probes (model 253 Irrometer Company, Riverside, CA). The data were analyzed using a one-way ANOVA. A significance level of ≤5% was accepted. The statistical analysis was performed using Statgraphics Plus 5.1 software (StatPoint Technologies, Warrenton, VA).

Picture 1. Experimental plot, pot-in-pot (PIP) and above ground pot (AGP) and datalogger in the center of the plot.

2.3 Growth and development

On four occasions during the experiment (March, June, September, and December), the main stem base diameter, plant height and length of main shoots were measured. At the end of the experiment, leaf area and dry weight (DW) of root and shoot was determined in six plants per treatment. The leaf area was determined with a LI-3100C (LI-COR Biosciences, Lincoln, NE). To calculate the DW, shoot and root were introduced in clearly identified envelopes and placed in a natural convection bacteriological stove (model 2002471, JP Selecta SA, Barcelona, Spain) at 60 ° C until constant weight was reached. Before introducing the roots in the stove, roots were washed with pressurized water using a hose with flat tip before being introduced in a dryer. Finally, the DW was determined by weighing with a GRAM ST series precision balance (sensitivity of 10 mg and up to 1200 g, Gram Precision SL, Barcelona, Spain). The index shoot DW/root DW (S/R) was determined, separating shoots and roots.

2.4 Soil matric potential (SMP) and temperature

The soil matric potential (SMP) was registered using eight watermark probes and four substrate temperature probes per treatment to perform the SMP corrections due to temperature (Thompson et al., 2006). The devices were connected to the datalogger and multiplexer, which were programmed to register data every minute and to save the hourly mean value. The watermark and temperature probes were installed in random pots, in a southerly orientation and 5 cm deep. SMP was estimated using the equation of Shock et al. (1998), which is the best way of fitting the studied interval, as described by Thompson et al. (2006).

2.5 Leaf water potential and gas exchange

Leaf water potential (Ψ_l) was determined using a pressure chamber (Soil Moisture Equipment Corp; Santa Barbara, Cal.) according to Scholander et al. (1965). The stomatal conductance (g_s) and net photosynthesis (P_n) were measured using a portable photosynthesis system (LI-6200, Licor, Inc., Lincoln, Neb.). All measurements were taken at midday in six plants per treatment the following months: March, June, September, and December.

2.6 Measurements of leaf color and SPAD

The color and SPAD measurements were made for 12 plants of each treatment at the end of experiment. For the determination of both, representative plant leaves were chosen, taken from south-facing mid-height and mature. The color was determined with a shot in the middle of the leaf blade with a Minolta CR10 colorimeter (Konica Minolta Sensing, Inc., Osaka, Japan) that calculated the color coordinates (CIELAB): lightness (L), tone (hue angle, H) and saturation (chrome, C). The SPAD was measured using the same criteria as for color but with a SPAD-502 chlorophyll meter (Konica Minolta Sensing, Inc., Osaka, Japan). For each measurement the average of three shots was determined.

3. Results and discussion

3.1 SMP and plant water relations

The mean monthly environmental temperatures during the experimental period ranged between 10 °C and 27 °C, and DLI ranged between 7 and 38 mol $m^{-2}day^{-1}$ (Fig. 1A and 1C). Mean monthly maximum values were 16 °C to 33 °C and maximum PPFD 680 to 1717 $\mu mol \cdot m^{-2} \cdot s^{-1}$, respectively (Fig. 1A and 1C), and mean monthly minimum temperatures varied between 6 °C and 21 °C (Fig. 1A). The registries were nearly the same reported by Miralles et al. (2009).

The mean monthly substrate temperatures in all the experimental months were similar in PIP and AGP, ranging between 17 °C and 31 °C. The AGP system showed higher mean monthly maximum substrate temperatures than PIP (Fig. 3B), with the thermal differences between both systems around 8 °C. Young and Bachman (1996) and Ruter (1993) described how, on the hottest days of summer, PIP substrates for different species were 2.3 °C and 6 °C lower, respectively, than AGP temperatures. As shown in figure 3B, PIP moderated substrate temperature increases from June to September, preventing mean monthly maximum temperatures >34 °C, unlike in AGP, where 43 °C was reached.

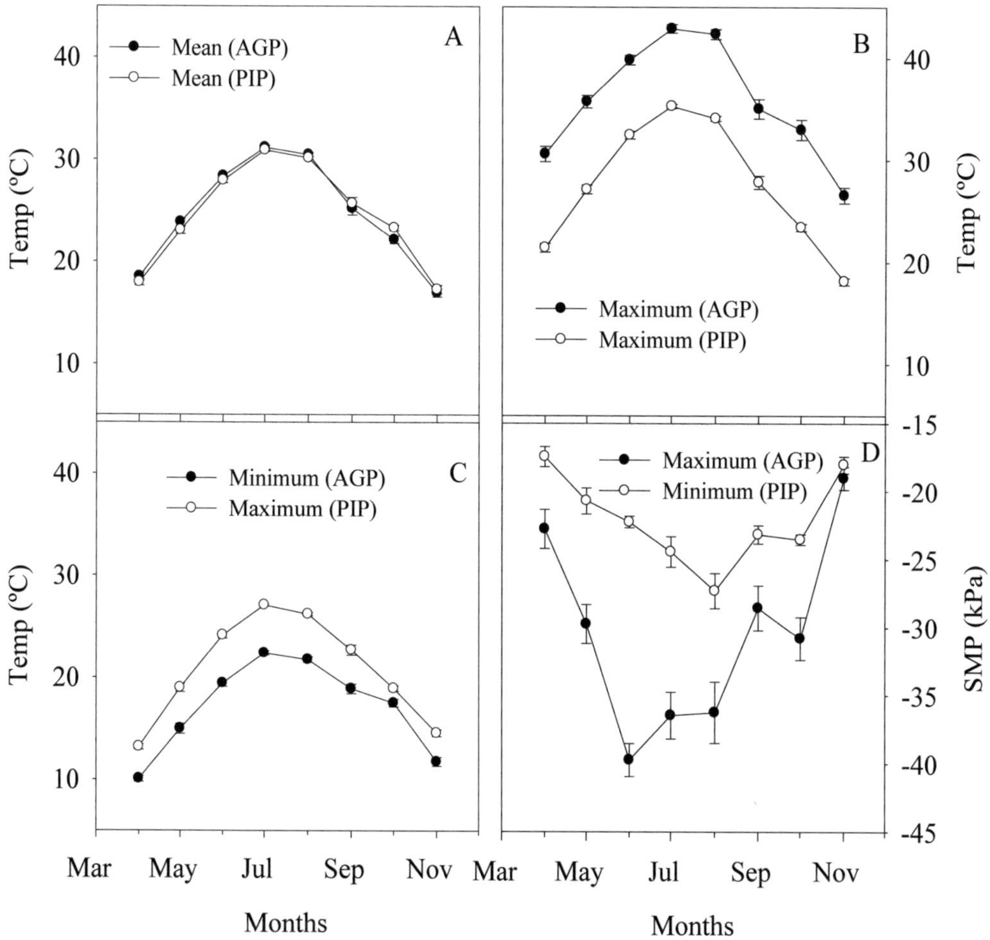

Fig. 3. Mean monthly temperature (A), mean monthly maximum temperature (B), mean monthly minimum temperature (C), and mean monthly minimum soil matric potencial (SMP) (D) evolution in substrate of PIP and AGP treatments. Error bars are standard errors (n = 4 for temperature and n=8 for SMP).

Mean monthly minimum substrate temperatures showed the opposite behavior to maximum temperatures (Fig. 3C), with PIP reaching higher temperatures than AGP. The thermal differences between both systems ranged from 1 °C to 5 °C, although the temperature differences between both systems were lower than the corresponding maximum values. Young and Bachman (1996) and Ruter (1993) found that, on the coolest winter days, PIP substrates were 1.1 °C and 3 °C warmer, respectively, than the corresponding AGP values. This behavior can be explained by the ground effect, which slowed the temperature loss at night. Miralles et al. (2009) confirmed during a one year experiment that PIP significantly moderated low and high substrate temperatures, particularly when temperatures were at their most extreme, as well as London et al. (1998).

Mean monthly temperatures were similar in both systems (Fig. 3A) because AGP reached higher daily temperatures than PIP but lower temperatures at night the one compensating the other.

Mean monthly minimum soil matric potential (SMP) was greater in PIP compared with AGP except in December which became similar (Fig. 3D). The greater differences were found in summer. The greater water demanding conditions increased water demands in summer in *R. alaternus,* while in winter due to plant growth stop, these differences in SMP disappeared. Mean monthly maximum SMP were not significantly different between treatments and mean monthly SMP had intermediate values between the minimum and the maximum (data not shown). Miralles et al. (2009) on its previous study with *M. communis* found a different behavior. In this case no differences were found from the beginning of the experiment (March) to August. In September and October, the mean monthly minimum SMP values were more negative in AGP and no more differences were found until February were PIP showed again higher mean monthly minimum SMP until the end of the experiment in May. The absence of differences the first months were related to low plant growth, and the SMP differences at the end of the experiment were related to the higher water consumption of plants in AGP following growth activation during the winter-spring transition. This may have been caused by higher maximum substrate temperature, together with more developed *M. communis* in the AGP system. In our experiment, *R. alaternus* plants grew more than *M. communis* plants and plants cropped in PIP grew more than AGP plants (Table 1). However, g_s was greater in AGP plants after summer (Fig. 4B) what would explain a greater water consumption in the pot, what produced lower SMP registries than PIP plants. Besides, substrate evaporation in *R. alaternus* was also greater than *M. communis* due to its plant architecture, which opposite to *M. communis,* it has a main shoot what leave the substrate surface expose to wind and with low shading level.

Miralles et al. (2009) described four periods of ten representative days (one per season) for *M. communis.* For the summer (Fig. 5), the high number of oscillations in daily minimum SMP is due to greater substrate drying; however, the differences between both systems were barely significant. These low differences in summer SMP between PIP and AGP were explained by greater evaporation because of the higher radiation that the AGP pots received, and the higher transpiration in PIP influenced by higher stomatal conductance.

In autumn (Fig. 6), when the irrigation frequency was lower, AGP reached more negative SMP values than PIP, possibly because transpiration rates leveled out due to similar stomatal conductance levels. Moreover, after summer, some roots from PIP plants entered the air chamber between the two pots of the PIP system, which may mean that transpired water did not come totally from the substrate, as occurred in AGP (Miralles et al., 2009). These differences between PIP and AGP agree with experiments performed by Martin et al. (1999) using *Acacia smallii* and *Cercidium floridum* in which AGP needed extra irrigation, as well as programmed irrigation, to keep moisture tensions for all rooting substrates between -0.005 and -0.01 MPa; AGP needed 5.3 L weekly per pot, and PIP needed 3.2 L per pot.

In November, December, and January, the mean monthly minimum SMP was similar in both systems, which could be a consequence of lower plant growth due to a decrease in temperature and solar radiation. Daily minimum SMP during a representative winter period (Fig. 7) showed less negative values, which were very similar in both systems, reflecting very low irrigation frequency. These registers showed that AGP reached more negative SMP before PIP, which suggests that PIP has lower irrigation requirements (Miralles et al., 2009).

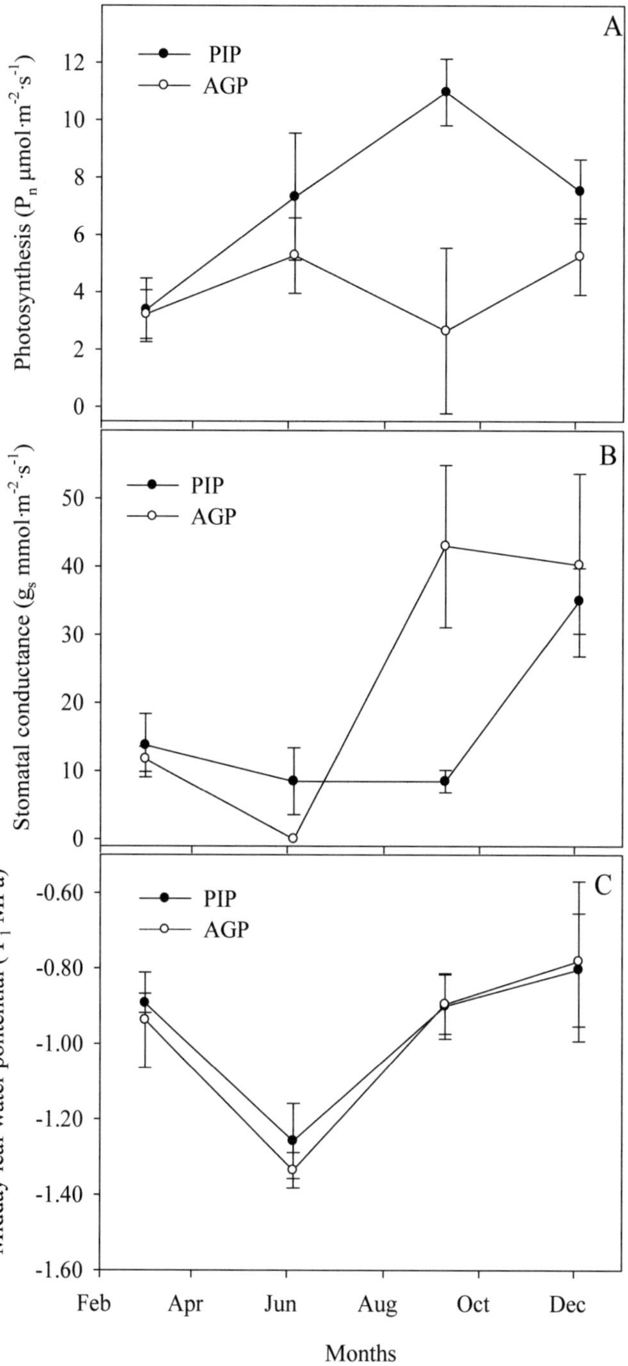

Fig. 4. Net photosynthesis (P_n) (A), stomatal conductance (g_s) (B) and leaf water potential (Ψ_1) (C). Error bars are standard errors ($n = 6$)

In February, as the temperature began to increase, the mean monthly minimum SMP in AGP became more negative than the corresponding values in PIP, and in spring, with this season's better environmental conditions for plants, these differences increased. This is reflected in the results shown in figure 8 (spring), where a greater number of SMP variations as a result of increasing water needs can be appreciated. Furthermore, AGP clearly reached more negative values than PIP, whose substrate conditions remained better. Some records in AGP reached SMP < -100 kPa (Fig. 8), which could have caused water stress. Nevertheless, such values were isolated, and the average leaf water potential, in general terms, pointed to no water stress (Miralles et al, 2009). Indeed, in well developed plants, no leaf water potentials under -1.0 MPa were recorded in either system, the values being greater than those recorded for leaf water potential registered in other experiments with *M. communis* plants subjected to moderate water stress (Vicente et al., 2006).

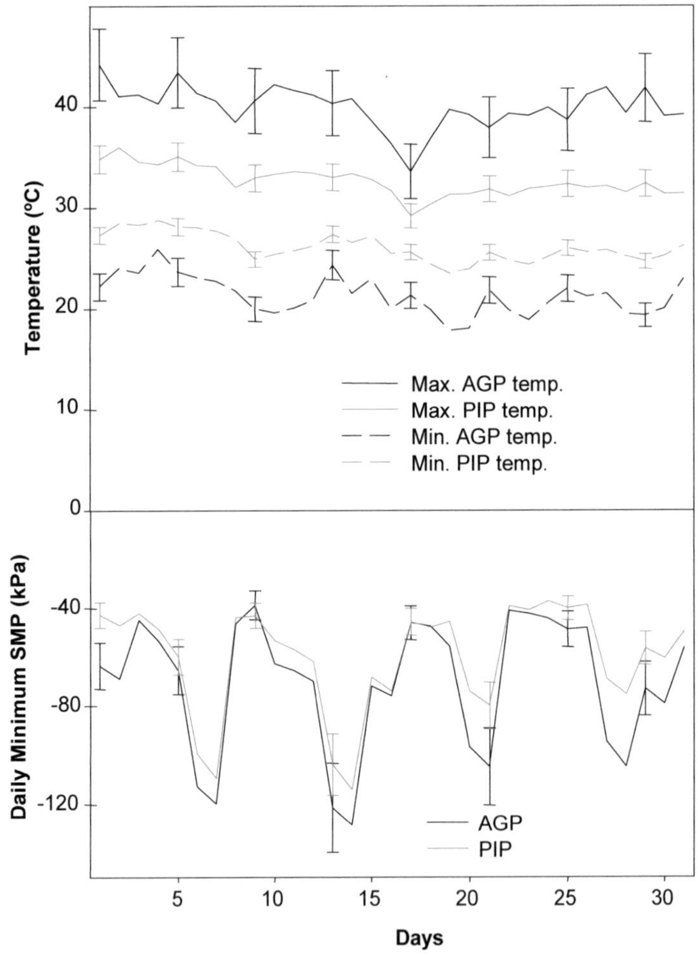

Fig. 5. Representative 31-day period for summer, showing daily maximum and minimum substrate temperature in PIP and AGP systems and daily minimum substrate SMP registers in PIP and AGP systems. Error bars are standard errors (n = 4 in temperature, and n = 6 in SMP). For clarity, only every fifth standard error value is shown. (Miralles et al., 2009)

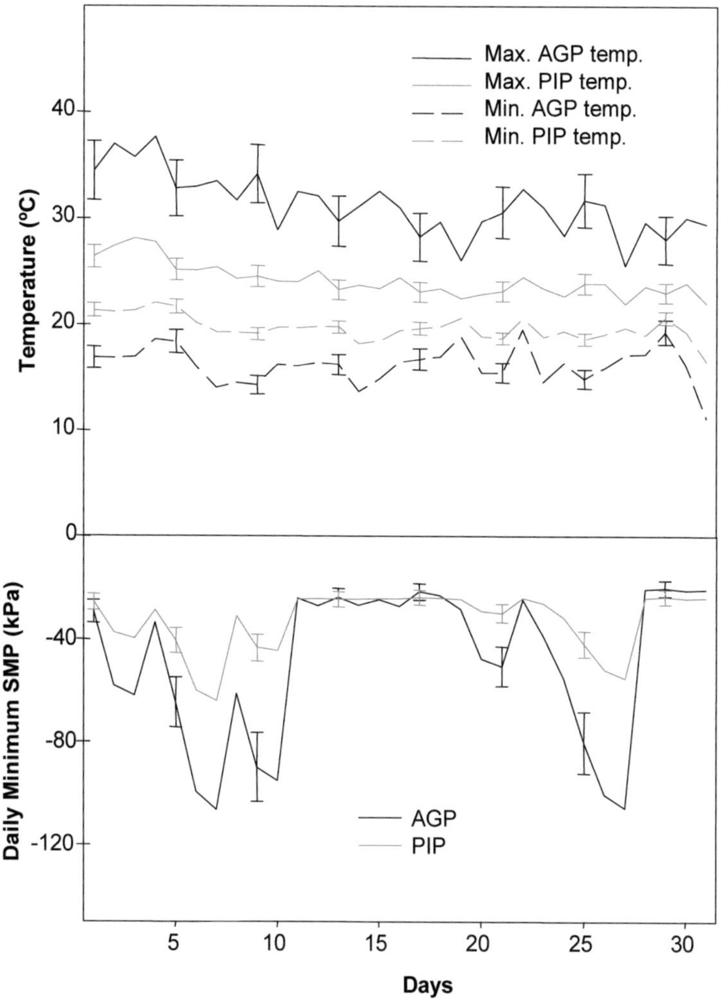

Fig. 6. Representative 31-day period for autumn, showing daily maximum and minimum substrate temperature in PIP and AGP systems and daily minimum substrate SMP registers in PIP and AGP systems. Error bars are standard errors (n = 4 in temperature, and n = 6 in SMP). For clarity, only every fifth standard error value is shown. (Miralles et al., 2009)

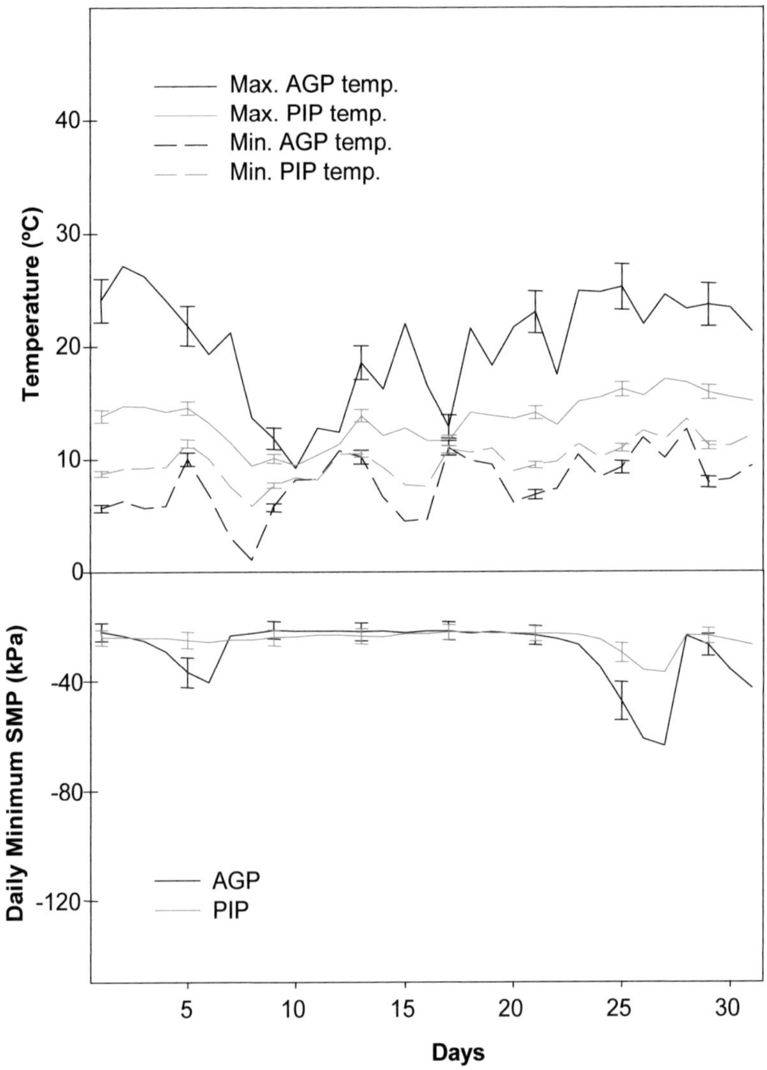

Fig. 7. Representative 31-day period for winter, showing daily maximum and minimum substrate temperature in PIP and AGP systems and daily minimum substrate SMP registers in PIP and AGP systems. Error bars are standard errors (n = 4 in temperature, and n = 6 in SMP). For clarity, only every fifth standard error value is shown. (Miralles et al., 2009)

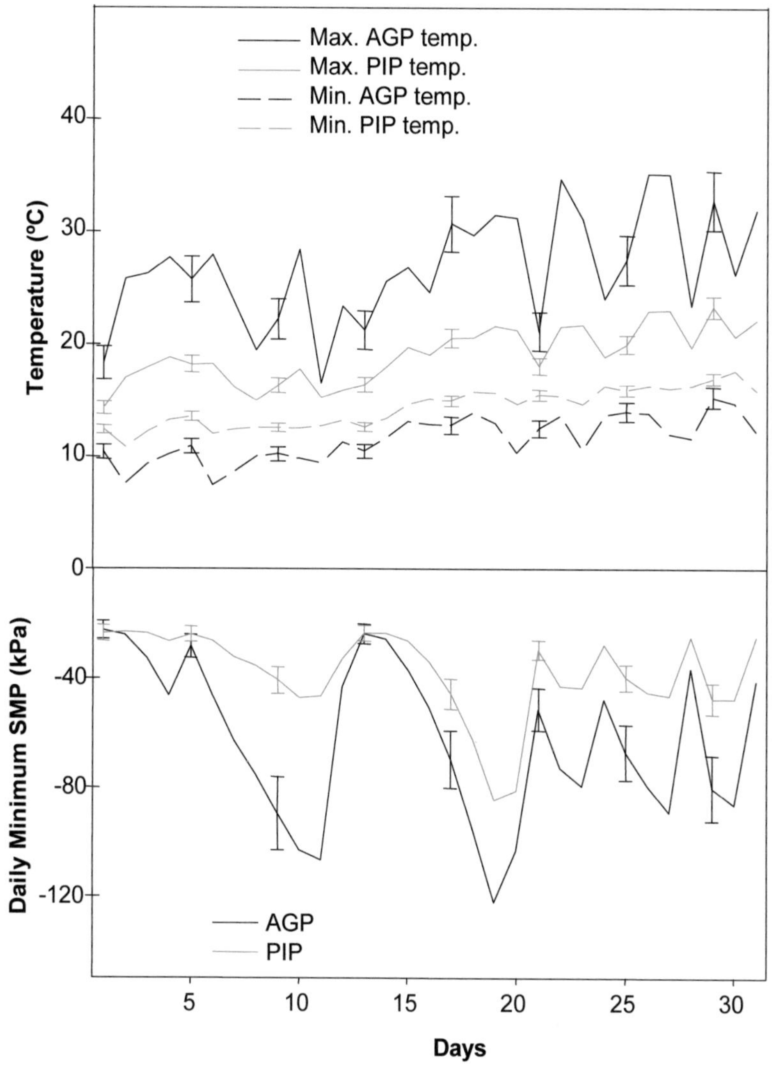

Fig. 8. Representative 31-day period for spring, showing daily maximum and minimum substrate temperature in PIP and AGP systems and daily minimum substrate SMP registers in PIP and AGP systems. Error bars are standard errors (n = 4 in temperature, and n = 6 in SMP). For clarity, only every fifth standard error value is shown. (Miralles et al., 2009)

At the beginning of cultivation (April) until the beginning of the summer, plants in PIP and AGP showed similar gas exchange values (Fig. 4A and 4B), which can explain the lack of difference in growth (Fig. 9). In autumn, P_n was greater in PIP plants and g_s in AGP plants (Fig. 4A and 4B), while in winter the differences disappeared. The Ψ_1 was not different between both treatments, but lower at the beginning of summer for all plants (Fig. 4C), so the differences observed in SMP (Fig. 3D) had no effect on Ψ_1, perhaps due to SMP were not too negative to produce water stress. Miralles et al. (2009) did not found Ψ_1 differences either, however, they registered SMP < -80 kPa in summer in AGP, so they explained that *M. communis* could have activated the messages sent by roots, perhaps of a chemical kind, which might have lowered stomatal conductance rates and, as a consequence, led to lower photosynthetic activity. Such behavior probably helped leaf water potential not to fall in these plants. Mendes et al. (2001) found a high sensitivity to stomatic closure in *M. communis* affected by water deficit. This fact, linked to high solar radiation in this period, could also influence this stomatal conductance behavior. Niinemets et al. (1999) described how stomatic sensitivity was intensified by high solar radiation.

3.2 Growth and development

Plant diameter evolution was greater in PIP plants than AGP plants since autumn, but both treatments had the greater diameter growth in summer (Fig. 9A). This greater diameter produced a final plant DW in this species in PIP (Table 1). Plant height and length of main shoots evolution in PIP and AGP was similar during the experimental period, increasing from 20 to 70 cm and 20 to 250 cm, respectively (Fig. 9B and 9C). Both, reduced their growth rate in autumn. Ruter (1993) observed no differences in shoot production in *Lagerstroemia indica* x *fauriei*. However, in *Magnolia* x *soulangiana*, the same author found that PIP plants had more shoots than their AGP counterparts. This effect in plant height was also found by Ruter (1993), although Miralles et al. (2009) reported that the height of PIP *M. communis* plants grew the same in both crop systems from transplantation in March to January, after which the AGP plants grew in height more than PIP plants until the end of the experiment, when they were 16% taller.

In our experiment, due to environmental conditions characterized by hot summers and mild winters (Fig. 1), it might be expected that, in summer, PIP plants would present higher growth because of the moderating effect on substrate temperature (Fig. 3B and 3C). Such behavior only occurred in plant diameter (Fig. 9A), while no differences were found in plant height or main shoot length (Fig. 9B and 9C). However, Miralles et al. (2009) did not have such effect, perhaps due to the physiological characteristics of *M. communis*, which rests at high temperatures (Brosse, 1979). Whatever the case, the influence of PIP on summer growth may depend on the species. For example, during the summer cultivation of *A. smallii*, Martin et al. (1999) observed that PIP increased plant height by 80% in contrast with AGP, but had no effect on *C. floridum*.

Table 1 shows a greater plant growth in PIP (shoot and root), which resulted in a 39% and 181% extra growth in shoot and roots, respectively, compared with AGP. In *L. indica* x *fauriei* and *Magnolia* x *soulangiana*, Ruter (1993) observed greater root dry weight (47% and 70%, respectively) when the plants were cultivated in PIP. Later works by the same author showed similar behavior in *Magnolia grandiflora* (Ruter, 1995). In *A. smallii*, Martin et al.

(1999) found a higher root dry weight in PIP (167 g) compared with AGP (97 g), although no differences for *C. floridum*. Furthermore, Young and Bachman (1996) recorded an increase of 26% in root dry weight when *Ilex* x *attenuate* was cultivated in PIP. The greater root growth was especially important inside the air chamber between pots what contributed to so high extra growth. Miralles et al. (2009) also reported greater root development in PIP (extra 14%), however, *M. communis* plants in PIP presented 11% less shoot dry weight compared with AGP shoot dry weight.

Parameter	AGP	PIP	Significance[a]
Root dry weight (g)	6.67	18.73	***
Shoot dry weight (g)	28.83	39.99	*
Plant dry weight (g)	35.50	58.72	**
Shoot DW/Root DW (S/R)	5.58	2.23	**
Leaf area (cm^2)	607.37	965.28	*
SPAD	66.41	67.27	ns
Lightness (L)	31.73	32.42	ns
Chrome (C)	21.27	21.98	ns
Hue angle (H)	107.98	107.78	ns

[a]Asterisk indicate statistically significant between means at *$P < 0.05$, **<0.005, ***<0.0005.
ns = Not significant.

Table 1. Final measurements of plant growth and biomass distribuitin, SPAD and Color.

The greater root growth in PIP reduced its S/R ratio compared with AGP, an effect also reported by Miralles et al. (2009). In contrast, Ruter (1993) described that *L. indica* x *fauriei* cultivation in PIP caused a substantial increase in the S/R ratio, although Martin et al. (1999) did not observe this difference in *A. smalli* or *C. floridum*. It has been suggested that a diminishing shoot/root ratio lowers the relative transpiration capacity, unlike water and nutrient absorption (Bernier et al., 1995). Mathers (2000) described that slight reductions in the shoot/root ratio gives plants greater water stress resistance during nursery production and transplantation, which accelerates plant establishment in the field (Owings, 2005). As mentioned by Guarnaschelli et al. (2006), these effects are more patent in drought conditions, so this change in biomass distribution in plants could improve survival after transplantation. In this study, the greater SMP of PIP plants along the experiment indicate that PIP plants water consumption in the pot was lower than AGP plants, so PIP plants should have lost less water from substrate evaporation, but they must have taken some water from the air chamber between pots, what would justify the large development of roots in it. Miralles et al. (2009) reported that, after two weeks without irrigation at the end of the experiment with *M. communis*, the PIP system led to 90% plant survival, compared with 62% for AGP (Picture 2), and he pointed that the presence of roots in the air chamber between the two pots could have affected the survival, because relative humidity was very high in this chamber (Fig. 2).

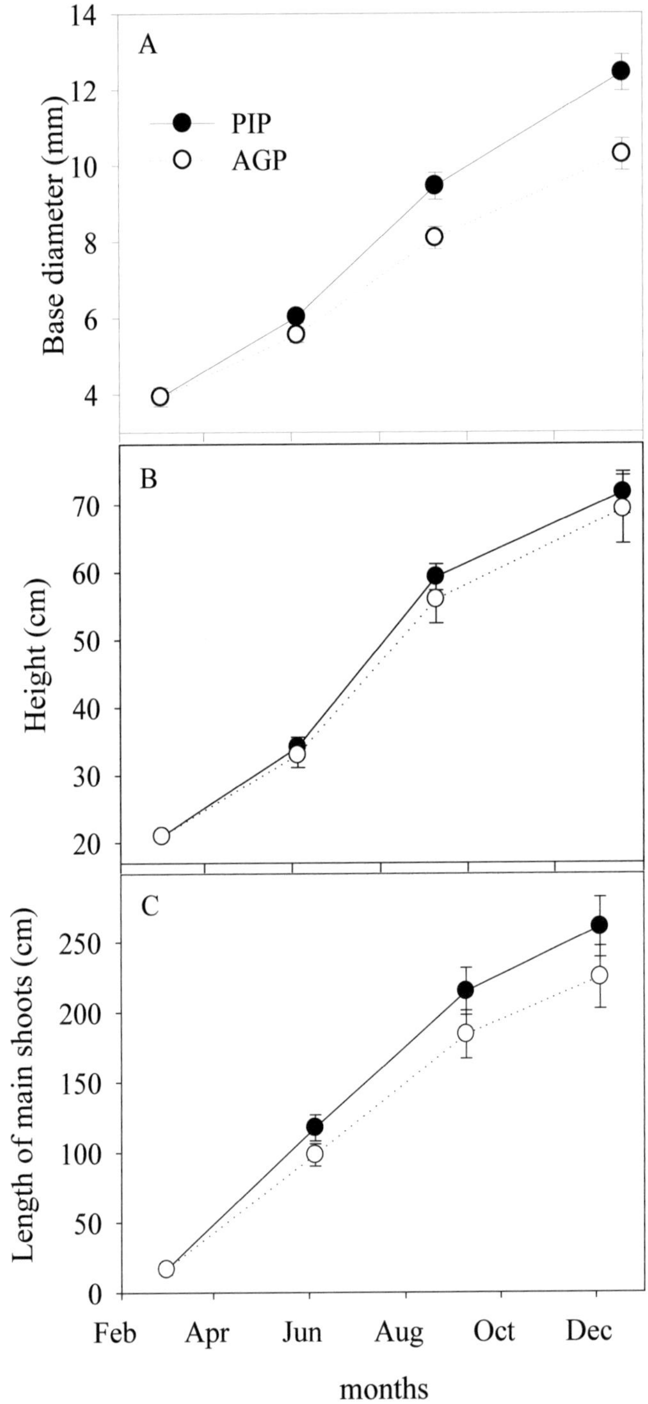

Fig. 9. Base diameter (A), plant height (B) and length of main shoots (C) evolution in PIP and AGP systems. Error bars are standard errors (n = 110).

The crop system did not affected leaf color and leaf SPAD, however, leaf area was greater in PIP (Table 1). So, since an aesthetic point of view, plants were quite similar, however those cropped in PIP had more foliage what supposed an advantage in this field.

Picture 2. Experimental plot after two weeks without irrigation at the end of the experiment.

4. Conclusion

The most relevant conclusions of this experiment are four. Firstly, the PIP system used in *R. alaternus* moderated extreme substrate temperatures. Secondly, PIP maintained a higher substrate SMP, which suggests lower irrigation needs than in AGP. Thirdly, PIP increased plant growth in terms of dry weight (more plant diameter) and also in leaf area (plant height, main shoot length and leaf color were not affected) what increased its aesthetic value. Lastly, the PIP system favored greater root development inside the air chamber between pots, which was translated into a lower shoot/root ratio, favoring taking some water from this chamber what maintained greater SMP in the pot substrate.

5. Acknowledgements

The authors thank R. Valdés and P. Cánovas for irrigating and supervising the plants while the experiment took place. This research was supported by CICYT projects (CICYT AGL2008-05258-CO2-1-AGR and CICYT AGL2008-05258-CO2-2-AGR), SENECA project

(08669/PI/08) and by the Consejería de Agricultura y Agua de la Región de Murcia, program (UPCT-CEBAS-IMIDA
2008).

6. References

Adrian, J. L.; Montgomery, C. C.; Behe, B. K.; Duffy, P. A. & Tilt, K. M. (1998). Cost comparisons of infield, above ground container and pot-in-pot production systems. *Journal of Environmental Horticulture*, 16, 65-68.

Bernier, P. Y.; Stewart, J. D. & Gonzalez, A. (1995). Effects of the physical properties of *Sphagnum* peat on water stress in container *Picea mariana* seedlings under simulated field conditions. *Scandinavian Journal of Forest Resources*, 10, 184-198.

Brosse, J. (1979). *Atlas of Shrubs and Lianes*. Paris, France, Bordas Editions.

Davidson, H.; Mecklenburg, R. & Peterson, C. (1988). *Nursery Management Administration and Culture*. Englewood, Cliffs, N.J.: Prentice Hall.

Fain, G. B.; Tilt, K. M.; Gilliam, C. H.; Ponder, H. G. & Sibley, J. L. (1998). Effects of cyclic micro irrigation and substrate in pot-in-pot production. *Journal of Environmental Horticulture*, 16, 215-218.

Franco, J. A.; Martínez-Sánchez, J. J.; Fernández, J. A. & Bañón, S. (2006). Selection and nursery production of ornamental plants for landscaping and xerogardening in semi-arid environments. *Journal of Horticultural Science and Biotechnology*, 8, 3-17.

Guarnaschelli, A. B.; Prystupa, P. & Lemcoff, J. H. (2006). Drought conditioning improves water status, stomatal conductance, and survival of *Eucalyptus globulus* subsp bicostata seedlings. *Annals Forest Science*, 6, 941-950.

Kuroyanagi, T. & Paulsen, G. M. (1988). Mediation of high-temperature injury by roots and shoots during reproductive growth of wheat. *Plant Cell Environmental*, 11, 517-523.

Levitt, J. (1980). *Responses of Plants to Environmental Stresses*. New York, N.Y.: Academic Press.

London, J. B.; Fernandez, R. T.; Young, R. E. & Christenbury, G. D. (1998). Media temperatures and plant growth in above-ground and in-ground pot-in-pot container systems. *HortScience*, 33, 512.

Martin, C. A.; McDowell, L. B. & Shiela, B. (1999). Below-ground pot-in-pot effects on growth of two southwest landscape trees was related to root membrane thermostability. *Journal of Environmental Horticulture*, 17, 63-68.

Mathers, H. M. (2000). Overwintering container nursery stock: Part. 1. Acclimatation and covering. *The Buckeye* (Dec.): 14, 16, and 18.

Mathers, H. M. (2003). Summary of temperature stress issues in nursery containers and current methods of protection. *HortTechnology*, 13, 617-624.

Mendes, M. M.; Gazarini, L. C. & Rodrigues, M. L. (2001). Acclimatation of *Myrtus communis* to contrasting Mediterranean light environments: Effects on structure and chemical composition of foliage and plant water relations. *Environmental and Experimental Botany*, 45, 165-178.

Miralles, J.; Nortes, P.; Sánchez-Blanco, M. J.; Martínez-Sánchez, J. J. & Bañón S. (2009). Above Ground and PIP Production Systems in *Myrtus communis*. *Transactions of the ASABE*, 52, 93-101

Neal, C.A. (2010). Crabapple and lilac growth and root-zone temperatures in northern nursery production systems. *HortScience*, 45, 30-35.

Niinemets, Ü.; Söber, A.; Kull, O.; Hartung, W. & Tenhumen, J. D. (1999). Apparent controls on leaf conductance by soil water availability and via light-acclimation on foliage structural and physiological properties in a mixed deciduous, temperature forest. *International Journal of Plant Science*, 160, 707-721.

Owings, A. D. (2005). Pot-in-pot nursery production. *Louisiana Agriculture*, 48, 20.

Parkerson, C.H. (1990). P & P: A new field-type nursery operation. *Proceedings of International Plant Production Society*, 40, 417-419.

Ruter, J. M. (1993). Growth and landscape performance of three landscape plants produced in conventional and pot-in-pot production systems. *Journal of Environmental Horticulture*, 11, 124-127.

Ruter, J. M. (1995). Growth of southern magnolia in pot-in-pot and above-ground production systems. In: *Proceedings of SNA Research Conference*, 40, 138-139. Atlanta, Ga.: Southern Nursery Association.

Ruter, J. M. (1997). The practicality of pot-in-pot. *American Nurseryman*, 8, 32-37.

Ruter, J. M. (1998a). Pot-in-Pot production and cyclic irrigation influence growth and irrigation efficiency of 'Okame' cherries. *Journal of Environmental Horticulture*, 16,159-162.

Ruter, J. M. (1998b). Fertilizer rate and pot-in-pot production increase growth of Heritage River birch. *Journal of Environmental Horticulture*, 16, 135-138.

Ruter, J. M. & Ingram, D. L. (1992). High root-zone temperatures influence RuBisCo activity and pigment accumulation in leaves of 'Rotundifolia' holly. *Journal American Society of Horticultural Science*, 117, 154-157.

Scholander, P.; Hammel, H. T.; Bradstreet, E. D. & Hemmingeen, E. A. (1965). Sap pressure in vascular plants. *Science*, 148, 339-346.

Shock, C. C., Barnum, J. M. & Seddigh, M. (1998). Calibration of watermark soil moisture sensors for irrigation management. In: *Proceedings of International Irrigation Show*, 139-146. Falls Church, Va.: Irrigation Association.

Thompson, R. B.; Gallardo, M.; Agüera, T.; Valdez, L. C. & Fernández, M. D. (2006). Evaluation of the Watermark sensor for use with drip-irrigated vegetable crops. *Irrigation Science*, 24, 185-202.

Tilt, K.M.; Williams, J.D.; Montgomery, C.C.; Behe, B.K. & Gaylor, M.K. (1994). Pot-in-pot production of nursery crops and Christmas trees, Auburn University, *Agricultural and Natural Resources Circle*, 893.

Vicente, M. J.; Martínez-Sánchez, J. J.; Conesa, E.; Bañón, S.; Navarro, A. & Sánchez-Blanco, M. J. (2006). Studio della modificazione dello sviluppo e dello stato idrico di piantine di *Myrtus communis* causati da deficit idrico e paclobutrazol. In: *Proc. III Convegno Nazionale Piante Mediterranee: Le Piante Mediterranee nelle Scelte Strategiche per l'Agricoltura e l'Ambiente*, 194. Bari, Italy: Università degli Studi di Bari.

Young, R. E. & Bachman, G. E. (1996). Temperature distribution in large, pot-in-pot nursery containers. *Journal of Environmental Horticulture*, 14, 170-176.

Zhu, H.; Krause, C. R.; Derksen, R. C.; Brazee, R. D.; Zondag, R. & Fausey, N. R. (2004). Real-time measurement of drainage from pot-in-pot container nurseries. *Transactions of the ASAE*, 47, 1973-1979.

Zinati, G. (2005). Pot-in-pot nursery production system: What you need to know before establishment. Pub. No. FS519. New Brunswick, N.J.: *Rutgers Cooperative Research and Extension*. Available at:
http://njaes.rutgers.edu/pubs/publication.asp?pid=FS519. Accessed 17 January 2008.

7

Towards a New Ecophysiological Approach to Understand Citrus Crop Yield Under Abiotic Stresses Mirroring in the Brazilian Savanna Genetic Resources

Marcelo Claro de Souza[1] and Gustavo Habermann[2]
[1]*Programa de Pós-Graduação em Ciências Biológicas (Biologia Vegetal) – Departamento de Botânica, Univ Estadual Paulista (UNESP), Rio Claro,*
[2]*Univ Estadual Paulista (UNESP), Departamento de Botânica, IB, Rio Claro*
Brazil

1. Introduction

Under the threat of unavoidable global warming and its consequences, citrus production has important impacts on water consumption in (irrigated) orchards and nurseries. Knowledge of agriculture indicates that more than 60% of variability in crop productivity may be explained by fluctuations in climate. More than 55% of the causes of crop losses can be identified as soil water deficits. In this context, areas for citrus production in Brazil are located in São Paulo and Minas Gerais states, but the most productive regions are the northwestern and the northern regions of São Paulo state, where frequent and intense droughts, high air temperatures and vapor pressure deficits occur. In these regions there are more incidences of citrus diseases, such as citrus variegated chlorosis and citrus sudden death. These factors have induced some citrus growers to irrigate their orchards. Field experiments have shown that irrigation may increase citrus production by more than 50% on an area basis, compared to non-irrigated areas. Considering extensive land areas for citrus in Brazil, irrigation systems may become very expensive. Alternatively, citrus groves have migrated to the south of São Paulo state, where soil water deficits and air temperatures are low. However, high temperatures enhance citrus growth and fruit production. Thus, the southern region, with approximately 5.5 million new citrus plants, might be interesting for plant water balance and disease attenuation, but not for carbon gain, and consequently, production. In fact, the northwestern and the northern regions were still responsible for the largest production in 2007/2008. To support the citrus industry, with 190 million productive plants, greenhouse- and screenhouse-protected nurseries use drip fertirrigation to produce high quality varieties and cultivars of scions grafted on specific rootstocks; screenhouse-protected systems also prevent plants from diseases and their vectors. But there are indications of over irrigation in nurseries, suggesting waste of water. It has also been reported that water may not be a renewable natural resource, and that water consumption is increasing, as the human population enlarges. The original habitat of citrus species is believed to be shaded environments of forest understories in southeastern Asia. On the other hand, citrus land areas in São Paulo state were originally occupied by the Brazilian savanna, or *Cerrado*. In this review, we re-visit and discuss the ecophysiology of sunlight and water use by citrus species, and present some aspects of the ecophysiological responses of native cerrado species, which could mirror, encourage or at least provoke physiological reflections on strategies to be used in citrus breeding programs of the 21st century.

2. General characteristics of the Brazilian savanna

The cerrado is a mosaic of biomes (Batalha, 2011) that occupies approximately 21% of the Brazilian territory (Figure 1). The cerrado is considered to be one of the last agricultural frontiers in the world (Borlaug, 2002). In the cerrado areas, a warm and rainy season is observed from October to March and a dry and cold season from April to September. It rains approximately 1500mm per year in the cerrado, and air temperatures may range between 22 and 27°C (Klink & Machado, 2005).

Fig. 1. Geographical distribution of the cerrado among Brazilian states.

From the native cerrado area (2 million Km^2) almost 50% has been destroyed and transformed into agricultural land areas.

Amongst other areas intended for conservation in the world, the cerrado is one of the 25 critical hot spots because of its high biological diversity and an increasing human population pressure that the cerrado has recently undergone (MMA, 2002).

Changes in the cerrado also brought several environmental effects, such as habitat fragmentation, biodiversity extinction, presence of invasive species, soil erosion, changes in the fire regimes, unbalance in the carbon cycle and probably regional climate changes (Klink & Machado, 2005). In a short term perspective, this reduction in the cerrado biodiversity might directly or indirectly affect the functioning of the ecosystem (Chapin III et al., 2000).

3. Climate and genetic origin center of *citrus* and cerrado woody plants

The native habitat of citrus species is believed to be the shaded environments of forest understory in southeastern Asia (Davies & Albrigo, 1994) (Figure 2). Such information is the backbone of discussions we intend to develop in the present text, since citrus plantations in São Paulo and Minas Gerais states in Brazil are currently occupying lands that were originally occupied by the native Brazilian savanna, or *Cerrado*. Consequently, the savanna vegetation, which is believed to have been selected and adapted in the early Cretaceous (Ratter et al., 1997) or in the Holocene (Ledru, 2002) is partially comprised of trees and shrubs showing physiological responses and plant development behavior that are considerably different from citrus plants.

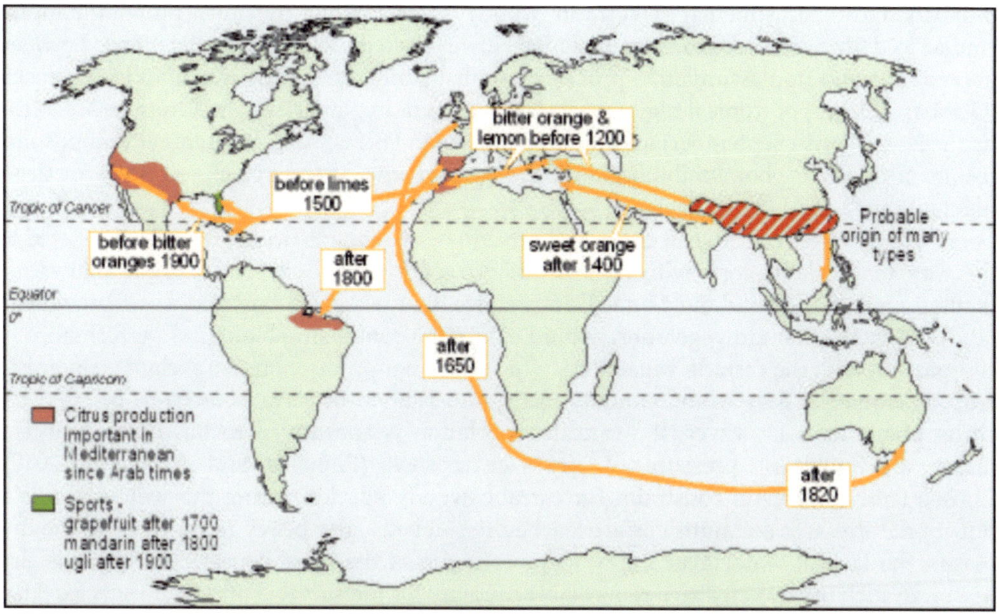

Fig. 2. Citrus origin center. Map adapted from information published by Davies & Albrigo (1994). Observation: This map was not originally drawn by the authors of this chapter

Despite their native shaded habitat in southeastern Asia, citrus plants are considered to be well acclimatized to tropical and subtropical sunny conditions (Ribeiro and Machado, 2007). Evidence suggests that citrus plants exhibit dynamic rather than chronic photoinhibition, given by the recovery of the maximum quantum efficiency of photosystem II, PSII, (Fv/Fm) at the end of the day, or when the photosynthetic photon flux density (PPFD) decreases (Ribeiro and Machado, 2007). Therefore, such reductions of Fv/Fm observed for citrus plants (Jifon and Syvertsen, 2003; Ribeiro and Machado, 2007) seem to be related to photoprotection, reducing the energy pressure on the PSII and avoiding injuries to the photosynthetic apparatus, which if not avoided would lead to photo oxidation.

However, a possible reason that citrus plants make use of the photoprotective mechanisms in tropical regions may be related to their native habitats in southeastern Asia, which suggests that the citrus plant is a shade species, or at least moderately adapted to shaded environments. Citrus leaves show photosynthetic responses that saturate at 600 $\mu mol\ m^{-2}\ s^{-1}$ of PPFD (Vu et al., 1986; Habermann et al., 2003; Ribeiro and Machado, 2007), which can be considered a low-light saturation of photosynthesis, typical of shade species (Gvinish, 1988). Tropical and subtropical areas receive high solar radiation loads, and daily-integrated global solar radiation may reach values of 35 $MJ\ m^{-2}\ day^{-1}$ or even higher values (Ribeiro and

Machado, 2007). Then, both citrus plants that are cultivated in tropical and subtropical areas and cerrado woody plants that are maintained as preserved sites in these regions have to cope with high irradiation loads, although cerrado woody plants are already adapted to such conditions. In fact, photochemical responses of cerrado woody plants are quite variable and dependent on species, because of the high plant diversity in cerrado ecosystems. Some case studies show that cerrado species have diurnal adjustments of non-photochemical energy dissipation in PSII to reduce the risk of photoinhibition when irradiances are high (Mattos, 1998; Lemos-Filho, 2000; Lemos-Filho et al., 2004; Franco et al., 2007). Therefore, photochemical adjustments in cerrado woody plants allow dynamic photoinhibitions, similar to daily adjustments in citrus plants. However, in citrus plants such balance between assimilative and non-assimilative processes occur mainly during the warm and wet season (October - March) of tropical regions, whereas for cerrado plants the correlation between the dry season (April - September) and photo damages to PSII is more frequent, although some results of dynamic photoinhibition in the wet season have already been described for these plants (Franco and Lüttge, 2002).

Therefore, differences between citrus and cerrado woody plants do not seem to be related to the way these plants cope with high irradiance loads. Then, one could argue that citrus plantations in São Paulo and Minas Gerais states, that occupy devastated areas previously occupied by the cerrado vegetation, would exhibit no contrasting biological performance in comparison with the cerrado vegetation, which does not produce any important economical crop. It would be easy to inaccurately justify devastation of cerrado areas to be used for citrus plantations. However, the cerrado vegetation is considered to have been selected under an important pressure of dry/wet seasons (Pinheiro and Monteiro, 2010). Consequently, the main constraint for cerrado woody species during the wet season is a nutritional stress, as soil nutrients are leached to soil beds after heavy rains, while in the dry season the lack of water availability to roots becomes the main concern. In fact, the dry season in such regions creates a remarkable pressure for the survival of cerrado species (Hao et al., 2008; Habermann et al., 2011a; Habermann and Bressan, 2011).

Under an ecological point of view, and considering the low-light saturation of citrus photosynthesis and its shallow root system, it can be said that citrus plants present a forest-like ecophysiological behavior. Moreover, only 20% of citrus leaves can be considered to be sun-exposed at the external canopy layer (Cohen and Fuchs, 1987), and only 10% of the total external PPFD can reach deep layers of a citrus canopy (Davies and Albrigo, 1994). Cerrado plant architectures, on the other hand, exhibit diverse strategies when it comes to sunlight penetration into the canopy. Despite the limited data concerning plant architecture in cerrado woody plants, their strategies range from vertical leaf orientation, which may not be related to leaf heat avoidance (Habermann et al., 2011b) to evergreen/deciduousness behaviors (Franco et al., 2005).

Although citrus and cerrado woody plants may show similar photochemical parameters, their overall ecophysiological strategies are quite distinct, which might represent hidden consequences for land use, which has not been studied so far.

4. Root system, water absorption/use strategies in citrus and cerrado plants

It has been long discovered by the citrus industry that the use of citrus rootstocks and scions is the best way to produce fruits in different tropical and subtropical regions from all over the world (Davies & Albrigo, 1994). Therefore, there are several species, varieties and cultivars of citrus rootstocks indicated for specific regions, with specific climates, under particular disease and pest pressures.

By the middle of the 20th century, the Brazilian citrus industry was found to be under the threat of a new disease, called Citrus Tristeza Virus (CTV). Some observations made by Dr. Sylvio Moreira and Dr. Victória Rossetti suggested that orange plants that had been previously grafted on 'Rangpur' lime (*Citrus limonia* L.) rootstocks had escaped that disease, which was later confirmed, as evidenced by the 'Rangpur' lime resistance to CTV. Since then, 'Rangpur' lime has been a widely used citrus rootstock species for orange production in Brazil (Azevedo et al, 2006). But by the beginning of the year 2000, another threat, called citrus sudden death, was discovered to affect plants grafted on the 'Rangpur' lime rootstock. Aside from being a threat to 'Rangpur' lime-grafted orange plants, the citrus sudden death seems to be more prevalent among plants cultivated in warm regions, such as the northern and northwestern regions of São Paulo state (Jesus Junior & Bassanezi, 2004). Citrus Variegated Chlorosis (CVC), caused by the bacterium *Xylella fastidiosa* (Rossetti et al., 1990), is another important citrus disease and, although it does not limit the usage of any specific rootstock, it has influenced the migration of citrus groves to the south of São Paulo state. In this region, the incidence of CVC seems to be low due to mild environmental stresses that do not allow a fast bacterial spread within the plant xylem (Gomes et al., 2003; Habermann & Rodrigues, 2009).

The above description is a typical example of how horticulture, citrus breeding and phytopathological research programs intervene for the success of citrus industry. Therefore, the exploration of soil resources, mainly water (see next section for nutrients), by citrus plants has always depended on the "available" rootstocks species, cultivars and varieties as genetic resources. On the other hand, cerrado woody species have never undergone human intervention, except devastation and use of land for (citrus!) plantations. Cerrado plants have been facing biotic and abiotic stresses for millions of years, and adaptive pressures have always been applied to them (Pinheiro and Monteiro, 2010).

Interestingly, the 'Rangpur' lime is considered to be the best rootstock species for regions subjected to severe environmental stresses, such as the northern and northwestern regions of São Paulo state. Using one-year-old 'Valência' sweet orange plants, grafted on 'Rangpur' lime rootstock, it was observed that the 'Rangpur' lime's roots grow approximately 40 cm within 60 days (Magalhães Filho et al., 2008). On the other hand, using 40-day-old plants of a cerrado woody species (*Styrax ferrugineus*) it was observed that the main root reaches 60 cm in length within 75 days (Habermann & Bressan, 2011) (Figure 3). Therefore, the length root growth for both plants is very similar. However, the cerrado plants were only 40-days-old, whereas the citrus plants showed this root growth rate capacity in plants that had been cultivated for one year, as required by standard horticulture procedures. These representative case studies show that the long and deep root that was observed for *S. ferrugineus* is common for many cerrado species (Rawistscher 1948; Franco 1998), whereas for citrus plants such root behavior is only observed for the 'Rangpur' lime rootstock, in comparison with other citrus rootstocks (Magalhães Filho et al., 2008). Thus, in nature, cerrado species show a fast root growth rate to reach deep soil layers, allowing them to uptake water from deep water sources during the dry season (April-September).

Therefore, considering that disease limitations (CTV, CVC and citrus sudden death) and seasonal water deficit determine the specific rootstock to be used for citrus cultivations in the central, northern and northwestern regions of São Paulo state (Habermann & Rodrigues,

2009), and also that citrus plantations in São Paulo and (in the south of) Minas Gerais states are nowadays occupying lands that were previously occupied by the native cerrado vegetation, it would be valuable to investigate the plant biology, and to a further extent, the plant physiology of cerrado woody species. Indeed, Haridasan (2008) highlights the importance of understanding different aspects of plant biology of several species from the cerrado vegetation.

Fig. 3. Morphological details of the root system of plants of *Styrax ferrugineus* (on the left), *S. camporum* (in the midlle), and *S. pohlii* (on the right) cultivated in rhizotrons (1 m long).

Since severe wet and dry seasonal events may have been one of the most important pressures for the new savanna vegetation that formed during the Tertiary and Quaternary geological eras (Pinheiro & Monteiro, 2010), it is expected that cerrado species present very good adaptations to cope with these pressures. In general, cerrado species present long and deep root systems (Rawistscher 1948; Franco 1998; Habermann & Bressan, 2011), low specific leaf areas (SLA) and thick leaves, especially in evergreen species (Franco et al., 2005). These characteristics make cerrado species very resilient to the seasonal water deficit, and some analysis (Prado et al., 2004) performed with data obtained from field observations demonstrated that there was no severe water stress during the dry season for 22 cerrado woody species.

Although efficient scion/rootstock combinations may increase the water use efficiency in citrus plants (Ribeiro & Machado, 2007), low stomatal conductance affects the CO_2 assimilation rates by decreasing carbon availability at the carboxylation sites. Such effects may be caused by the low values of soil temperatures (Magalhães Filho et al., 2009), soil water availability (Ribeiro & Machado, 2007) and also by high vapor pressure deficits

(Habermann et al., 2003) observed during the dry season in subtropical areas (Ribeiro & Machado, 2007; Habermann & Rodrigues, 2009). Significant reductions in the water use efficiency have been reported for irrigated field-grown citrus plants during the dry season (Ribeiro, 2006), which confirm that other factors, other than soil water availability, may affect the photosynthesis of citrus plants (Ribeiro & Machado, 2007). These effects may have significant consequences for fruit production, when considering different regions, with different micro-climates and different disease pressures and their physiological damages for plants (Habermann & Rodrigues, 2009).

Therefore, although cerrado woody species are (obviously) never studied under the plant production perspective, it is clear that such species are more suitable, or well adapted to deal with environmental stresses that occur in the areas that are currently cultivated with citrus plants in São Paulo and Minas Gerais states.

5. Soil fertility and aluminum "toxicity" for cerrado and citrus plants

Aluminum (Al^{3+}) is the third most abundant element on Earth, after oxygen (O_2) and silicon (Si^{4+}). Aluminum in the soil (Al^{3+}) constitutes a major limitation to crops (Rengel, 1992; Hartwig et al., 2007), mainly in tropical regions, causing imbalances in nitrogen, sulphur and carbon cycles (Bolan and Hedley, 2003; Tang and Rengel, 2003).

When the soil pH drops to values below 5, Al^{3+} becomes soluble in the soil solution and, if absorbed by plants, it inhibits the root growth, limiting water and nutrient uptakes (Kochian et al., 2004). It is well established that Al^{3+} inhibits the root cell expansion/elongation and cell divisions nearby (Kochian et al., 2004). Aluminum is mostly associated with the cell apoplast, but some associations between Al^{3+} and the symplastic cell environment have also been reported (Kochian et al., 2004). In addition, Al^{3+} is reported to disrupt the cytoskeletal dynamics, interacting with both microtubules and actin filaments, which may affect calcium (Ca^{2+}) homeostasis, playing an important role in toxicity (Kochian et al., 2004).

On the other hand, for cerrado species, Al^{3+} does not seem to negatively affect the absorption of other cations, such as Ca^{2+}, magnesium (Mg^{2+}) and potassium (K^{+}) (Haridasan, 1982; Medeiros and Haridasan, 1985). The reason for this lack of aluminum toxicity in cerrado plants is still unknown, and the metabolic pathways involved in this process are still unclear.

The cerrado areas present well drained and deep soils, with deep water tables, and these soils are rich in aluminum (Haridasan, 2008). When compared with soils that are used for agriculture, soils from cerrado areas are considered to be poor in nutrients (Table1). However, cerrado plants are well adapted to these low fertile soils, and a high plant biodiversity is supported by these soils (Castro et al., 1999). In cerrado areas, one may find around 210 species per hectare, including trees, shrubs and grasses (Gardner, 2006).

Approximately 70% of citrus groves from São Paulo state occur in areas where root growth restrictions have been demonstrated due to high Al^{3+} contents in the soil (Pereira et al., 2003). Therefore, a high level of knowledge about mineral nutrition and the discovery of mechanisms involved in aluminum toxicity avoidance in cerrado species would be very important to figure out an alternative research strategy to "construct" transgenic plants of citrus rootstocks that could become insensitive to the toxic Al^{3+} in the soil.

Citrus plants absorb great amounts of Mg^{2+} and Ca^{2+} (Oliveira, 1986), and are very sensitive to Al^{3+}, which is highly available in acidic soils (Malavolta & Violante Neto, 1988; Table 1). In citrus leaves, aluminum affects photosystem II in thylakoid membranes, reducing the photochemical performances of the photosynthetic apparatus, causing leaf chlorosis and also reducing the dry mass accumulation and plant development (Jiang et al., 2008). On the other hand, cerrado species show an opposite response pattern. A lack of Al^{3+} for the roots of cerrado species can reduce photosynthetic rates, causing leaf chlorosis and necrosis, and can even reduce the dry mass accumulation. Moreover, two aluminum hyperaccumulator cerrado species use chloroplasts as a sink for Al^{3+}, with no apparent signs of toxicity (Andrade et al., 2011).

Soil	pH	OM	P (resin)	Al^{+3}	H+Al	K^+	Ca^{2+}	Mg^{2+}	BS	V%
	($CaCl_2$)	(g dm^{-3})	(mg dm^{-3})	(mmolc dm^{-3})						(%)
Unfertilized*	4.0	12.0	1.0	9.0	47.0	0.3	1.0	1.0	2.0	4.0
Fertilized**	5.2	14.3	65.7	0.0	17.7	2.6	36.7	14.0	53.3	74.3

OM – organic matter, BS – base saturation, V% - fertility rate
*adapted from Habermann & Bressan, 2011
**adapted from Corá et al., 2005

Table 1. Fertility parameters, and macro- and micronutrient contents in contrasting soils from different cerrado conditions (fertilized and non-fertilized soils) in São Paulo state, Brazil.

Aluminum hyperaccumulator plants are mainly woody species that absorb and maintain, in their (leaf) tissues, more than 1000 mg Kg^{-1} of Al^{3+}, exceeding the Al^{3+} presence in the soil or in the non-accumulating species growing nearby (Haridasan, 1982, Jansen et al., 2002). In 1971 Professor Goodland suggested that cerrado plants accumulate aluminum (Haridasan, 2000), resulting in plants that exhibited what he called olygotrophyc schleromorphism. This author suggested that the scleromorphism in cerrado species could be caused by aluminum toxicity, since the symptoms of aluminum toxicity are similar to the deficiency caused by malnutrition (Haridasan 2000). Based on results reported by Hutchinson (1943), Chenery (1948 a, b) and Webb (1954), who had also identified aluminum accumulation syndromes in plants from the Australian and other savannas, Goodland suggested that the Vochysiaceae and Melastomataceae plant families from the cerrado could also be considered aluminum accumulators (Haridasan 2000).

Currently, it is known that the most important hyperaccumulator species in the cerrado are: *Miconia ferruginata, M. burchellii, M. albicans, M. fallax, M. pohliana* (Melastomataceae); *Symplocos platyphylla* (Symplocaceae); *Palicourea rigida, P. squarrosa, Coussarea* spp., *Faramea cyanea, Rudgea viburnioides* (Rubiacea); *Vochysia thyrsoidia, V. rufa, V. haenhiana, V. elliptica, V. tucanorum, V. divergens, Qualea dichotoma, Q. Grandiflora, Q. Multiflora, Q. parviflora, Calisthene major, Salvertia convalhariadora* (Vochysiaceae) (Haridasan, 1982, 1987, 1988; Haridasan & Araújo, 1987).

Vochysia thyrsoidea and *Miconia fallax* cultivated without Al^{3+} in the soil show chlorotic and necrotic leaves, and severe reduction in the growth rate (Haridasan, 1988; Haridasan, 2000; Haridasan, 2008) (Figure 4 and 5) and *Styrax camporum* shows reductions in the root development [(Habermann, G. 2010, unpublished data) Figure 6]; however, when these species are transplanted to an Al^{3+}-rich soil, or Al^{3+}-rich nutritive solution, the plants recover well, suggesting that some cerrado species are incapable of surviving in calcareous soil, because aluminum might be an essential nutrient for these plants.

Fig. 4. Seedlings of *Vochysia thyrsoidea* growing in an acid and Al^{3+}-rich soil, showing healthy green leaves (top); and in a calcareous soil, showing chlorotic leaves (bottom) (pictures originally published by Haridasan, M. in Braz. J. Plant Physiol, v. 20, p. 183-195, 2008).

Fig. 5. Seedlings of *Miconia albicans*, growing in acid cerrado soil (left) (healthy green leaves) and alkaline calcareous soil (right) (chlorotic leaves and stunted growth) (picture originally published by Haridasan, M. in Braz. J. Plant Physiol, v. 20, p. 183-195, 2008).

Fig. 6. Seedlings of *Styrax camporum* cultivated in a nutritive solution without aluminum showing dark necrotic roots (on the left) and with 20 mg L^{-1} of aluminum showing normal development of roots (on the right) (Habermann, G. 2010).

Not only does toxic Al^{3+} in the soil negatively affect citrus crop production, but also a low soil pH may provoke serious problems to the nutritional plant balance. Most citrus groves in São Paulo and Minas Gerais states show soil pH around 5.0 (Table 1). Correcting soil pH can enhance production from 46 to 61 kg per 'Valência' sweet-orange tree (Anderson, 1987). Notwithstanding, soil pH corrections are very expensive and yearly-mandatory.

6. Conclusions

We tried to show that many studies about the plant biology and ecophysiology of cerrado species have been and are still being conducted, but a few, if any, are investigating native plants and their use as genetic resources for citrus plants. In addition, it was highlighted that since cerrado plants are adapted to the natural stresses that occur in the areas that are nowadays occupied by citrus plantations, they may show several (adapted) responses to deal with such environmental stresses. Thus, it is about time that citrus research and plant breeding biotechnology start looking at the plant biology of cerrado woody plants. As far as we know, this research strategy is practically neglected, not only for citrus plants, but for many other crops.

In the future, under global warming effects, it will be too late to overcome the consequences of not having used native plants as genetic resources. Irrigation, which is extremely expensive, will be the last and only option for orchards. Brazil, which is the largest producer of citrus fruit, may lose its productive capacity by not having researched its native genetic resources, which have been and are being destroyed even before we begin to understand and make good use of them.

7. References

Anderson, C.A. (1987). Fruit yields, tree size, and mineral nutrition relationships in 'Valencia' orange trees as affected by liming. *Journal of Plant Nutrition,* v.10, p.1907-1916.

Andrade, L.R.M., Barros, L.M.G., Echevarriac, G.F., Amaral, L.I.V., Cottab, M.G., Rossatto, D.R., Haridasan, M., Franco, A.C. (2011) Al-hyperaccumulator Vochysiaceae from the Brazilian Cerrado store aluminum in their chloroplasts without apparent damage. *Environmental and Experimental Botany,* v.70, p.37-42.

Azevedo, F.Al., Mourão Filho, F.A.A.; Mendes, B.M.J.I,., Almeida, W.A.B., Schinor, E.H., Pio, R., Barbosa, J.M., Gonzalez, S.G., Carrer, H., Lam, E. (2006). Genetic transformation of Rangpur lime (*Citrus limonia* Osbeck) with bO (bacterio-opsin) gene and its initial evaluation for *Phytophthora nicotianae* resistance. *Plant Molecular Biology Reporter,* v. 24, p. 185-196.

Batalha, M. (2011). O cerrado não é um bioma. Biota Neotropica, v. 11, no. 1.

Bolan, N.S., Hedley, M.J. (2003). Role of carbon, nitrogen and sulfur cycles in soil acidification. In: Rengel Z, ed. *Handbook of soil acidity.* New York, NY, USA: Marcel Dekker, 29 - 56.

Borlaug, N.E.(2002). Feeding a world of 10 billion people: the miracle ahead. In: Bailey, R. (ed.). *Global warming and other eco-myths.* p. 29-60.

Castro, A.A.J.F., Martins, F.R., Tamashiro, J.Y., Shepherd, G.J. (1999). How rich is the flora of the Brazilian cerrados? *Annals of the Missouri Botanical Garden,* v.86, p.192-224.

Chenery, E.M. (1948a). Aluminium in the plant world. Part I. General survey in the dicotyledons. *Kew Bulletin,* v.3, p.173-183.

Chenery, E.M. (1948b). Aluminium in plants and its relation to plant pigments. *Annals of Botany,* v.12, p.121-136.

Chapin III, F.S., Zavaleta, E.S., Eviner, V.T, et al(2000). Consequences of changing biodiversity. *Nature,* v.405, p.234 - 242.

Cohen, S., Fuchs, M. (1987). The distribution of leaf area, radiation, photosynthesis and transpiration in a Shamouti orange hedgerow orchard. Part I. Leaf area and radiation. *Agric. For. Meteorol.* v.40, p.123-144.

Corá, J.E., Silva, G.O., Martins Filho, M.V. (2005). Manejo do solo sob citros. In: Dirceu de Mattos Júnior; José Dagoberto De Negri; Rose Mary Pio; Jorgino Pompeu Júnior. (Org.). *Citros.* Campinas: Instituto Agronomico de Campinas e Fundag, 1 ed., p. 347-368.

Davies, F.S.; Albrigo, L.G. (1994). *Citrus.* Wallingford: CAB International, 254p.

Franco, A.C., Matsubara, S., Orthen, B. (2007). Photoinhibition, carotenoid composition and the coregulation of photochemical and non-photochemical quenching in neotropical savanna trees. *Tree Physiology,* v.27, p.717-725.

Franco, A.C., Bustamante, M., Caldas, L.S., Goldstein, G., Meinzer, F.C., Kozovits, A.R.Rundel, P., Coradin, V.T.R. (2005). Leaf functional traits of Neotropical savanna trees in relation to seasonal water deficit. *Trees,* v.19, p.326-335.

Franco, A.C., Lüttge, U. (2002). Midday depression in savanna trees: coordinated adjustments in photochemical efficiency, photorespiration, CO2 assimilation and water use efficiency. *Oecologia,* v.131,p.356 - 365.

Franco A.C. (1998). Seasonal patterns of gas exchange, water relations and growth of Roupala montana, an evergreen savanna species. *Plant Ecology*, v.136, p. 69 - 76.

Gardner, T. (2006). Tree-grass coexistence in the Brazilian cerrado: demographic consequences of environmental instability. *J. Biogeogr.* v.33, p.448-463.

Givnish, T.J. (1988). Adaptation to sun and shade: a whole plant perspective. *Australian Journal of Plant Physiology*, v.15, p.63 - 92.

Gomes, M.M.A., Lagôa, A.M.M.A., Machado, E.C., Medina, C.L. (2003). Abscisic acid and indole-3-acetic acid contets in orange trees infected by Xylella fastidiosa and submitted to cycles of water stress. *Plant Growth Regul.* v.39, p.263 - 270.

Habermann, G., Bressan, A.C.G. (2011). Root, shoot and leaf traits of the congeneric *Styrax* species may explain their distribution patterns in the cerrado sensu lato areas in Brazil. *Functional Plant Biology*, v.38, p.209-218.

Habermann, G., Ellsworth, P.F.V., Cazoto, J.L., Simão, E., Bieras, A.C. (2011a) Comparative gas exchange performance during the wet season of three Brazilian *Styrax* species under habitat conditions of cerrado vegetation types differing in soil water availability and crown density. *Flora*, v.206, p.351-359.

Habermann, G., Ellsworth, P.F.V., Cazoto, J.L., Feistler, A.M., Silva, L., Donatti, D.A., Machado, S.R. (2011b). Leaf paraheliotropism in *Styrax camporum* confers increased light use efficiency and advantageous photosynthetic responses rather than photoprotection. *Environmental and Experimental Botany*, v.71, p.10-17.

Habermann, G., Rodrigues, J.D. (2009). Leaf gas exchange and fruit yield in sweet orange trees as affected by citrus variegated chlorosis and environmental conditions. *Scientia Horticulturae*, v. 122, p.69-76.

Habermann, G., Machado, E.C., Rodrigues, J.D., Medina, C.L. (2003). Gas exchange rates at different vapor pressure deficits and water relations of 'Pera' sweet orange plants with citrus variegated chlorosis (CVC). *Scientia Horticulturae*, v.98, p.233-245.

Hao, G.Y., Hoffmann, A., Scholz, F.G., Bucci, S.J., Meinzer, F.C., Franco, A.C., Cao, K., Goldstein, G. (2008). Stem and leaf hydraulics of congeneric tree species from adjacent tropical savanna and forest ecosystems. *Oecologia*, v.155, p.405-415.

Haridasan, M. (2008). Nutritional adaptations of native plants of the cerrado biome in acid soils. *Brazilian Journal of Plant Physiology*, v.20, p.183-195.

Haridasan, M. (2000). Nutrição mineral de plantas nativas do cerrado. *Revista Brasileira de Fisiologia Vegetal*, v. 12, p. 54-64.

Haridasan, M. (1988). Performance of *Miconia albicans* (Sw.) Triana, an aluminum acculmulating species in acidic and calcareous soils. *Communications in Soil Science and Plant Analysis*,v.19, p.1091-1103.

Haridasan, M. (1982). Aluminium accumulation by some cerrado native species of central Brazil. *Plant and Soil*, v.65, p.265-273.

Haridasan, M., Araújo, G.M.(1987). Aluminium accumulating species in two forest communities in the cerrado region of central Brazil. *Forest Ecology and Management*, v. 24, n. 1, p. 15-26.

Hatwing, I., Oliveira, A.C., Carvalho, F.I.F., Bertan, I., Silva, J.A.G., Schidt, D.A.M., Valério, I.P., Maia, L.C., Fonseca, D.N.R., Reis, C.E.S. (2007). Mecanismos associados à tolerância ao alumínio em plantas. *Semina Ciências Agrárias*, v.28,p.219-228.

Hutchinson, D.E. (1943). The biochemistry of aluminium and certain related elements. *Quarterly Reviews of Biology*, v.18, p.1-29, 123-153, 242-262, 331-363.

Jansen, S., Broadley, M.R., Robbrecht, E., Smets, E. (2002). Aluminum Hyperaccumulation in Angiosperms: A Review of Its Phylogenetic Significance. *The Botanical Review*, v.68, p. 235-269.

Jesus Junior, W.C., Bassanezi, R. (2004). Análise da Dinâmica e Estrutura de Focos da Morte Súbita dos Citros. *Fito. Bras.*, v.29, p.399-405.

Jiang, H.X., Chen, L.S., Zheng, J.G., Han, S., Tang, N., Smith, B.R. (2008). Aluminum-induced effects on Photosystem II photochemistry in Citrus leaves assessed by the chlorophyll a fluorescence transient. *Tree Physiology*, v.28, p.1863 - 1871.

Jifon, J.L., Syvertsen, J.P. (2003). Moderate shade can increase net gas exchange and reduce photoinhibition in citrus leaves. *Tree Physiology*, v.23, p.119-127.

Klink, C.A., Machado, R.B. (2005). A conservação do Cerrado brasileiro. *Megadiversidade*, v.1, p. 147-155.

Kochian, L.V., Hoekenga, O.A., Piñeiros, M.A. (2004). How do crop plants tolerate acid soils? Mechanisms of aluminum tolerance and phosphorous efficiency. *Annu. Rev. Plant Biol.*, v.55, p.459 - 493.

Ledru, M.P. (2002). Late Quaternary history and evolution of the Cerrado as revealed by palynological records. *In The Cerrados of Brazil*, eds Oliveira PS, Marquis RJ (Columbia Univ Press, New York), pp 33 - 50.

Lemos Filho, J.P., Goulart, M.F., Lovato, M.B. (2004). Chlorophyll fluorescence parameters in populations of two legume trees: *Stryphnodendron adstringens* (Mart.) Coville (Mimosoideae) and *Cassia ferruginea* (Schrad.) Schrad. ex DC. (Caesalpinoideae). *Revista Brasileira de Botânica*, v.27, p.527-532.

Lemos Filho, J.P. (2000). Fotoinibição em três espécies do cerrado (*Annona crassifólia, Eugenia dysenterica* e *Campomanesia adamantium*) na estação seca e na chuvosa. *Revista Brasileira de Botânica*, v.23, p.45-50.

Ma, J.F., Ryan, P.R., Delhaise, E. (2001). Aluminium tolerance in plants and the complexing role of organic acids. *Trends Plant Sci*, v.6, p.273 - 278.

Magalhães Filho, J.R., Machado, E.C., Machado, D.F.S.P., Ramos, R.A., Ribeiro, R.V. (2009). Variação da temperatura do substrato e fotossíntese em mudas de laranjeira 'Valência'. *Pesquisa Agropecuária Brasileira*, v.44, p.1118-1126.

Magalhães Filho, J.R., Amaral, L.R., Machado, D.F.S.P., Medina, C.L., Machado, E.C. (2008). Deficiência hídrica, trocas gasosas e crescimento de raízes em laranjeira 'Valência' sobre dois tipos de porta-enxerto. *Bragantia*, v.67, p.75-82.

Malavolta, E., Violante Neto, A. (1988). Nutrição mineral, calagem, gessagem e nutrição dos citros. In: DONADIO, L.C. (Ed.). *Produtividade de citros*. 2.ed. Jaboticabal: Funep, p.233-284.

Mattos, E.A. (1998). Perspectives in comparative ecophysiology of some Brazilian vegetation types: leaf CO2 and H2O exchange, chlorophyll a fluorescence and carbon isotope discrimination. In: Scarano, F.R., Franco, A.C. *Ecophysiological Strategies of Xerophytic and Amphibious Plants in the Neotropics*, Series Oecologia Brasiliensis, PPGE-UFRJ, Rio de Janeiro, Brazil, v.4, p.1 - 22.

Medeiros, R.A., Haridasan, M. (1985). Seasonal variations in the foliar concentrations of nutrients in some aluminium accumulating plants of the cerrado region of central Brazil. *Plant and Soil*, v.88, p.433-436.

MMA (2002). *Biodiversidade brasileira: avaliação e identificação de áreas prioritárias para conservação, utilização sustentável e repartição de benefícios da biodiversidade brasileira.* Brasília: MMA/SBF, 404p.

Oliveira, J.B. (1986). Solos para citricultura no Estado de São Paulo. *Laranja*, v.7, p.337-351.

Pereira, W.E., Siqueira, D.L., Puiatti, M., Martinez, C.A., Salomão, L.C.C., Cecon, P.R. (2003). Growth of citrus rootstocks under aluminium stress in hydroponics. *Scientia agricola*, v.60, p.31-41.

Pinheiro, M.H.O., Monteiro, R. (2010). Contribution to the discussions on the origin of the cerrado biome: Brazilian savanna. *Brazilian Journal of Biology and Technology*, v.70, p.95-102.

Prado, C.H.B.A., Wenhui, Z., Rojas, M.H.C., Souza, G.M. (2004). Seasonal leaf gas exchange and water potential in a woody cerrado species community. *Braz. Journal of Plant Physiology*, v.16, p.7 - 16.

Ratter, J.A.; Ribeiro, J.F.; Bridgewater, S. (1997). The Brazilian Cerrado vegetation and threats to its biodiversity. *Ann Bot.*, v.80, p.223 - 230.

Rawistscher, F. (1948). The water economy of the campos cerrados in the Southern of Brazil. *Journal of Ecology*, v.36, p.237 - 268.

Rengel, Z. (1992). Role of calcium in aluminium toxicity. *New Phytologist*, v.121, p.499-513.

Ribeiro, R.V., Machado, E.C. (2007). Some aspects of citrus ecophysiology in subtropical climates: re-visiting photosynthesis under natural conditions. *Brazilian Journal of Plant Physiology*, v.19, p.393-411.

Ribeiro, R.V. (2006). *Variação sazonal da fotossíntese e relações hídricas de laranjeira 'Valência'.* Piracicaba, Universidade de São Paulo. PhD thesis.

Rossetti, V., Garnier, M., Bové, J.M., Beretta, M.J.G., Teixeira, A.R., Quaggio, J.A., Negri, J.D. (1990). Présence de bactéries dans le xylème d'orangers atteints de chlorose variégée, une nouvelle maladie des agrumes au Brésil Comptes Rendus de l'Academie des Sciences, Ser. III, v. 310, p.345 - 349.

Tang, C., Rengel, Z. (2003). Role of plant cation /anion uptake ratio in soil acidification. In: Rengel, Z, ed. Handbook of soil acidity. New York, NY, USA: Marcel Dekker, 57 - 81.

Vu, J.C.V., Yelenosky, G., Bausher, M.G. (1986). CO2 exchange rate, stomatal conductance and transpiration in attached leaves of Valencia orange. *HortScience*, v.21, p.143-144.

Webb, L.J. (1954). Aluminium accumulation in the Australian-new Guinea flora. *Australian Journal of Botany*, v.2, p.176-196.

8

Antioxidant Enzyme Activities as a Tool to Discriminate Ecotypes of *Crithmum maritimum* L. Differing in Their Capacity to Withstand Salinity

Ben Hamed Karim[1], Magné Christian[2] and Abdelly Chedly[1]
[1]*Centre de Biotechnologie de Borj Cédria/Laboratoire des Plantes Extrêmophiles*
[2]*Université de Brest/LEBHAM, EA 3877,*
Institut Universitaire Européen de la Mer
[1]*Tunisia*
[2]*France*

1. Introduction

Salt stress causes a range of adverse effects in plants, mainly ionic disorders, osmotic stress and nutritional imbalance. A common feature of these effects is the production of reactive oxygen species (ROS) (Ashraf and Foolad, 2007). Thus, salt stress causes stomatal closure, which reduces the CO_2/O_2 ratio inside leaf tissues and inhibits CO_2 fixation (Hernández *et al.*, 1999). As a consequence, an over reduction of the photosynthetic electron transport chain occurs, which causes the generation of ROS such as singlet oxygen ($^1O_2$), superoxide anion ($O_2^{\cdot -}$), hydrogen peroxide (H_2O_2), and hydroxyl radical (•OH). It is now widely accepted that ROS are responsible for various stress-induced damages to macromolecules and ultimately to cellular structure (Mittler, 2000; Ashraf, 2009). Plants are equipped with a set of non-enzymatic scavengers and of antioxidant enzymes that act in concert to alleviate cellular damage under oxidative stress conditions (Foyer and Noctor, 2000). Superoxide dismutase (SOD) reacts with the superoxide radical at almost diffusion-limited rates to produce H_2O_2 (Scandalios, 1993). H_2O_2 is scavenged by peroxidases, especially ascorbate peroxidase (APX), and catalase (CAT). CAT has been found predominantly in leaf peroxisomes where it functions chiefly to remove H_2O_2 formed in photorespiration or in β-oxidation of fatty acids in the glyoxysomes (Dat *et al.*, 2000). APX, which uses ascorbic acid as a reductant in the first step of the ascorbate-glutathione cycle is the most important plant peroxidase involved in H_2O_2 detoxification (Foyer and Halliwell, 1976; Noctor and Foyer, 1998). ROS are also scavenged non-enzymatically by hydrophilic antioxidants, such as ascorbate and glutathione (GSH), most of them being found in photosynthetic tissues.
In the last two decades, a number of reviews have concentrated much on the role of various antioxidant enzymes and metabolites in plant salt tolerance (Jitesh *et al.*, 2006; Ashraf, 2009). Due to considerable variations in the mechanisms of defense against ROS among plant species, it is difficult to generalize the involvement of this phenomenon in salt tolerance

(Ashraf and Harris, 2004). A number of experimental approaches have been adopted to understand the link between salt tolerance and antioxidant activities. Several studies compared crop species, genera and cultivars differing in salt tolerance, and showed considerable variations in the production of antioxidant in response to salt stress. Moreover, increased levels of antioxidant enzymes and/or metabolites have been correlated to the degree of salt tolerance in a number of plant species including wheat, rice, maize, cotton, tomato and potato (Ashraf, 2009). However, such relationship could not be found in other plants such as cowpea (Cavalcanti *et al.*, 2004), *Arabidopsis* (Katsuhara *et al.*, 2005) and strawberry (Turhan *et al.*, 2008).

Recently, many authors addressed investigations on naturally salt tolerant plants (or halophytes) in which the mechanisms of salt tolerance are fully developed and functional to make them productive under stress. Jithesh *et al.* (2006) reviewed the regulation of antioxidant enzymes under salt stress in halophytes, especially those of mangroves. They concluded that antioxidant enzymes protect halophytes from deleterious ROS production during salt stress. The antioxidant response in halophytes, like in glycophytes, varies among species and genera. Most studies reported an induction in the activities of antioxidant enzymes and an accumulation of antioxidant metabolites, in response to salinity (Cherian *et al.*, 1999; Takemura *et al.*, 2000; Parida *et al.*, 2004). Moreover, some halophytes like *Thelungiella halophila* exhibited high levels of SOD expression even in unstressed conditions (Taji *et al.*, 2004). Thus, the SOD activities in mangrove *Rhizophora stylosa* were more than 40 times higher than those of the glycophyte pea. High concentration of ascorbate was also observed in the leaves of that plant. This robust antioxidant activity is considered to be effective in protecting mangroves from excess irradiance (which generates ROS) under natural conditions (Cheeseman *et al.*, 1997).

Higher antioxidant defences in leaves compared to those of roots accounted for the survival of the halophyte *Crithmum maritimum* L. under high salinity conditions (Ben Hamed *et al.*, 2007). In the same species, Ben Amor *et al.* (2005) showed an increased total SOD activity in shoot tissues in presence of 50 mM NaCl stress while SOD activity in root tissues decreased upon salt treatment (up to 200 mM). These findings showed that the activities of antioxidant enzymes under salt stress may vary depending on the tissues. It may also depend on the conditions of treatments and on the species from which ecotypes were collected. In the present work, we examined the effects of salt stress on two ecotypes of *Crithmum maritimum* grown under the same controlled conditions. The effects of salt stress were investigated on growth, lipid peroxidation, electrolyte leakage, hydrogen peroxide levels and antioxidative enzyme activity. The differential response of that system might contribute to better understand the mechanism of salinity resistance and, in turn, might lead to the development of elite ecotypes able to withstand high salinity levels in the soils.

2. Material and methods

2.1 Plant material and growth conditions

Seeds of two *Crithmum maritimum* (Apiaceae) ecotypes were collected from coastal sites near Tabarka (North-West of Tunisia, humid bioclimatic stage) and Korbous (North-East of Tunisia, semi-arid bioclimatic stage). Seeds were sown in pots (4 seeds per pot) filled with 3 kg of inert sand and irrigated with distilled water until germination. Then, seedlings were daily watered with Hewitt solution (Ben Hamed *et al.*, 2007). Thirty four day-old seedlings were partitioned into 3 lots, which were treated with 0, 100, and 300 mM NaCl, respectively (the salt concentrations were daily stepwise increased with 50 mM NaCl). The experiment was conducted in a growth room under controlled conditions : 15-25°C temperature and 70-90% relative humidity, 8-16 h night-day photoperiod, and 440 µmol m^{-2} s^{-1} photosynthetic active radiations (PAR). Plants were harvested at the beginning of salt treatment (initial harvest) and after 50 days. For plant growth, plants were separated into leaves, stems and roots. Samples from fully expanded leaves were liquid N_2 frozen and stored at -80 °C for enzyme activity analysis.

2.2 Plant growth

Plant shoot and root dry weights (DW) were measured after 96 h at 80°C. The relative growth rate (RGR) was also calculated.

$$\text{RGR (day}^{-1}) = \text{M}/\underline{\text{M}}\,\Delta\text{t}$$

Where Δ is the difference between the values at the final and initial harvests, *t* the salt treatment duration (days), and $\underline{M}$ the logarithmic mean of M, the whole plant dry weight (g):

$$\underline{\text{M}} = \Delta\text{M}/\Delta \ln \text{M})$$

2.3 H_2O_2 determination

H_2O_2 was extracted by homogenizing plant leaves with phosphate buffer (50 mM, pH 6.8) including 1 mM hydroxylamine. The homogenate was centrifuged at 6000 *g* for 25 min. The H_2O_2 concentration in the supernatant was then measured colorimetrically (Chaparzadeh *et al.*, 2004). Aliquot of the extract was mixed with 0.1% titanium chloride in 20% (v/v) H_2SO_4 and the mixture was centrifuged at 6000 *g* for 15 min. The intensity of yellow color of the supernatant was measured at 410 nm. H_2O_2 concentration was calculated using the extinction coefficient 0.25 $mM^{-1}cm^{-1}$, and expressed on the basis of fresh weight (FW).

2.4 Lipid peroxidation

The extent of lipid peroxidation was estimated by determining the concentration of malondialdehyde (MDA) (Hagege *et al.*, 1990). Leaf materials were homogenized in 0.1% (w/v) TCA solution. The homogenate was centrifuged at 15000 *g* for 10 min and 1 ml of the supernatant obtained was added to 4 ml of 0.5% (w/v) TBA in 20% (w/v) TCA. The mixture was incubated at 90°C for 30 min, and the reaction was stopped by placing the reaction tubes in an ice water bath. Then, the samples were centrifuged at 10000 *g* for 5 min, and the absorbance of the supernatant was read at 532 nm. The value for non-specific absorption at 600 nm was subtracted. The concentration of MDA was calculated from the extinction coefficient 155 $mM^{-1}cm^{-1}$.

2.5 Electrolyte leakage

Fresh leaf samples were cut into discs of uniform size (5 mm diameter) and placed in test tubes containing 10 ml of double distilled water. The tubes were incubated in a water bath at 32°C for 2 h and the initial electrical conductivity of the medium (EC1) was measured. Samples were heated at 100°C for 20 min to release all electrolytes, cooled to 25°C and the final electrical conductivity (EC2) was measured (Sairam *et al.*, 2002). The electrolyte leakage (EL), expressed in % of total electrolytes, was calculated by using the formula: EL= (EC1/ EC2) x 100.

2.6 Enzyme extractions

All of the following operations were performed at 4°C. Fresh leaf and root samples (0.5 g) were rapidly extracted in a pre-chilled mortar with 10% (w/v) PVP (polyvinylpyrrolidone) in 50 mM K-phosphate buffer (pH 8), containing 0.1 mM EDTA (ethylenediaminetetraacetic acid), 1 mM DTT (dithiothreitol), and 0.5 mM PMSF (phenyl-methyl-sulphonyl-fluoride). For ascorbate peroxidase extraction, 20 mM sodium ascorbate was added to the medium to maintain the enzyme active during the extraction procedure. The homogenate was centrifuged at 12000 *g* for 30 min. and the supernatant was dialysed for 2 h against the extraction buffer containing only 5 mM sodium ascorbate. Three replicates per treatment were used. The supernatants were collected and protein concentration was determined according to Bradford (1976), using bovine serum albumin as a standard.

2.7 Enzyme assays

Total superoxide dismutase (SOD, EC 1.11.1.5) activity was assayed according to Scebba *et al.* (1999). Increasing volumes (5, 10, 20, and 40 µl) of leaf and root crude extracts were added to the reaction mixture at a final volume of 3 ml. The reaction mixture contained 50 mM potassium phosphate buffer (pH 7.8), 0.1 mM EDTA, 13 mM L-methionine, 2 µM riboflavin and 75 µM NBT (nitroblue tetrazolium). The reaction was started by exposing the mixture to cool white fluorescent light for 15 min. The blue colour developed was measured spectrophotometrically at 560 nm. The volume of sample causing 50% inhibition of colour development (compared to the blank where sample was replaced by extraction buffer) contained one unit of SOD activity and the specific activity of plant extracts was expressed as units per mg of protein.

Total catalase (CAT, EC 1.11.1.6) activity was measured spectrophotometrically according to the method of Lück (1965), by monitoring the decline in absorbance at 240 nm due to H_2O_2 consumption. The 3 ml reaction mixture contained 66 mM sodium phosphate buffer (pH 7.0) and 30% H_2O_2 (the absorbance should be about 0.5 at 240 nm and with a 1 cm light path). The reaction was initiated by adding an appropriate dilution of the shoot or root crude extract to this solution. CAT activity was expressed as units per mg of protein, one unit being the amount of enzyme which liberates half the peroxide in 100 s at 25°C (Lück, 1965).

Total peroxidase (POD, EC 1.11.1.7) activity was determined spectrophotometrically by measuring the oxidation of *o*-dianisidine (3,3'-dimethoxybenzidine) at 460 nm (Ranieri *et al.*, 2000). The reaction mixture contained 20 mM phosphate buffer (pH 5.0), 1 mM dianisidine, 3 mM H_2O_2 and 50 µl of extract. POD specific activity was expressed as units (µmol of dianisidine oxidized per minute) per mg of protein.

Total ascorbate peroxidase (APX, EC 1.11.1.11) activity was measured spectrophotometrically according to Nakano and Asada (1981) by following the decline in absorbance at 290 nm due to ascorbate oxidation. The oxidation rate of ascorbate was estimated between 1 and 60 s after starting the reaction with the addition of H_2O_2. The 1 ml reaction mixture contained 50 mM HEPES-NaOH (pH 7.6), 0.22 mM ascorbate, 1 mM H_2O_2, and an aliquot of enzyme extract. Corrections were made for the low, non-enzymatic oxidation of ascorbate by H_2O_2 and for the oxidation of ascorbate in the absence of H_2O_2. APX activity was expressed as units (µmol of ascorbate oxidized per minute) per mg of protein.

Total glutathione reductase (GR, EC 1.6.4.2) activity was determined by following the rate of GSSG-dependent oxidation of NADPH, through the decrease in the absorbance at 340 nm (Di Baccio *et al.*, 2004). The assay mixture (1 ml final volume) was composed of 0.4 M potassium phosphate buffer (pH 7.5), 0.4 mM Na_2EDTA, 5.0 mM GSSG and 100 µl of crude extract. The reaction was initiated by the addition of 2.0 mM NADPH. Corrections were made for the background absorbance at 340 nm, without NADPH. Activity was expressed as units (µmol of NADPH oxidized per minute) per mg of protein.

2.8 Native gel electrophoresis and SOD activity staining

Samples of crude *C. maritimum* leaf extracts were analyzed by electrophoresis in 10% (w/v) polyacrylamide slab gel, at pH 8.9 under non-denaturating conditions.
Staining for SOD activity was carried out as described by Beauchamp and Fridovich (1971). The gel was first soaked in 50 mM sodium phosphate, (pH 7.5) containing 4.8 mM 3-(4,5-dimethylthiazol-2-yl)-2,5-diphenyltetrazolium bromide (MTT) in darkness for 20 min, followed by soaking in 50 mM sodium phosphate (pH 7.5) containing 0.4% (v/v) N, N, N′, N′-tetramethylethylenediamine (TEMED) and 26 µM riboflavin, and subsequently illuminated under white light for 10 min. The three types of SOD, namely Fe-SOD, Mn-SOD and Cu-Zn SOD, were identified using inhibitors: Mn-SOD was visualised by its insensitivity to 5 mM H_2O_2 and 2 mM KCN, while Cu-Zn SOD was sensitive to 2 mM KCN and Fe-SOD was inhibited by 5 mM H_2O_2 (Navari-Izzo *et al.*, 1998).

2.9 Statistical analysis

Analysis of variance (ANOVA) using AV1W MSUSTAT program with orthogonal contrasts and mean comparison procedures was performed to detect significant differences between treatments. Mean separation procedures were carried out using the multiple range tests with Student's least significant difference (LSD) ($P \leq 0.05$).

3. Plant growth parameters

Growth of both ecotypes was estimated through dry mass production and RGR. The response of Tabarka plants to salinity was different to that of Korbous ecotype: whereas growth decreased under low (100 mM) and significantly under high (300 mM NaCl) salinity in Korbous plants (Fig. 1A and B), it increased by 40% under 100 mM NaCl in Tabarka plants.
In the following sections, we will try to check if the different physiological behavior of the ecotypes of *C. maritimum* observed mainly under low salinity (100 mM NaCl) is concomitant with different or similar antioxidant responses.

3.1 Oxidative stress parameters

Lipid peroxidation, as estimated by the MDA content, electrolyte leakage and H_2O_2 content are considered as the most useful indicators to detect oxidative disturbances in plants. In Korbous ecotype, MDA level (Fig. 2A), H_2O_2 concentration (Fig. 2B) and electrolyte leakage (Fig. 2C) increased in the leaves of salt treated plants. Conversely, all these parameters were higher in control plants of Tabarka ecotype.

3.2 Activity of antioxidant enzymes

Superoxide dismutase (SOD) is the first enzyme that eliminates superoxide radicals in plant cells. Total SOD activity in Korbous plants was unaffected by NaCl exposure but it increased in the leaves of salt-treated Tabarka ecotype (Fig. 3A). Individual activities of SOD

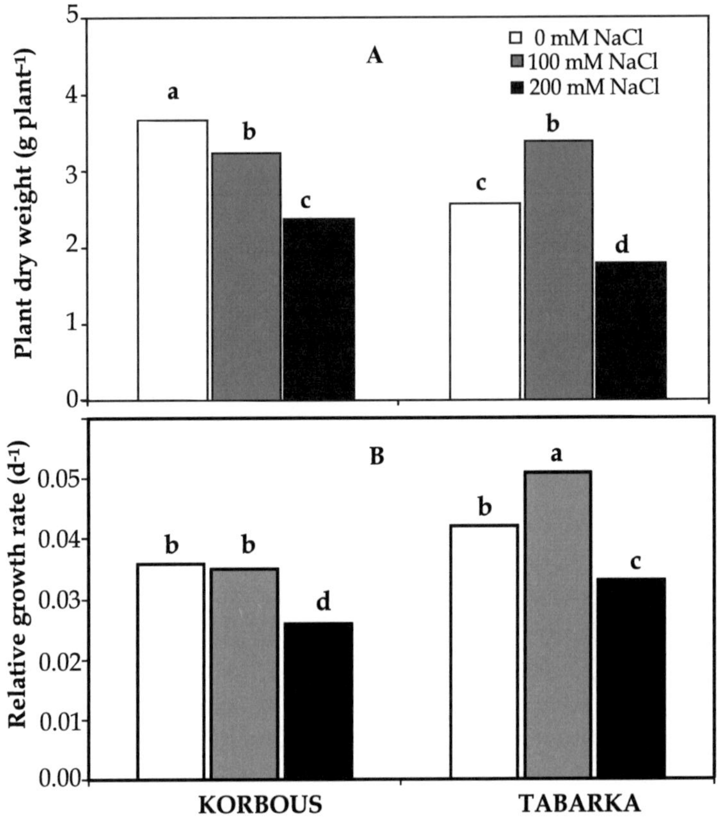

Fig. 1. Effect of NaCl on plant dry weight (**A**) and relative growth rate (**B**) of *Crithmum maritimum* ecotypes. The results are the means of 10 replicates. Means followed by different letters are significantly different at $P \leq 0.05$ according to student's LSD test.

isoforms in response to NaCl was examined after polyacrylamide gel electrophoresis (Fig. 3 B). The gel showed that plants from Korbous subjected to salt maintained steady activities of Mn and Fe SOD isoforms. Conversely, each isoform was stimulated by NaCl treatment in Tabarka ecotype and a new isoform (type Cu-Zn) was induced. Cu Zn, Mn and Fe SOD isoforms were identified in the presence of specific inhibitors (Figure 3 C).

SOD reaction on $O_2^{\cdot-}$ results in the production of H_2O_2, which needs to be controlled. The enzyme that are well known to play major roles in the detoxification of H_2O_2 are catalase (CAT), dianisidine (POD) and ascorbate peroxidases (APX) and Glutathione reductase (GR). In the absence of salt, the CAT activity in Tabarka plant was significantly higher than that of Korbous (Fig. 4A). The CAT activity did not change in plants from Korbous treated by 100 mM, while it increased in salt treated plants from Tabarka. POD activities increased under salinity conditions compared to control plants in both ecotypes (Fig. 4B). Also, we noticed higher control POD activities in Korbous ecotype. The activity of APX was significantly higher in Tabarka, both in the absence and presence of NaCl (Fig. 4C). Salt treatment did not affect enzyme activities in the two ecotypes. Under control conditions, the activity of GR was 50% higher in Korbous than in Tabarka plants (Fig. 4D). Salt-treated plants from Korbous exhibited a slight (-10%) decrease in GR activity, while a strong increase (+80%) in GR activity was observed in Tabarka ecotype.

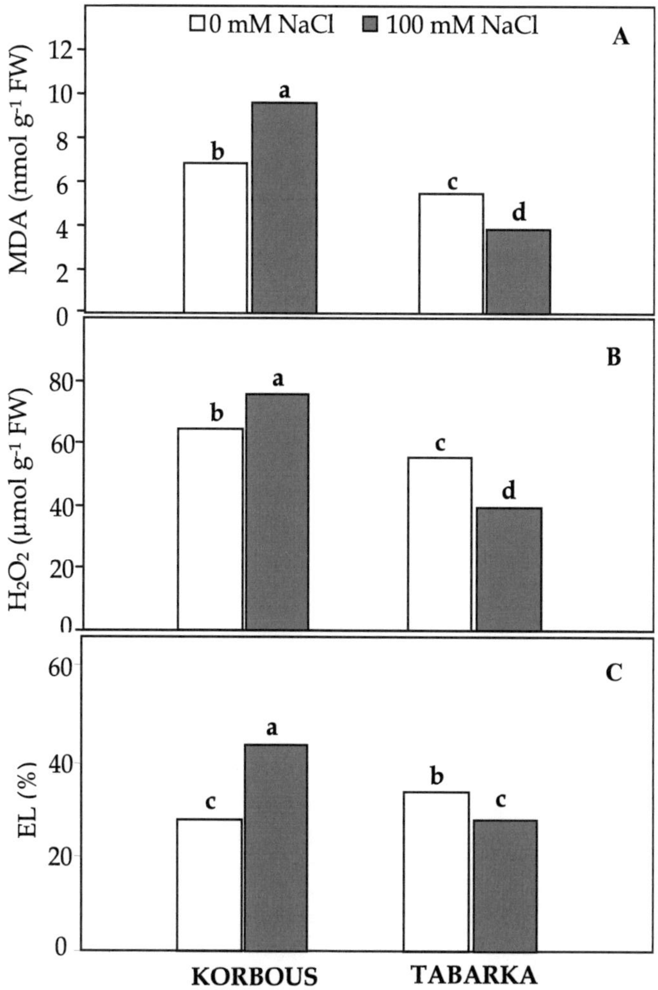

Fig. 2. Effect of NaCl on malondialdehyde levels (A), hydrogen peroxide (H_2O_2) concentration (B), and % of electrolyte leakage (C) in leaves of *Crithmum maritimum* ecotypes. The results are the means of 5 replicates. Means followed by different letters are significantly different at $P \leq 0.05$ according to student's LSD test.

4. Discussion

The two ecotypes of *Crithmum maritimum* exhibited strongly divergent responses to salinity. Tabarka ecotype showed an halotolerant behavior, as shown by stimulation of leaf biomass production and leaf expansion under 100 mM NaCl. Conversely, leaf growth and expansion of Korbous ecotype were significantly reduced under salt treatment, even at low NaCl concentration. These results showed an intraspecific variability in the response to salt stress of *C. maritimum*, a perennial Apiaceae, which is considered as a promising crop species for oil production (Zarrouk *et al.*, 2004). The same conclusion was drawn for another oilseed halophyte, *Cakile maritima* (Ben Amor *et al.*, 2006; Ksouri *et al.*, 2007; Megdiche *et al.*, 2009).

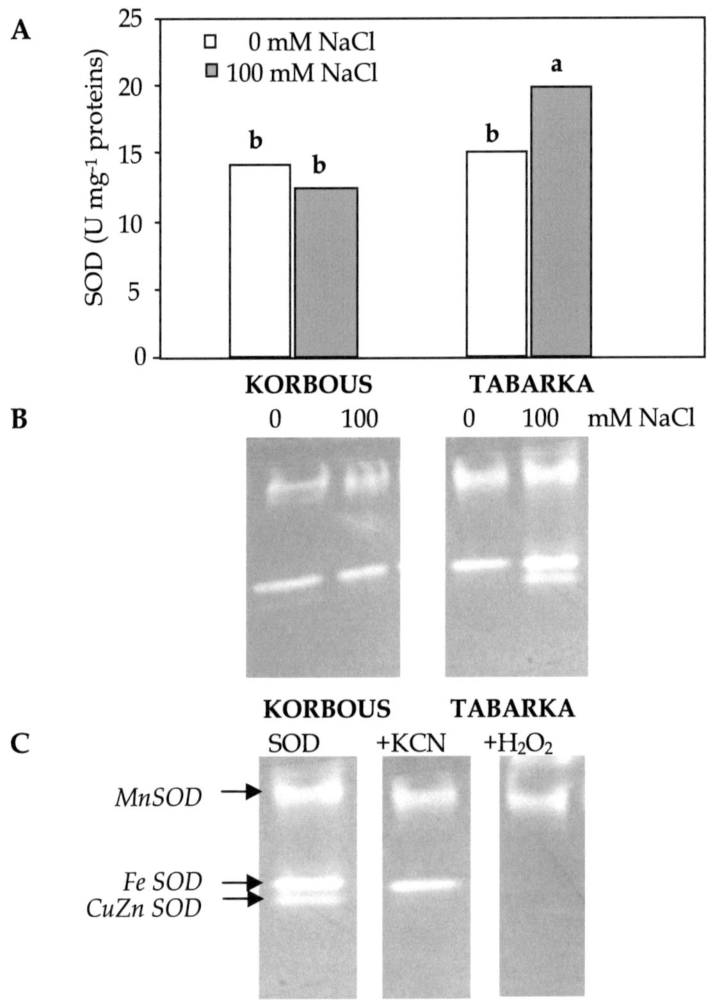

Fig. 3. Effect of NaCl on the total activity of superoxide dismutases (**A**) and in gel activities of SOD isoforms (**B**). Differentiation of CuZn, Mn and Fe SOD in the crude extract of leaves of salt-treated plants of *C. maritimum* (ecotype Tabarka) (**C**).The results in (A) are the means of 5 replicates. Means followed by different letters are significantly different at $P \leq 0.05$ according to student's LSD test.

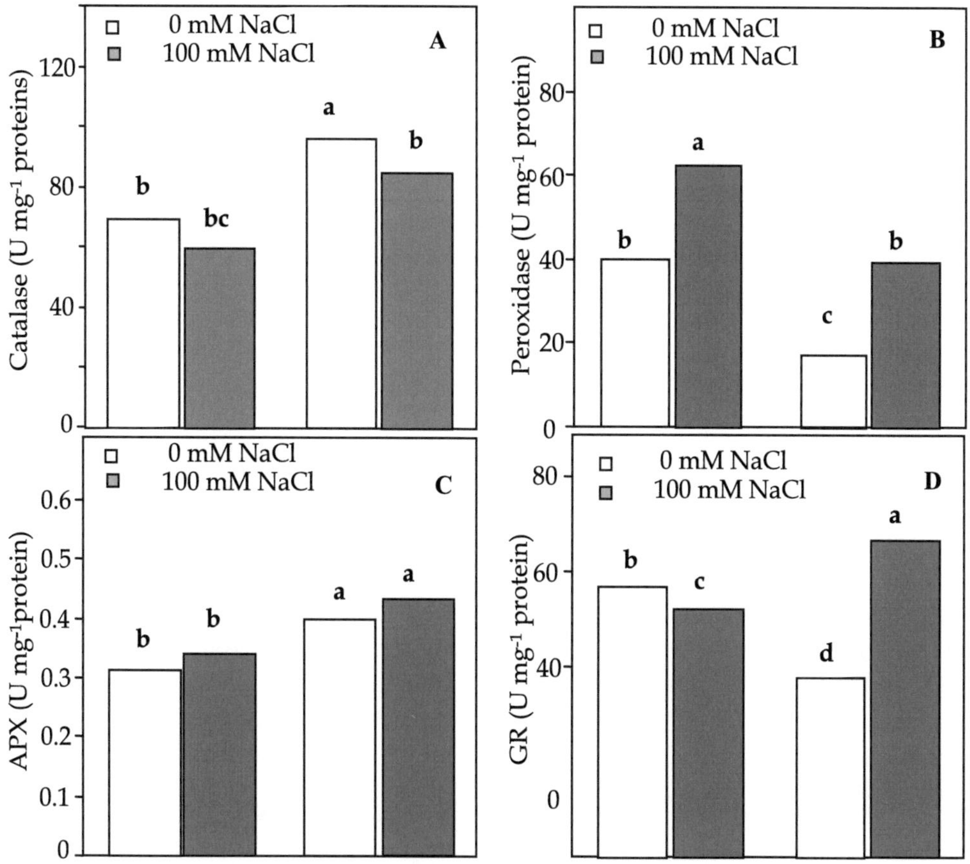

Fig. 4. Effect of NaCl on the total activity of catalase (CAT), dianisidine peroxidase (POD), ascorbate peroxidase (APX) and glutathione reductase (GR). Total CAT (**A**). POD (**B**). APX (**C**). GR (**D**). The results are the means of 5 replicates. Means followed by different letters are significantly different at $P \leq 0.05$ according to student's LSD test.

Salt stress is usually accompanied by the formation of various reactive oxygen species (ROS). These are toxic and may result in a variety of metabolic disturbances including photosynthetic component damages, protein and enzyme inactivation, and membrane permeabilization because of lipid peroxidation (Price and Hendry, 1991; Mittler, 2002; Meloni *et al.*, 2003). MDA level, H_2O_2 concentration and electrolyte leakage are considered as reliable indicators of the oxidative stress resulting from several abiotic constraints (Shulavev and Oliver, 2006). Our results showed variability of these indicators in leaves of the two ecotypes exposed to salt stress. The significant increases in MDA and H_2O_2 concentration and electrolyte leakage in Korbous leaves at 100 mM NaCl, compared to the lower MDA and H_2O_2 concentrations and electrolyte leakage in Tabarka leaves under the same conditions suggest that the latter is better protected against oxidative damage under salt stress. Another explanation would be that unlike Korbous, Tabarka leaves escape from the applied stress. For instance, the two accessions could differ in their capacity to compartmentalize salt in their leaf cells, or to maintain the leaf water equilibrium. This point deserves further investigation.

The maintaining of leaf growth in Tabarka plants might be the result of the presence of more efficient antioxidative systems. The activities of antioxidant enzymes are determinant in the functioning of these systems. Under NaCl stress, the up-regulation of antioxidative enzymes characterized the response of the more tolerant ecotype. These enzymes include SOD and POD. SOD plays a significant role in protecting living cells against the toxicity of active O_2 species due to their capacity to scavenge superoxide (Scandalios, 1993). The effects of salinity stress on SOD activity has been reported in a number of plant species, lines and varieties. Thus, considerable variations in the activity of SOD in response to salinity appeared at the inter-specific or intra-specific level (Ashraf, 2009). In the present study, the activity of this enzyme increased in the leaves of the more tolerant ecotype (Tabarka), in the presence of NaCl, but no significant change in SOD activity was detected in the more sensitive ecotype (Korbous). Such a differential response to salt stress was reported between cultivars of potato (Rahnama and Ebrahimzadeh, 2005), rice (Dionisio-Sese and Tobita 1998), millet (Srenivasulu *et al.*, 2000), wheat (Sairam *et al.*, 2002), tomato (Shalata and Tal, 1998), strawberry (Turhan and Gulen Erics, 2008), cotton (Gossett *et al.*,1994) and beet (Bor *et al.*, 2003). In plants, scavenging of superoxides occurs by differential regulation of SOD isoforms (Wang et al., 2004). Studies on regulation of individual isoforms of SOD are therefore important because their role during salt stress could be elucidated (Jithesh *et al.*, 2006). However, all the previous reports reflect the effect of NaCl stress on total SOD activity. Most of the studies in halophytes have revealed increased activities of Mn SOD and Fe SOD isoforms localized in mitochondria and chloroplasts, respectively (Parida *et al.*, 2004; Wang *et al.*, 2004). This is not surprising because the immediate targets of salt stress are the chloroplasts and mitochondria. Superoxide radicals and other ROS are formed in the chloroplasts during photosynthesis, and could be translocated in mitochondria during oxygen reduction. However, superoxide radicals could also leak towards the cytosol, resulting in the induction of the cytosolic Cu-Zn SOD. Accordingly, our study revealed an increased activity of Cu-Zn SOD isoform in Tabarka ecotype in the presence de NaCl and suggested that superoxides were mainly over-produced in cytosol under salinity. Similar results were reported for the halophyte *Bruguiera gymnorrhiza* in presence of NaCl, mannitol and abscissic acid (Takemura *et al.*, 2000).

H_2O_2, which is toxic and produced by the activity of SOD to prevent cellular damage, must be eliminated by conversion to H_2O in subsequent reactions. CAT and APX regulate H_2O_2 level in plants (Mittler, 2002). Our data showed that CAT activities in leaves of both ecotypes decreased at 100 mM NaCl while APX activity did not change. Contrarily to our results, Shalata and Tal (1998), Bor *et al.* (2003), Vaidyanathan *et al.* (2003) and Sekmen *et al.* (2007) observed salt-induced activities of CAT and APX in salt-tolerant tomato, sugar beet, rice and plantain, respectively. It could be that, in our case, little H_2O_2 is formed in the leaves of salt-treated *C. maritimum*, so that CAT and APX activations are not required to detoxify H_2O_2. Another explanation could be the involvement of POD in H_2O_2 elimination since POD are known to participate in H_2O_2 scavenging. Salt tolerant plants often exhibit a stimulation of POD activities under salt stress (Bor *et al.*, 2003, Meloni *et al.*, 2003; Ben Amor *et al.*, 2006). Here, we obtained the same results with *Crithmum maritimum* plantlets. PODs are therefore probably the most important enzymes in the detoxification of H_2O_2 in that species. However, our results show that POD activity is constitutively higher in the more sensitive ecotype. Some studies reported that high levels of POD activity in condition of saline stress are correlated with a more reduced growth of plants and it appeared that POD activity is involved in inhibiting the growth of aerial parts, rather than in protecting tissues against the accumulation of H_2O_2 (Lin and Kao, 2000). Such growth reduction has been attributed to the hemicellulose ferruloylation and the insolubilization of hydroxyproline rich glycoprotein (HPRG), those reactions being catalyzed by POD (Dionisio-Sese and Tobita, 1998). A similar role of POD has been proposed in cowpea plants subjected to NaCl (200 mM), where induction of POD activity in leaf tissues could be involved in stopping growth through the activation of tissue lignification (Cavalcanti *et al.*, 2004).

GR, the last enzyme of ascorbate-glutathione cycle, catalyzes NADPH-dependent reduction of oxidized glutathione. GR is important for plant protection against oxidative stress (Foyer and Halliwell, 1976). Although 100 mM NaCl treatment remarkably enhanced GR activity of Tabarka ecotype, slightly decreased that of Korbous ecotype. Because decreased GR activity enhances plant sensitivity to environmental stress (Aono *et al.*, 1995), the salt induced decrease in GR activity found in leaves of Korbous plants may explain the impairment of leaf growth and loss of membrane integrity. In the same way, oxidative stress in Tabarka appears to be prevented by the strong activity of SOD, POD and the ascorbate-glutathione cycle, as shown by the increased GR activities under 100 mM NaCl stress. The induction of GR by NaCl could increase the NADP / NADPH ratio, promoting the availability of NADP to accept electrons from the chain of electron transfer and limiting ROS levels in chloroplasts (Ben Amor *et al.*, 2006).

5. Conclusion

In the present study, plant growth and leaf antioxidant activities were analyzed in two Tunisian ecotypes of the halophyte *Crithmum maritimum* (Tabarka and Korbous, sampled from humid and semi-arid bioclimatic stages, respectively) under controlled salt stress.

Both ecotypes responded differently to low salt treatment. Korbous ecotype behaved like a facultative halophyte, with growth decreasing as NaCl levels rised in the medium. Conversely, biomass production in Tabarka ecotype was stimulated by 100 mM NaCl.

Our results showed also that intraspecific variability in *C. maritimum* salt tolerance coincides with differential antioxidant responses.

As a whole, our findings suggest the capacity of a plant ecotype or species to induce antioxidative enzymes in response to salt may greatly contribute to its ability to sustain growth.

6. Acknowledgements

This work was supported by the Tunisian Ministry of Higher Education and Scientific Research (LR10CBBC02), by the Tunisian-French "Comité Mixte de Coopération Universitaire" (CMCU) network # 08G0917.

7. References

Aono, M.; Saji, H.; Fujiyama, K.; Sugita, M.; Kondo, N. & Tanaka, K. (1995). Decrease in activity of glutathione reductase enhances paraquat sensitivity in transgenic *Nicotiana tabacum*. *Plant Physiology* 107, 645–648.

Ashraf, M. (2009). Biotechnological approach of improving plant salt tolerance using antioxidants as markers. *Biotechnolgy Advances* 27, 84-93.

Ashraf, M. & Foolad, M.R. (2007). Roles of glycine betaine and proline in improving plant abiotic stress resistance. *Environmental and Experimental Botany* 59, 206-216.

Ashraf, M. & Harris, P.J.C. (2004). Potential biochemical indicators of salinity tolerance in plants. *Plant Science* 166, 3-16.

Beauchamp, C. & Fridovich, I. (1971). Superoxide dismutase: improved assays and applicable to acrylamide gels. *Analytical Biochemistry* 44, 276-287.

Ben Amor, N.; Ben Hamed, K.; Debez, A.; Grignon, C. & Abdelly, C. (2005). Physiological and antioxidant responses of the perennial halophyte *Crithmum maritimum* to salinity. *Plant Science* 168, 889–899.

Ben Amor, N.; Jiménez, A.; Megdiche, W.; Lundqvist, M.; Sevilla, F. & Abdelly, C. (2006). Response of antioxidant systems to NaCl stress in the halophyte *Cakile maritima. Physiologia Plantarum* 126, 446-457.

Ben Hamed, K.; Castagna, A.; Elkahoui, S.; Ranieri, A. & Abdelly, C. (2007). Sea fennel (*Crithmum maritimum* L.) under salinity conditions: a comparison of leaf and root antioxidant responses. *Plant Growth Regulation* 53, 185-194.

Bor, M.; Özdemir, F. & Turkan, I. (2003). The effect of salt stress on lipid peroxidation and antioxidants in leaves of sugar beet *Beta vulgaris* L. and wild beet *Beta maritima* L.. *Plant Science* 164, 77-84.

Bradford, M. (1976). A rapid and sensitive method for the quantification of microgram quantities of protein utilizing the principle of protein-dye binding. *Analytical Biochemistry* 72, 248-254

Cavalcanti, F.R.; Oliveira, J.T.A.; Martins-Miranda, A.S.; Viegas, R.A. & Silveira, J.A.G. (2004). Superoxide dismutase, catalase and peroxidases activities do not confer protection against oxidative damage in salt-stressed cowpea leaves. *New Phytologist* 163, 563-571.

Chaparzadeh, N.; D'Amico, M.L.; Khavari-Nejad, R.A.; Izzo, R. & Navarri-Izzo, F. (2004). Antioxidative response of *Calendula officinalis* under salinity conditions. *Plant Physiology and Biochemistry* 42, 695-701.

Cheeseman, J.M.; Herenedeen, L.B.; Cheeseman, A.T. & Clough, B.F. (1997). Photosynthesis and photoprotection in mangroves under field conditions. *Plant & Cell Environment* 20, 579-588.

Cherrian, S.; Reddy, M.P. & Pandya, J.B., (1999). Studies on salt tolerance of *Avicennia marina* (Forsk.) Vierh.: effect of NaCl on growth, ion accumulation and enzyme activity. *Indian Journal of Plant Physiology* 4, 266-270.

Dat, J.; Vandenabeele, S.; Vranova, E.; Van Montagu, M.; Inzé, A.D. & Van Breusegem, F. (2000). Dual action of the active oxygen species during plant stress responses. *Cell Molecular and Life Science* 57, 779-795.

Di Baccio, D.; Navari-Izzo, F. & Izzo, R. (2004). Seawater irrigation: antioxidant defense responses in leaves and roots of a sunflower (*Helianthus annuus L.*) ecotype. *Journal of Plant Physiology* 161, 1359-1366.

Dionisio-Sese, M.L. Tobita, S. (1998). Antioxidant responses of rice seedlings to salinity stress. *Plant Science* 135, 1–9.

Foyer, C.H. & Halliwell, B. (1976). The presence of glutathione and glutathione reductase in chloroplasts: a proposed role in ascorbic acid metabolism. *Planta* 133, 21-25.

Foyer, C.H. & Noctor, G. (2000). Oxygen processing in photosynthesis: regulation and signalling. *New Phytologist* 146, 359-388.

Gossett, D.R.; Millhollon, E.P. & Lucas, M.C. (1994). Antioxidant response to NaCl stress in salt-tolerant and salt-sensitive cultivars of cotton. *Crop Science* 34, 706-714

Hagege, D.; Nouvelot, A. ; Boucaud, J. & Gaspar, T. (1990). Malondialdehyde titration with thiobarbiturate in plant extracts: avoidance of pigment interference. *Phytochemical Analysis* 1, 86-89.

Jithesh, M.N.; Prashanth, S.R.; Sivaprakash, K.R. & Parida, A.K. (2006). Antioxidative response mechanisms in halophytes: their role in stress defence. Journal of Genetics 85, 237-254.

Katsuhara, M.; Otsuka, T. & Ezaki, B. (2005). Salt stress-induced lipid peroxidation is reduced by glutathione S-transferase, but this reduction of lipid peroxides is not enough for a recovery of root growth in *Arabidopsis*. *Plant Science* 169, 369-373.

Ksouri, R.; Megdiche, W.; Debez, A.; Falleh, H.; Grignon, C. & Abdelly, C. (2007). Salinity effects on polyphenol content and antioxidant activities in leaves of the halophyte *Cakile maritima*. *Plant Physiology and Biochemistry* 45, 244-249.

Lin, C.C. & Kao, C.H. (2000). Effect of NaCl on H_2O_2 metabolism in rice leaves. *Plant Growth Regulation* 30, 151-155.

Lück, H. (1965). Catalase, In: *Methods of Enzymatic Analysis*, Bergmeyer, H.U. (Ed.), pp 885-888, Academic Press, New York.

Megdiche, W.; Ben Amor, N.; Debez, A.; Hessini, K.; Ksouri, R. & Abdelly, C. (2009). Physiological and biochemical traits involved in the genotypic variability to salt tolerance of Tunisian *Cakile maritima*. *African Journal of Ecology* 47, 774-783.

Meloni, D.A.; Oliva, M.A.; Martinez, C.A. & Cambraia, J. (2003). Photosynthesis and activity of superoxide dismutase, peroxidase and glutathione reductase in cotton under salt stress. *Environmental and Experimental Botany* 49, 69-76.

Mittler, R. (2002). Oxidative stress, antioxidant and stress tolerance. *Trends in Plant Science* 7, 405-410.

Nakano, Y. & Asada, K. (1981). Hydrogen peroxide is scavenged by ascorbate-specific peroxidase in spinach chloroplast. *Plant & Cell Physiology* 22, 867-880.

Navari-Izzo, F.; Quartacci, M.F.; Pinzino, C.; Dalla vecchia, F. & Sgherri, C.L.M. (1998). Thylakoid-bound and stromal antioxidative enzymes in wheat treated with excess copper. *Physiologia Plantarum* 104, 630-638.

Noctor, G. & Foyer, C.H. (1998). Ascorbate and glutathione: keeping active oxygen under control. *Annual Review of Plant Physiology and Plant Molecular Biology* 49, 623-647.

Parida, A.K. & Das, A.B. (2005). Salt tolerance and salinity effects on plants. *Ecotoxicology and Environmental Safety* 60, 324-349.

Parida, A.K.; Das, A.B. & Mohanty, P. (2004). Investigations on the antioxidative defence responses to NaCl stress in a mangrove, *Bruguiera parviflora*: differential regulations of isoforms of some antioxidative enzymes. *Plant Growth Regulation* 42, 213–226.

Price, A. and Hendry, G. (1991). Iron catalyzed oxygen radical formation and its possible contribution to drought damage in nine native grasses and three cereals. *Plant Cell Environment* 14, 477-481.

Rahnama, H. & Ebrahimzadeh, H. (2005). The effect of NaCl on antioxidant enzyme activity in potato seedlings. *Biologia Plantarum* 49, 93-99.

Ranieri, A. ; Petacco, F. ; Castagna, A. & Soldatini, G.F. (2000). Redox state and peroxidase system in sunflower plants exposed to ozone. *Plant Science* 159, 159-168.

Sairam, R.K.; Rao, K.V.; Srivastava, G.C. (2002). Differential response of wheat genotypes to long term salinity stress in relation to oxidative stress, antioxidant activity and osmolyte concentration. *Plant Science* 163, 1037–1046.

Scandalios, J.G. (1993). Oxygen stress and superoxide dismutase. *Plant Physiology* 101, 7-12.

Scebba, F.; Sebastiani, L. & Vitagliano, C. (1999). Protective enzymes against activated oxygen species in wheat (*Triticum aestivum* L.) seedlings: Responses to cold acclimation. *Journal of Plant Physiology* 155, 762-768.

Sekmen, A.H.; Türkan, I. & Takio, S. (2007). Differential responses of antioxidative enzymes and lipid peroxidation to salt stress in salt-tolerant *Plantago maritima* and salt-sensitive *Plantago media. Physiologia Plantarum* 131, 399-411.

Shalata, A. & Tal, M. (1998). The effect of salt stress on lipid peroxidation and antioxidant in the leaf of the cultivated tomato and its wild salt-tolerant relative *Lycopersicon pennellii. Physiologia Plantarum* 104, 169-174.

Shulavev, V. & Oliver, D.J. (2006). Metabolic and proteomic markers for oxidative stress. New tools for reactive oxygen species research. Plant Physiology 141, 367-372.

Sreenivasulu, N.; Grimm, B.; Wobus, U. & Weschke, W. (2000). Differential response of antioxidant compounds to salinity stress in salt tolerant and salt sensitive seedlings of foxtail millet (*Setaria italica*). *Physiologia Plantarum* 109, 435-442.

Taji, T.; Seki, M.; Sakuri, T.; Kobayashi, M.; Ishiyama, K.; Naruska, Y.; Zhu, J.K. & Shinosaki, K. (2004). Comparative genomics in salt tolerance between *Arabidopsis thaliana*-related halophyte salt cress using *Arabidopsis* microarray. *Plant Physiology* 135, 1697-1709.

Takemura, T.; Hanagata, N.; Sugihara, K.; Baba, S.; Karube, I. & Dubinsky, Z. (2000). Physiological and biochemical responses to salt stress in the mangrove, *Bruguiera gymnorrhiza. Aquatic Botany* 68, 15-28.

Turhan, E. & Gulen Erics, A. (2008). The activity of antioxidative enzymes in three strawberry cultivars related to salt-stress tolerance. *Acta Physiologia Plantarum* 30, 201-208.

Vaidyanathan, H.; Sivakumar, P.; Chakrabarty, R. & Thomas, G. (2003). Scavenging of reactive oxygen species in NaCl-stressed rice (*Oryza sativa* L.) - differential response in salt-tolerant and sensitive varieties. *Plant Science* 165, 1411-1418.

Wang, B.; Luttge, U. & Ratajczak, R. (2004). Specific regulation of SOD isoforms by NaCl and osmotic stress in leaves of the C_3 halophyte *Suaeda salsa* L.. *Journal Plant Physiology* 161, 285-293.

Zarrouk, M.; El Almi, H.; Ben Youssef, N.; Sleimi, N.; Ben Miled, D.; Smaoui, A. & Abdelly, C. (2004). Lipid composition of seeds of local halophyte species: *Cakile maritima, Zygophyllum album* and *Crithmum maritimum*. In: Cash Crop Halophytes: Recent Studies, Lieth, H. (Ed.), Kluwer Academic Publishers, pp 121-126.

9

Use of Finite Element Method to Determine the Influence of Land Vehicles Traffic on Artificial Soil Compaction

Biris Sorin-Stefan and Vladut Valentin
"Politehnica" University of Bucharest, INMA Bucharest
Romania

1. Introduction

It is known that the soil is an heterogenous, polyphasic, disperse, structured and porous material, whose non-linear behaviour at the interaction with the rolling bodies of land vehicles is difficult to be modeled mathematically, being necessary to use simplifications and idealizations. The compaction phenomenon of agricultural soil can be defined as an increase of its dry density and the closer packing of solid particles or reduction in porosity (McKyes, 1985), which can result from natural causes, including rainfall impact, soaking and internal water tension (Gill & Vandenberg, 1968; Arvidsson, 1997). Knowing the behaviour of soil under the action of such rolling bodies is an important element, because, by optimizing the pressures applied on the soil, the negative effects of soil compaction can be diminished, both at the surface and in depth.

Agricultural soil is a particullary important case, being the development environment of plant roots, for which artificial compaction is very important. Soil compaction reduces water permeability, favoring water runoff on soil surface, causing erosion and preventing proper restoration of soil moisture. Soil aeration is also reduced, with direct consequences on the metabolic processes occuring in plant roots. Another negative effect for agriculture is the increase caused by compaction of mechanical resistance, thus the development of roots being delayed. All those mentioned effects may reduce the quantity and quality of agricultural products (Koolen & Kuipers, 1983).

Between land vehicles traffic and soil compaction, between soil compaction and the parameters characterizing the development environment of plants and between this environment and the level of crop production, there are direct qualitative relations, of cause-effect type. This raises the following two problems: the soil may be too compact to be actually used for agricultural production, requiring the prevention and reduction of this phenomenon; on the other hand, soil may not be compact enough to be used for road construction, dams or building foundations. In this last case, the problem is to obtain the maximum degree of compaction, with minimum effort (Gill & Vandenberg, 1968).

Compaction phenomenon must be described mathematically, by equations that take into account the forces causing it. From this point of view, there are two categories of forces: in the first category enter the mechanical forces generated by the traffic of land vehicles. These forces are applied for short periods of time and can be measured relatively easily (Gill &

Vandenberg, 1968). Second category is that of forces generated by natural phenomena. For example, drying has as effect soil compaction. Forces in this category are acting for long periods of time, are difficult to define and hard to measure. Estimating the degree of compaction can only be made by using precise equations, which describe soil behaviour at compaction. In the attempt to developing these equations, outstanding scientific efforts have been made.

Soil degradation is a worldwide topic of actuality (Keller, 2004). The European Union noticed that it is necessary to protect soil and has identified soil compaction as one of the most important threats for soil, which leads to its degradation (COM, 2002). Soil compaction in an environmental problem (Pagliai et al., 2004). It is also one of the causes of erosion and land flooding (Horn et al., 1995; Soane & Ouwerkerk, 1995; Gieska et al., 2003). In addition, it contributes directly or indirectly to crop growth and pesticides leaching in ground water and to the emission of nitrogen oxides in the atmosphere (Lipiec & Stepniewski, 1995). Efforts to improve depth compaction by deep loosening are expensive and often ineffective. Therefore, soil compaction must be prevented. It is believed that the risk of unwanted changes in soil structure can be descreased by limiting the applied mechanical stresses (Dawidowski et al., 2001), and by limiting the precompression tensions.

The most important factors, which have a significant influence in the process of artificial compaction of agricultural soil, are: the type of soil, moisture content of the soil, intensity of external load, area of the contact surface between the soil and the tyre or track, shape of the contact surface, and the number of passes (Biriş, 2003).

The impact of land vehicles on soil properties can be simulated using compaction models (Bailey et al., 1986; O'Sullivan et al., 1999), which are an important instrument in the development strategy for soil compaction prevention. The traffic of land vehicles on soil in unappropriate moisture conditions is one of the most important causes of soil degradation by artificial compaction (Trautner, 2003). The value of soil moisture must be taken into account when referring to the traffic of a land vehicle which develops soil works, crops maintenance, harvest or transport. It is known that, for proper processing of agricultural soil, its normal moisture in the processed layer must be 18-22 % (Biris et al., 2007; Biris et al., 2009). Too dry soil compacts less, deforms less, but the resistance to processing is higher. For optimum moisture soils, the area of maximum pressures is found in the plough area, while soil deformations are resonable, the resistance at mechanical processing being minimum. In case of too wet soil, maximum stresses are concentrated near the rolling bodies, soil strains are very high, shear stresses are high, and soil processing under these conditions is not recommended (Arvidsson, 1997).

Soil moisture influences the deforming pattern and the size of soil strains at the contact with the rolling bodies of land vehicles (Bakker et al., 1995; Way, 1995). Strains are very low for dry soils, but, under these conditions, agricultural works of soil processing are nor recommended, because friction forces between soil particles and working bodies are very large, leading to large mechanical resistances, and abrasion wear phenomena for these working bodies are emphasized. Transport works, traffic, phytosanitary treatments applying, harvesting, etc. is recommended to be performed at smaller moisture values.

At the present moment, one of the most advanced methods for modelling the phenomenon of stresses propagation into soil is the Finite Element Method (FEM), which is a numerical method for obtaining approximate solutions of ordinary and partial differential equations of this distribution (Biris et al., 2003; Biris et al., 2007, Biris et al., 2009; Britto & Gunn, 1987; Gee-Clough et al., 1994; Van den Akker, 2004).

This paper presents a model for prediction of the stress state in agricultural soil below agricultural tyres in the driving direction and perpendicular to the driving direction, which are different from one another, using the Finite Element Method. It was created a general model of analysis using FEM, which allows the analysis of equivalent stress distribution and the total displacements distribution in the soil volume, making evident both of the conditions in which the soil compaction is favoured and of the study of graphic variation of equivalent stress and the study of shifting in the depth of soil volume.
This work has theoretical and obvious practical importance, because it allows that, by running the program of analysis through Finite Elements Method, to determine in a short period of time how are distributed in the soil the stresses that generate soil compaction at the interaction with the rolling bodies of land vehicles, for all traffic conditions and for any physical and mechanical properties of soil, no matter what value its moisture has.

2. Modelling the soil artificial compaction

2.1 Modelling the soil stress spreading

Because the agricultural soil is not a homogeneous, isotropic, and ideal elastic material, the mathematical modelling of stress propagation phenomenon is very difficult. Many mathematical models of stress propagation into the soil under different traction devices are based on the Boussinesq equations, which describe the stress distribution under a load point (Figure 1) acting on a homogeneous, isotropic, semi-infinite, and ideal elastic medium (Hammel, 1994). Frohlich developed equations to account for stress concentration around the application point of a concentrated load for the problem of the half-space medium subjected to a vertical load (Kolen, 1983).

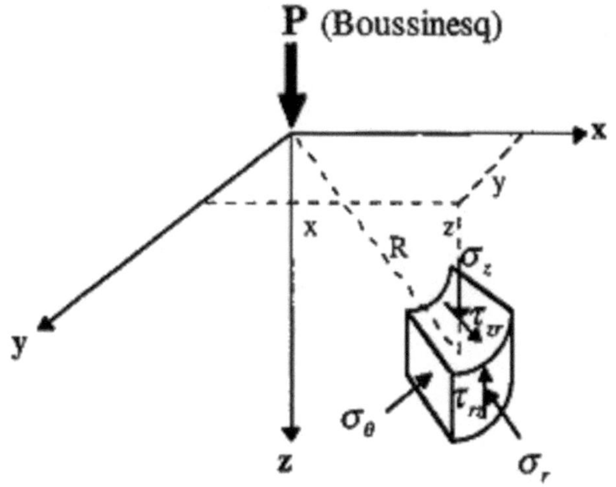

Fig. 1. Stress state produced by a concentrated vertical load (Upadhyaya, 1997)

Many models of dynamic soil behaviour are using elastic properties of soil, and when the soil is represented by a linear-elastic, homogenous, isotropic, weightless material, the elastic properties required to fully account for the behaviour of the material are: Young's modulus (E), shear modulus (G), and Poisson's ratio (υ).

For agricultural soils, the relationships between stresses and strains are measured on soil samples in the laboratory or directly in the field. The stress-strain relationships are given by constitutive equations (Gee-Clough, 1994).
The Drucker-Prager plasticity model can be used to simulate the behaviour of agricultural soil. The yield criteria can be defined as:

$$F = 3 \cdot \alpha \cdot \sigma_m + \bar{\sigma} - k = 0 \tag{1}$$

where a and k are material constants which are assumed unchanged during the analysis, σ_m is the mean stress and $\bar{\sigma}$ is the effective stress, a and k are functions of two material parameters (Φ and c) obtained from the experiments, where Φ is the angle of internal friction and c is the material cohesion strength.
Using this material model, the following considerations should be noted: strains are assumed to be small; problems with large displacements can be handled providing that the small strains assumption is still valid; the use of NR (Newton-Raphson) iterative method is recommended; material parameters Φ and c must be bounded in the following ranges: $90 \geq \Phi \geq 0$ and $c \geq 0$.
The required input parameters for the constitutive model of the agricultural soil of wet clay type are (Gee-Clough, 1994):

- Soil cohesion (c): 18.12 kPa
- Internal friction angle of soil (φ): 30°
- Soil density (γ_w): 1270 kg/m^3
- Poisson's ratio (υ_s): 0.329
- Young's modulus (E): 3000 kPa

The stress levels under a point load as shown in figure 1 are given in cylindrical coordinates as follows (Upadhyaya, 1997):

$$\sigma_z = \frac{3 \cdot P \cdot z^3}{2 \cdot \pi \cdot R^5} \tag{2}$$

$$\sigma_r = \frac{P \cdot z^3}{2 \cdot \pi} \cdot \left[\frac{3 \cdot z \cdot r^2}{R^5} - \frac{1 - 2 \cdot \upsilon}{R \cdot (R + z)} \right] \tag{3}$$

$$\sigma_\theta = \frac{P \cdot (1 - 2 \cdot \upsilon)}{2 \cdot \pi} \cdot \left[\frac{1}{R \cdot (R + z)} - \frac{z}{R^3} \right] \tag{4}$$

$$\tau_{rz} = \frac{3 \cdot P \cdot r \cdot z^2}{2 \cdot \pi \cdot R^5} \tag{5}$$

where P –is the point load, υ -Poisson's ratio, $\sigma_{z,r,\theta}$ –normal stress components, and τ_{rz} –shear stress component.
Figure 2 shows the stress state in soil, of an infinite cubic soil element, which can be written in a matrix, named the matrix of the stress tensors (Koolen, 1983; McKyes, 1985). Stresses acting on a soil element can be described by mechanical invariants, which are independent of the choice of reference axes. The invariants yields are (Keller, 2004):

$$I_1 = \sigma_1 + \sigma_2 + \sigma_3 = \sigma_x + \sigma_y + \sigma_z \tag{6}$$

$$I_2 = \sigma_x\sigma_y + \sigma_x\sigma_z + \sigma_y\sigma_z - \tau_{xy}^2 - \tau_{xz}^2 - \tau_{yz}^2 = \sigma_1\sigma_2 + \sigma_1\sigma_3 + \sigma_2\sigma_3 \tag{7}$$

$$I_3 = \sigma_x\sigma_y\sigma_z + 2\tau_{xy}\tau_{xz}\tau_{yz} - \sigma_x\tau_{yz}^2 - \sigma_y\tau_{xz}^2 - \sigma_{xy}^2 = \sigma_1\sigma_2\sigma_3 \tag{8}$$

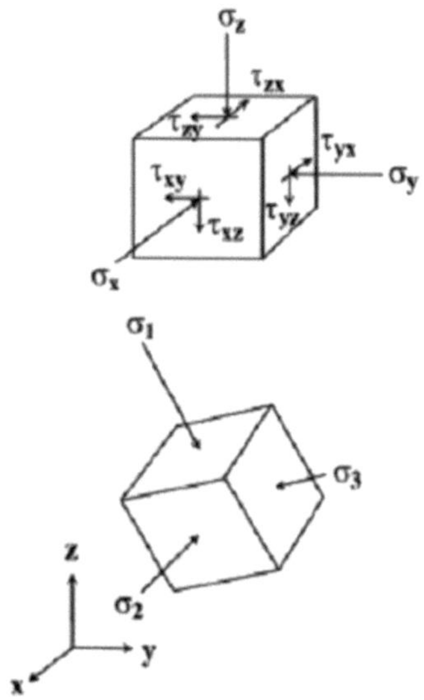

Fig. 2. Stress tensor components (Koolen, 1983)

It is useful to define the stress measures that are invariant. Such stress is the octahedral normal stress and the octahedral shear stress (Keller, 2004):

$$\sigma_{oct} = \frac{1}{3}(\sigma_1 + \sigma_2 + \sigma_3) = \frac{1}{3}I_1 \tag{9}$$

$$\tau_{oct} = \frac{1}{3}\sqrt{(\sigma_1 - \sigma_2)^2 + (\sigma_2 - \sigma_3)^2 + (\sigma_1 - \sigma_3)^2} = \sqrt{\frac{2}{9}\cdot(I_1^2 - 3I_2)} \tag{10}$$

The critical state soil mechanics terminology uses the mean normal stress p and the deviator stress q. If $p=\sigma_{oct}$ (Eq. 9), q is given as (Keller, 2004):

$$q = \frac{1}{\sqrt{2}}\sqrt{(\sigma_1 - \sigma_2)^2 + (\sigma_2 - \sigma_3)^2 + (\sigma_1 - \sigma_3)^2} = \sqrt{(I_1^2 - 3I_2)} \tag{11}$$

2.2 Utilization of incremental methods for studying the soil non-linear behaviour

The incremental methods are used to deal with material and geometrically non-linear problems. The basis of the incremental procedure is the subdivision of the load into many small increments. Each increment is treated in a piecemeal linear behaviour with the stiffness matrix evaluated at the start of the increment. The tangent stiffness, E_t (Figure 3) for

each element is calculated from the stress-strain curves according to the current stress level of that element. The Finite Element Method (FEM) is proving to be very promising into modelling this propagation phenomenon. In a FEM calculation when the coordinates are continually updated, the strain increment $d\epsilon$, has the mean of a ratio between an incremental length and the current length.
The relationship between ε and ϵ has the form (Gee-Clough, 1994):

$$\varepsilon = 1 - e^{-\epsilon} \tag{12}$$

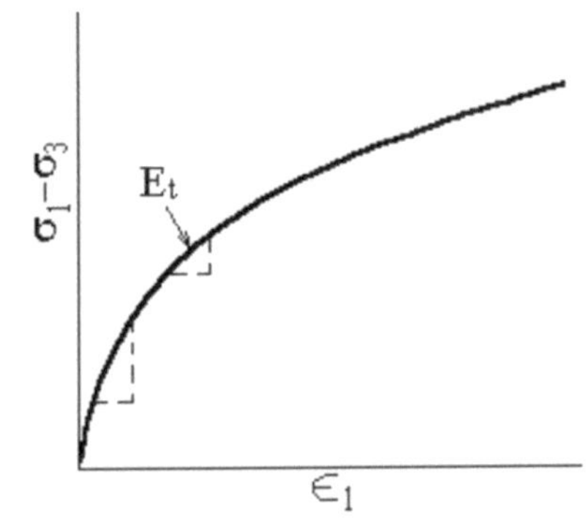

Fig. 3. Stress-strain curve for agricultural soil

According to the relationship between ε and ϵ the following revised stress-strain and tangent stiffness formulae were derived and used in the calculation (Gee-Clough, 1994):

$$\sigma_1 - \sigma_3 = \frac{1 - e^{-\epsilon_1}}{a + b \cdot (1 - e^{-\epsilon_1})} \tag{13}$$

$$E_t = \frac{1}{a} \cdot [1 - b \cdot (\sigma_1 - \sigma_3)] \cdot [1 - (b + a) \cdot (\sigma_1 - \sigma_3)] \tag{14}$$

For saturated soil under an un-drained condition, the volume change is generally considered to be negligible. But for FEM calculation purposes, it is common to assume a constant Poisson's ratio slightly less than 0,5 (Gee-Clough, 1994). In terms of the concept of the incremental method, for a soil with nonlinear properties when increments are very small, Hooke's law in which the Young's modulus, E_t, and Poisson's ratio, υ_t, are variables (depending on current stress and strain values) is valid. On this basis, for a plane strain problem, a formula for the volume modulus, K_t, can be derived:

$$K_t = \frac{d(\sigma_x + \sigma_y)}{d(\varepsilon_x + \varepsilon_y)} = \frac{E_t}{(1 - \nu_t - 2 \cdot \upsilon_t^2)} \tag{15}$$

where: ε_x, ε_y are strains in x and y directions; σ_x, σ_y are stresses in x and y directions.
If υ_t is constant, as E_t decreases (soil failure), K_t also decreases. This means that soil volume changes can be large. Assuming K_t is constant, and the initial values of E_t and υ_t are E_0 and υ_0, respectively, then the Poisson's ratio formula can be derived as in eq. (15) in which a

maximum υ_t and a minimum E_t may be specified to avoid the calculation problem (Gee-Clough, 1994):

$$\upsilon_t = 0{,}25 \cdot (\sqrt{9 - \frac{8 \cdot E_t}{E_0} \cdot (1 - \upsilon_0 - 2 \cdot \upsilon_0^2)} - 1) \quad (16)$$

2.3 Tyre deformation at running path interaction

Under the action of an external load (weight per wheel), a tire deforms as it is shown in figure 4. According to Hedekel's equation, tire deformation is given by the following relationship:

$$f = \frac{F}{2 \cdot \pi \cdot p_i \cdot \sqrt{R \cdot r}} \quad [\text{mm}] \quad (17)$$

where: F – vertical load acting on the wheel, [N]; p_i – air pressure inside the tire, [MPa]; R – free radius of the wheel, [mm]; r – radius of tire running path in cross section, [mm].

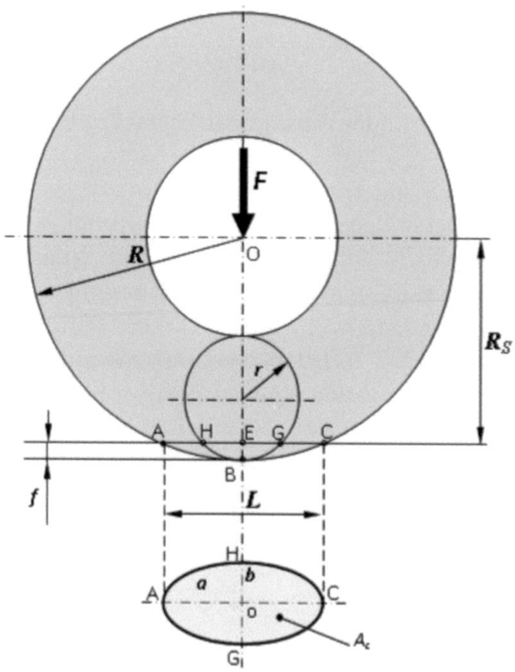

Fig. 4. Tire deformation under the action of an external load

Static tire radius is given by:

$$R_s = R - f \quad [\text{mm}] \quad (18)$$

and the length of the contact chord is:

$$L = 2 \cdot \sqrt{R^2 - R_s^2} \quad [\text{mm}] \quad (19)$$

Figure 5 shows the influence of tire pressure on the dimensional characteristics of the wheel (figure 4), respectively tire deformation (Eq. 17), static radius R_s (Eq. 18) and the length of contact chord L (Eq. 19), for the rear wheel.

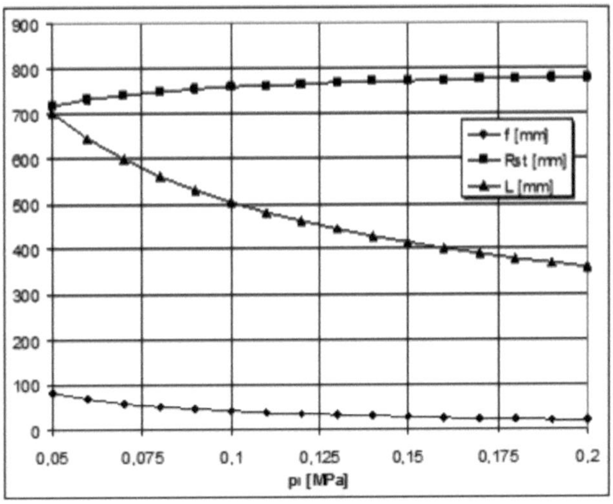

Fig. 5. Influence of tire pressure on the dimensional characteristics of the wheels

2.4 Contact surface between tyre and running path

The calculation of the contact surface between tyre and running path is rather complicate due to this interface complexity which depends on soil varaiable parameters and tyre parameters. It is usually necessary to make simplifying assumption of the true contact area. Surface of contact can be approximated by a circle, in case of rigid running paths and tyres with high pressure. In case of low inflation pressures, the more elliptical the contact area becomes. Low tire inflation pressure and high axle loads lead to high tire deflection and the contact area is no longer elliptical, but rectangular with curved ends (Upadhyaya & Wulfsohn, 1990). Figure 6 shows the theoretical shape of contact area between the soil and agricultural tyres. The pressure distribution along the width of tyre is described by a decay function (Keller, 2004):

$$p(y) = C \cdot \left[\frac{w(x)}{2} - y \right] \cdot e^{-\delta \cdot \left[\frac{w(x)}{2} - y \right]} ; 0 \le y \le \frac{w(x)}{2} \tag{20}$$

and the pressure distribution in the driving direction is described by a power-law function:

$$p(x) = p_{x=0,y} \cdot \{1 - [\frac{x}{\frac{l(y)}{2}}]^{\alpha}\} ; 0 \le x \le \frac{l(y)}{2} \tag{21}$$

where C, δ and a are parameters, $w(x)$ is the width of contact between the tyre and soil, $p_{x=0,y}$ is the pressure under the tyre centre and $l(y)$ is the length of contact between the tyre and soil.

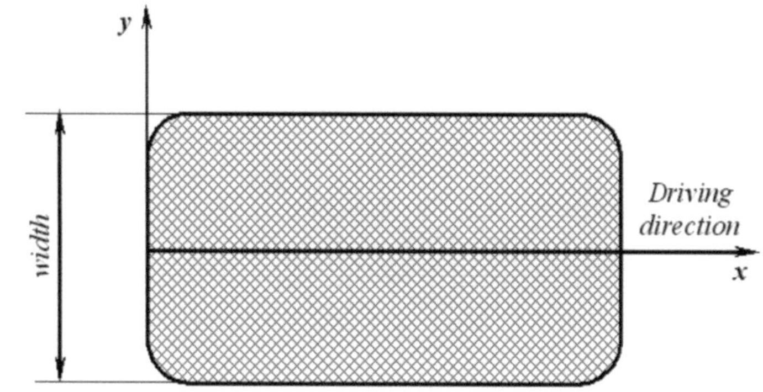

Fig. 6. Shape of the contact surface between the soil and the tyre

Figure 7 shows the vertical load distribution in the contact area beneath agricultural tyres for three considerations: the real distribution with measured values (left), a model with uniform load distribution (centre), and a better model with irregular load distribution (right).

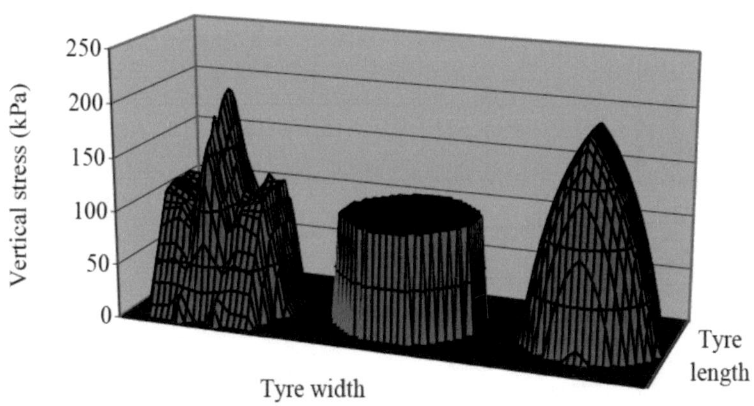

Fig. 7. Distribution of the vertical load in the contact area (Keller, 2004)

Equation (20) can describe different cases of pressure distribution, e.g. maximum pressure under the tyre centre or pressure under the tyre edge. The parameters C, δ and a are calculated from wheel load, tyre inflation pressure, recommended tyre inflation pressure at given wheel load, tyre width and overall diameter of the unloaded tyre. All these parameters are easy to measure or readily available from e.g. tyre catalogues.

2.5 FEM model for studying the distribution of soil strains and stresses

For the modelling using the finite element method it was considered a soil volume with the depth of 1 meter, the width of 3 meter and length of 4 meter (Figure 8) under the action of different tractors and harvester-threshers (Table 1). The structural nonlinear analysis was made on the ideal model, the soil being considered a homogeneous and isotropic material. It was used the COSMOS/M 2.95 Programme for FEM modelling.

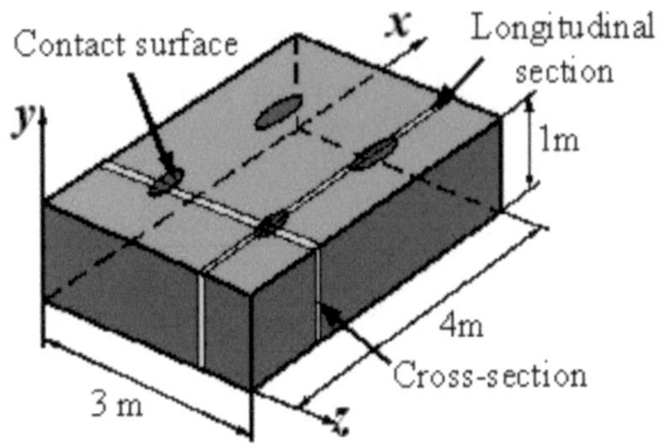

Fig. 8. Analyzed soil volume

Applicant	Soil interaction elements	Gauge [mm]	Mass (total / deck) [kg]		The active width for load [mm]	Pressure on the soil [kPa]
Romanian tractor U-445 (45 HP)	The front wheels	1500	1920	720	170	82.5
	The back wheels			1200	315	44.2
Romanian tractor U-650 (65 HP)	The front wheels	1600	3380	1170	180	110
	The back wheels			2210	367	57.3
Romanian Caterpillar SM-445 (45 HP)	Track	1300	2600		360	31
Harvester-thresher NH-TX66	The front wheels	2950	14000	11000	615	106.5
	The back wheels			3000	408	103.2
Romanian harvester-thresher Sema-140	The front wheels	2850	11033	9033	587	115
	The back wheels			2000	317.5	148.5

Table 1. The principal characteristics of the rolling devices used in modelling

The static radius of tire and length of contact chord are computed, thus providing the contact area between the wheel and the soil and the value of the pressure applied by the rolling body on the soil.

Figures 9, 10 and 11 show the results of FEM analysis in cross-section and figures 12, 13, 14 and 15 the results of FEM analysis in longitudinal section for two 45 HP tractors with tires and with caterpillar (U-445 and SM-445), respectively for two harvester-threshers (New Holland TX-66 and SEMA-140). These results are: the stresses distribution in soil and the graphical variation of stresses along the vertical-axial direction and along to the longitudinal direction.

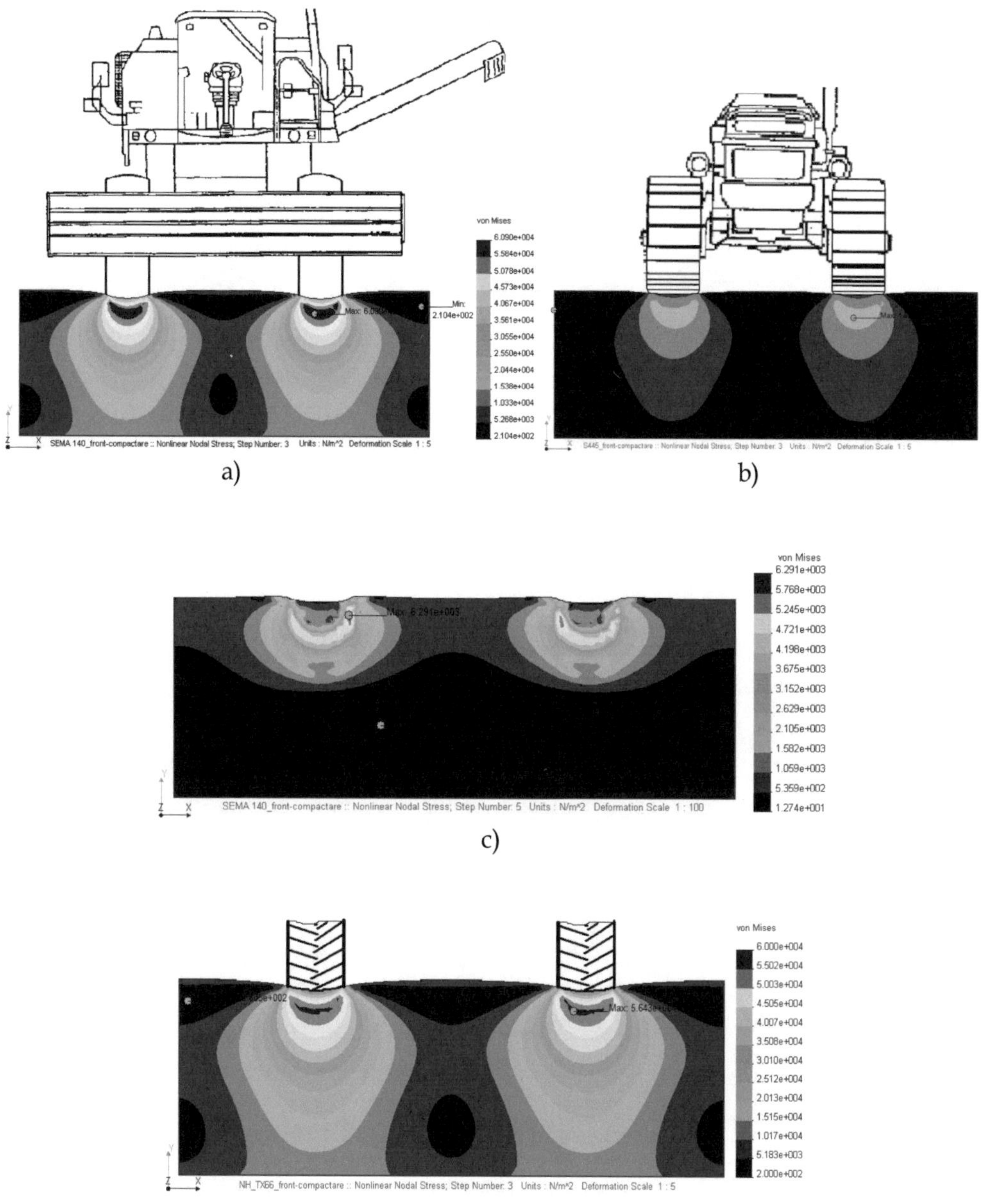

Fig. 9. Stresses distribution in cross-section for: a) SEMA 140 harvester-thresher, b)SM-445 caterpillar tractor, c) SEMA 140 harvester-thresher after the first transit, d) New Holland TX-66 harvester-thresher (Units: Pa)

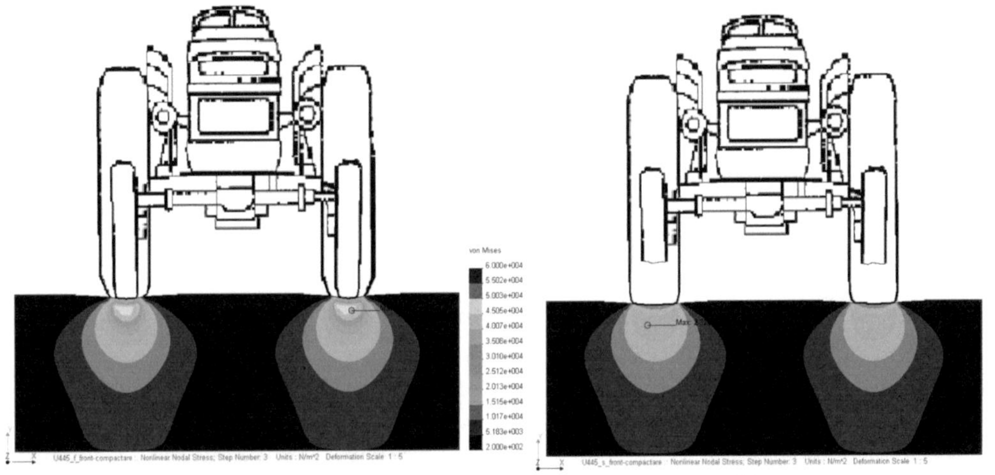

a) b)

Fig. 10. Stresses distribution in cross-section for: a) front wheels of U-445 tractor (U-445_f) (Units: Pa), b) back wheels of U-445 tractor (U-445_b) (Units: Pa)

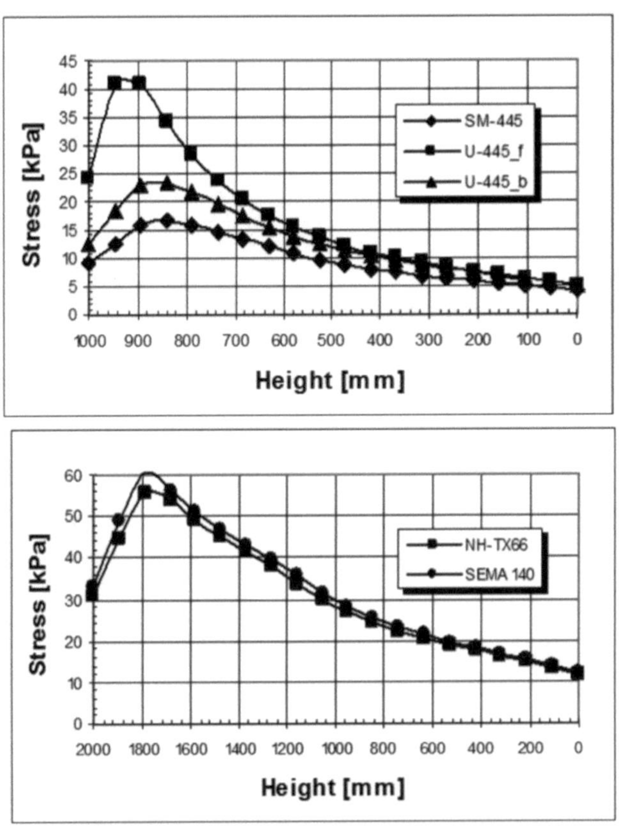

Fig. 11. Graphical distribution along the axial-vertical direction

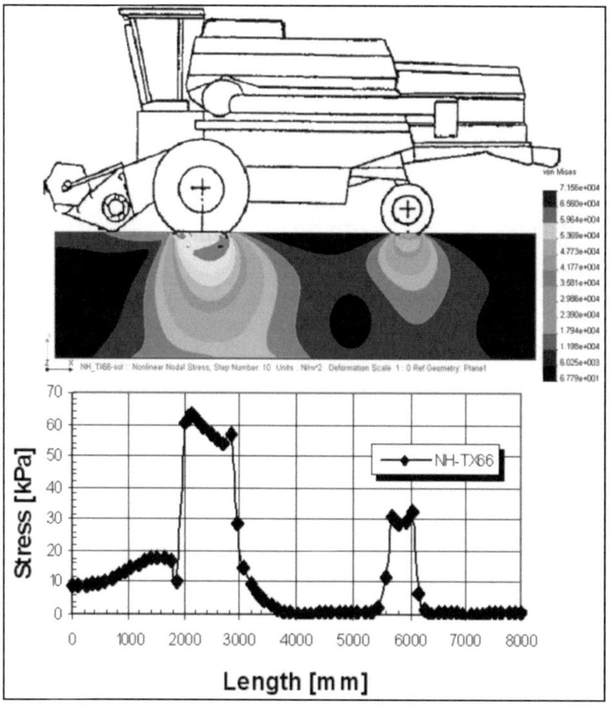

Fig. 12. Stresses distribution and graphical variation along the longitudinal direction to the top layer of the soil in longitudinal section for New Holland TX-66 harvester-thresher

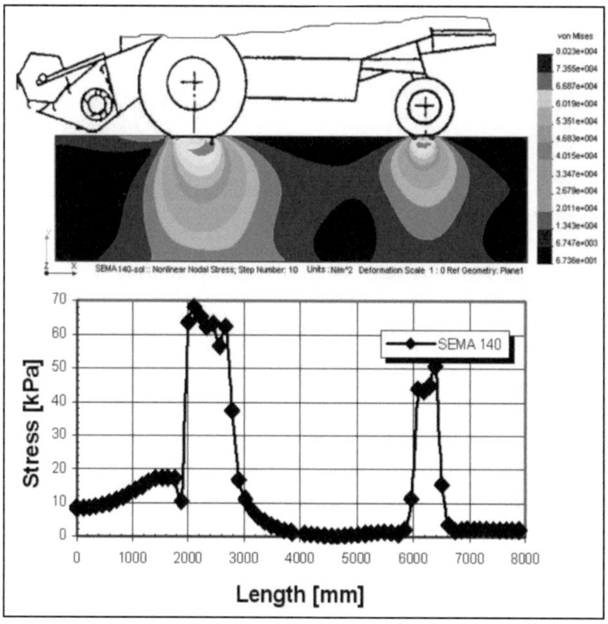

Fig. 13. Stresses distribution and graphical variation along the longitudinal direction to the top layer of the soil in longitudinal section for SEMA 140 harvester-thresher

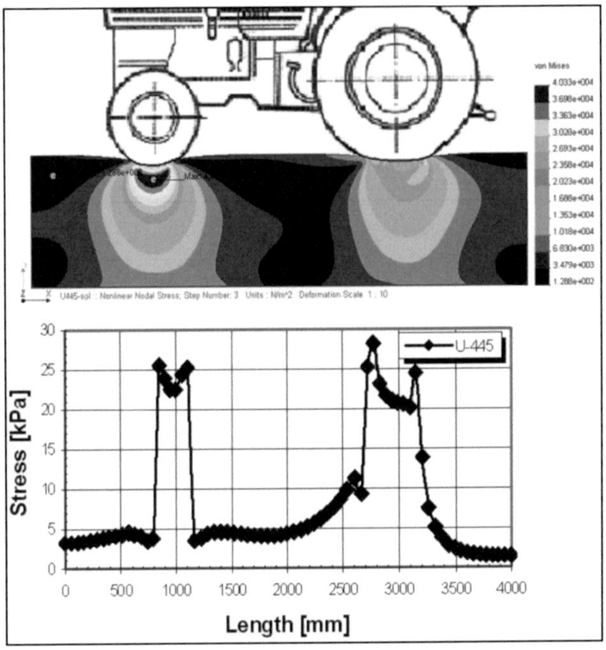

Fig. 14. Stresses distribution and graphical variation along the longitudinal direction to the top layer of the soil in longitudinal section for U-445 tractor, b) SM-445 caterpillar tractor

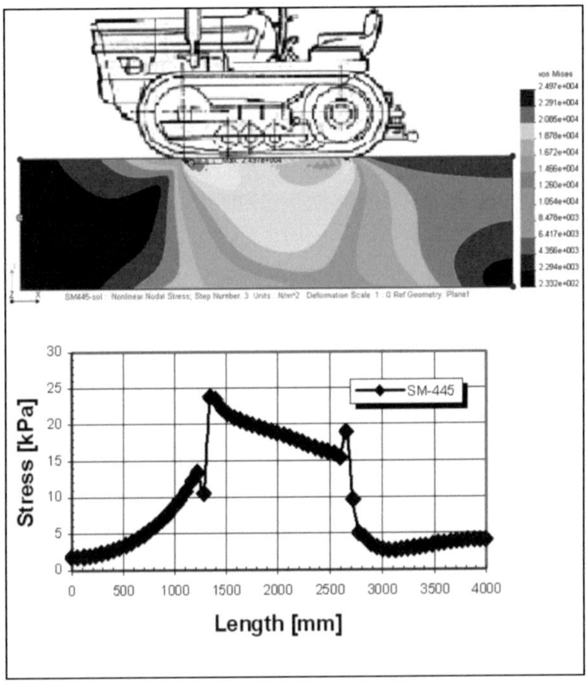

Fig. 15. Stresses distribution and graphical variation along the longitudinal direction to the top layer of the soil in longitudinal section for SM-445 caterpillar tractor

Figure 16 shows the results of FEM analysis in cross-section for a "1/2 symmetrical model" which consists in equivalent stresses distribution in agricultural soil under the action of a uniform load in the case of back wheel of U-650 tractor. Figure 17 shows the distribution of equivalent stresses in agricultural soil in cross-section for the same "1/2 symmetrical model" under the action of an un-uniform load (Decay function, Eq. 20) in the case of back wheel of U-650 tractor.

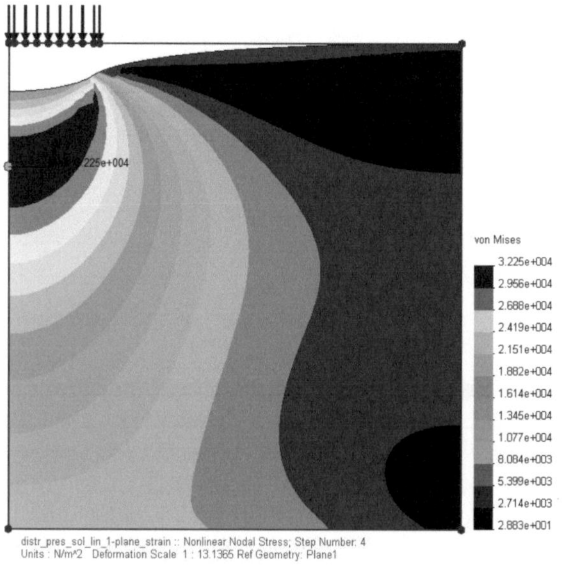

Fig. 16. Distribution of equivalent stresses for uniform load in the case of back wheel of U-650 tractor (Units: Pa)

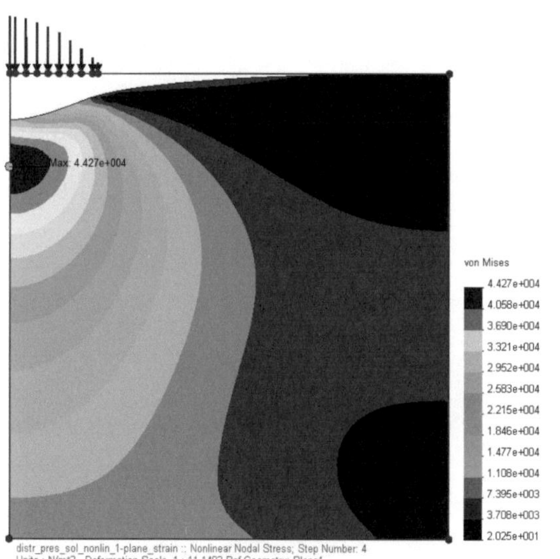

Fig. 17. Distribution of equivalent stresses for un-linear load in the case of back wheel of U-650 tractor (Units: Pa)

Figure 18 shows the graphical variation of equivalent stresses along the vertical-axial direction for the two cases of loading.

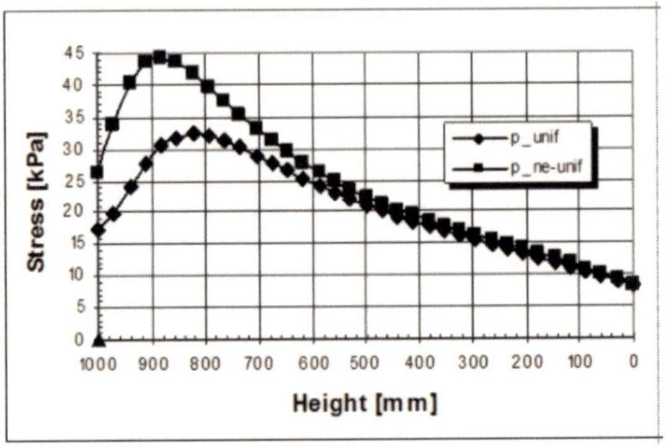

(p_unif: uniform load; p_ne-unif: un-linear load)

Fig. 18. Graphical variation of stresses along the vertical-axial direction

2.6 Laboratory tests for studying the stress and strain distribution into the soil

In order to check the model elaborated using FEM, laboratory tests were taken using a data acquisition system (Figure 19). The system was connected to Flexi Force Tekscan W-B201-L force sensors (Figure 20), vertically mounted in the soil, at 10 cm distance, in a metallic container with 1x1x1 m dimensions (Figure 21). The load on the wheel (of different types) in statically state was applied using the Hydropulse equipment (Figure 22).

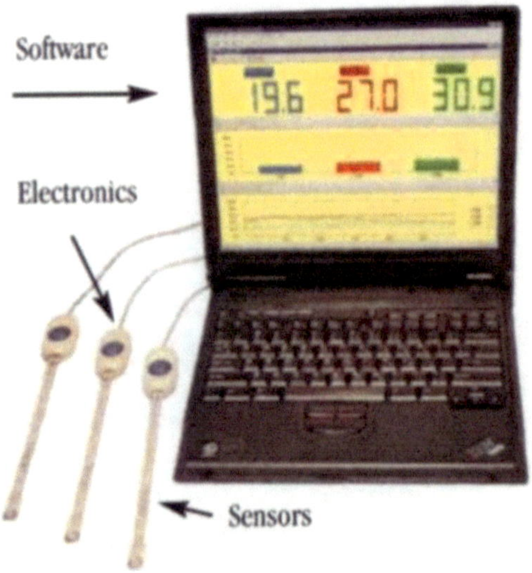

Fig. 19. Data acquisition system

Fig. 20. Flexi Force Tekscan W-B201-L force sensors

Fig. 21. Metallic container

Fig. 22. Test stand in Hydropulse laboratory

2.7 Comparative analysis of results measured and calculated, using FEM model

In figures 23 and 24 are comparatively presented the variation curves of the equivalent stresses with the points obtained by FEM calculus and by experimental tests for different depths along the tire's vertical axis in the case of the U-445 tractor.

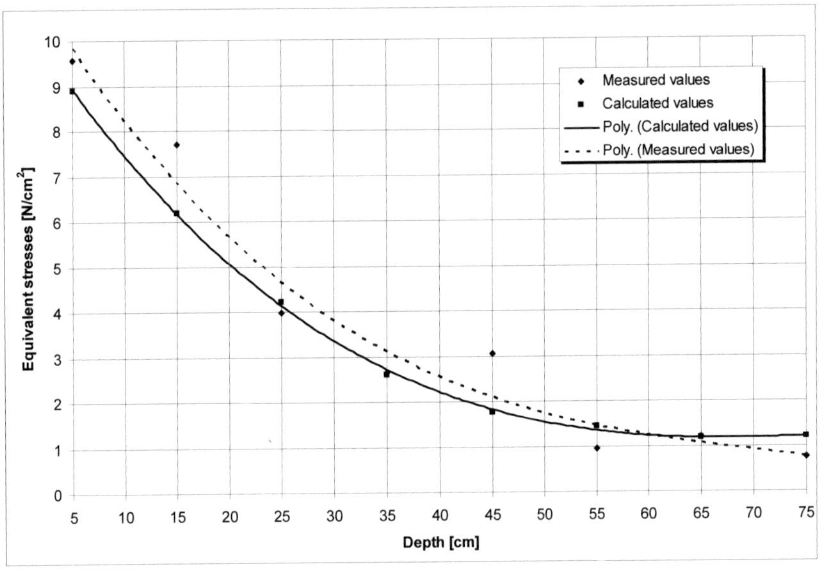

Fig. 23. Equivalent stresses calculated and measured for the front deck of U-445 tractor

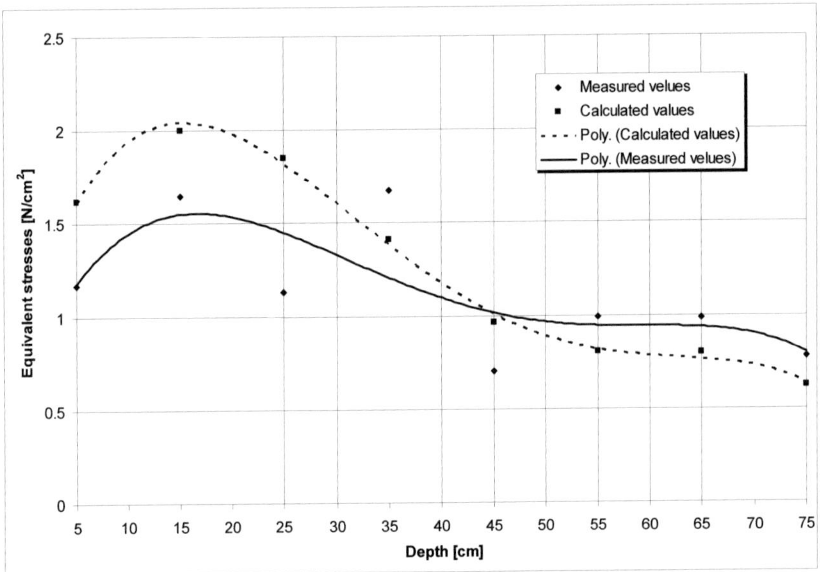

Fig. 24. Equivalent stresses calculated and measured for the back deck of U-445 tractor

3. Conclusions

The Finite Element Method is at the present the most advanced mathematical tool which can be used for the study of agricultural soil artificial compaction process. For mathematical modelling the soil is considered as a homogeneous and isotropic material, and the Drucker-Prager plasticity model can be used to simulate the behaviour of agricultural soil.

This study shows that from these analysed tractors and harvester-threshers, the highest artificial compaction of soil was caused by the front wheels of SEMA-140 harvester-thresher (see Figure 13), when the equivalent maximum stress in soil is approx. 60 kPa, and in the case of the front wheels of NH TX-66 harvester-thresher (Fig. 12), when the maximum equivalent stress is higher then 55 kPa. In these cases it is recommended to extend the contact area between the wheel and the soil.

In the case of the front wheels of U-445 tractor (see figure 10.a), the equivalent maximum stress in soil is approx. 42 kPa (Fig. 11). We can see that the equivalent maximum stress in soil in the case of analyzed caterpillar tractor (SM-445) is less than 20 kPa (Figure 11 and 15). This study represents a supplementary argument for using the caterpillar for the reduction of artificial soil compaction. The present researches are directed to using the rubber caterpillar, and also to using the reduce-pressure tyres with largest contact area with the soil.

We can see from the figures 16, 17, and 18, that the distribution of equivalent stresses in soil volume is strongly influenced by the loading distribution in the contact area.

As we can see in figure 23 and 24, between the calculated and measured results is a difference of 8% for the front wheel and 12% for the back wheel of U-650 tractor. There is a true development possibility of the pseudo-analytical procedures for the modelling of the stress propagation in agricultural soil, based on the work of Boussinesq, Fröhlich and Söhne, using the numerical calculus procedures, respectively the finite element method.

4. Acknowledgment

This work was supported by POSDRU based on POSDRU/89/1.5/S/62557 financing program.

5. References

Arvidsson, J. (1997). Soil compaction in agriculture - from soil stress to plant stress, Doctoral Thesis, *Agraria 41*, Swedish University of Agricultural Sciences, Uppsala, Sweden.

Bailey, A.C., Johnson, C.E. & Schafer R.L. (1986). A model for agricultural soil compaction, *Journal of Agricultural Engineering Research*, 33, 257-262.

Bakker, D.M., Harris, H.D. & Wong, K.Y. (1995). Measurement of stress path under agricultural vehicles and their interpretation in critical state space, *Journal of Agricultural Engineering*, 61, 247-260.

Biriş, S.Şt., Prunău, M., Vlăduţ, V. & Bungescu, S.T. (2003). Study of agricultural soil artificial compaction using the finite element method, *Scientific Bulletin of the „Politehnica" University of Timişoara, Transactions on Mechanics*, Tom 48(62), Fascicola 2, pg. 41-50.

Biriş, S.Şt., Maican, E., Faur, N., Vlăduţ, V. & Bungescu, S. (2007). FEM model for appreciation of soil compaction under the action of tractors and agricultural machines, *Proceeding of the 35th International Symposium „Actual Tasks on Agricultural Engineering"*, Croatia, Opatija, 19-23 february, pg. 271-279.

Biriş, S.Şt., Vlăduţ, V., Ungureanu, N., Paraschiv G. & Voicu Gh. (2009). Development and experimental testing of a FEM model for the stress distribution analysis in agricultural soil due to artificial compaction, *Agricultural Conspectus Scientificus*, Vol. 74, No. 1, pg. 21-29.

Britto, A.M. & Gunn, M.J. (1987). *Critical State Soil Mechanics via Finite Elements*, Ellis Horwood, Chichester, 488 pg.

COM (Commission of the European Communities), (2002). *Communication from the Commission to the Council, the European Parliament, the Economic and Social Committee and the Committee of the Regions – Towards a Thematic Strategy for Soil Protection*, Brussels, 35 pg.

Dawidowski, J.B., Morrison, J.E. & Snieg, M. (2001), Measurement of soil layer strength with plate sinkage and uniaxial confined methods. *Transactions of the ASAE*, 44, 1059-1064.

Gee-Clough, D., Wang, J. & Kanok-Nukulchai, W. (1994). Deformation and Failure in Wet Clay Soil: Part 3, Finite Element Analysis of Cutting of Wet Clay by Tines, *J. of Agric. Engng. Res.* 58, pg. 121-131.

Gieska, M., Van der Ploeg, R.R., Schweigert, P. & Pinter, N. (2003). Physikalische Bodendegradierung in der Hildesheimer Börde und das Bundes-Bodenschutzgesetz. *Berichte über Landwirtschaft*, 81, 485-511.

Gill, W.R. & Vandenberg, G.E. (1968). *Soil dynamics in tillage and traction*, U.S.A. Department of Agriculture, Handbook 316, USA, Washington D.C.

Hammel, K. (1994). Soil stress distribution under lugged tires, *Soil and Tillage Research*, 32, pg. 163-181.

Horn, R., Domzal, H., Slowinska-Jurkiewicz, A. & Van Ouwerkerk, C. (1995). Soil compaction process and their effects on the structure of arable soils and the environment, *Soil and Tillage Research*, 35, 23-36.

Keller, T. (2004). Soil compaction and soil tillage - studies in agricultural soil mechanics, Doctoral Thesis. *Agraria 489*, Swedish University of Agricultural Sciences, Uppsala, Sweden.

Koolen, A.J. & Kuipers, H. (1983). *Agricultural soil mechanics*, Advanced Series in Agricultural Sciences, Vol. 13. Springer, Heidelberg, 241 pg.

Lipiec, J. & Stepniewski, W. (1995). Effects of soil compaction ande tillage systems on uptake and losses of nutrients. *Soil and Tillage Research*, 35, 37-52.

McKyes, E. (1985). *Soil cutting and tillage*, Elsevier, Amsterdam-Oxford-New York-Tokyo.

O'Sullivan, M.F., Henshall, J.K. & Dickson J.W. (1999). A simplified method for estimating soil compaction, *Soil and Tillage Research*, 49, 325-335.

Pagliai, M., Vignozzi, N. & Pellegrini, S. (2004). Soil structure and the effect of management practices. *Soil and Tillage Research*.

Soane, B.D. & Van Ouwerkerk, C. (1995). Implication of soil compaction in crop production for the quality of the environment. *Soil and Tillage Research*, 35, 5-22.

Trautner, A. (2003). On Soil Behaviour During Field Traffic, Doctoral Thesis, *Agraria 372*, Swedish University of Agricultural Sciences, Uppsala, Sweden.

Upadhyaya, S.K. & Rosa, U.A. (1997). Prediction of traction and soil compaction, *Proceeding of 3rd International Conference on Soil Dynamics*. Tiberias, Israel, pg. 19-58.

Van den Akker, J.J.H. (2004). SOCOMO: a soil compaction model to calculate soil stresses and the subsoil carrying capacity, *Soil and Tillage Research*, 79, pg. 113-127.

Way, T.R., Bailey, A.C., Raper, R.L. & Burt, E.C. (1995). Tire lug height effect on soil stresses and bulk density, *Transactions of the American Society of Agricultural Engineers*, 38, 669-674.

10

The Influence of Water Stress on Yield and Related Characteristics in Inbred Quality Protein Maize Lines and Their Hybrid Progeny

Dagne Wegary[1], Maryke Labuschagne[2] and Bindiganavile Vivek[3]
[1]*Melkassa Agricultural Research Center, Nazareth,*
[2]*University of the Free State, Bloemfontein,*
[3]*CIMMYT – India, New Delhi*
[1]*Ethiopia*
[2]*South Africa*
[3]*India*

1. Introduction

Water stress is one of the factors most frequently limiting maize production, food security, and economic growth in sub-Saharan Africa. The unprecedented combination of climatic risk, declining soil fertility, the need to expand food production into more marginal areas as population pressure increases, high input costs, extreme poverty, and unavailability of credit systems, have resulted in small holder farmers in southern and eastern Africa producing maize in extremely low-input/low risk systems (Banziger and Diallo, 2004). As a consequence, crop yields are falling to very low levels and food insecurity is widespread amongst agricultural communities (Kamara et al., 2004). The development of maize germplasm able to tolerate water stress is crucial if the productivity of maize based farming systems is to be sustained or increased (Betran et al., 2003).

Maize genotypes perform differently under water stress conditions due to the existence of genetic variability for tolerance to stress (Bolanos and Edmeades, 1993; Lafitte and Edmeades, 1994; Banziger et al., 2000; 2006; Diallo et al., 2004). Betran et al. (2003) observed hybrids performing well under stress and suggested the possibility of combining stress tolerance and yield potential in tropical maize hybrids. Tolerance of maize to water stress is partly related to the development of the root system, which in turn influences water and nutrient uptake by crop plants (Moll et al., 1982; Kamara et al., 2004). In general, however, the amount of grain yields recorded from maize genotypes fall with the severity of water stress (Betran et al., 2003). Breeding strategies to develop stress tolerant maize inbred lines include screening and selection of inbreds under managed stress conditions, multi-location testing of progeny in a representative sample of the target environments, and selection under high plant populations (Beck et al., 1997). Additional information from adaptive secondary traits (ears per plant, anthesis-silking interval and leaf senescence) that show differential expression between optimal and stress conditions is genetically variable and is correlated with grain yield and is commonly used to increase selection efficiency (Bolanos and Edmeades, 1993; 1996; Banziger and Lafitte, 1997). When genetic variance and heritability for grain yield declines under water stress (Blum, 1988; Bolanos and Edmeades, 1996), variances and heritability of anthesis-silking interval and ears per plant remain stable across water stress levels or may even increase (Bolanos and Edmeades, 1996). Anthesis-silking interval and ears per plant have, therefore, been used in selection indices to increase selection efficiency for water stress tolerance (Bolanos et al., 1993; Bolanos and Edmeades, 1996).

The choice of the most effective breeding scheme and the rate of the genetic improvement are dependent upon the relative magnitude of various gene effects (Dhillon and Pollmer, 1978). The expression and genetic variation of grain yield and secondary traits in maize vary with stress level. Additive genetic effects were found to be more important for grain yield under water stress and well-watered conditions (Betran et al., 2003; Makumbi et al., 2004). Betran et al. (1999) reported that as water stress increases so does the importance of general combining ability (GCA) and additive genetic effects. Derera et al. (2008) reported the preponderance of additive effects for grain yield and ears per plant under water stress and the importance of both additive and non-additive effects in controlling grain yield under well-watered conditions. Both additive and non-additive gene effects are important for days to anthesis, silking and anthesis-silking interval under both water stress and non-water stress environments (Derera et al., 2008).

Determining of the mode of gene action controlling yield and secondary traits in QPM germplasm under water stress, and optimal conditions would help in devising a viable conventional breeding strategy to develop nutritionally enhanced cultivars adapted to stress and optimal environments. The aim of this study was to determine (i) the combining ability and (ii) modes of gene action for grain yield and related traits in QPM inbred lines under water stress, and optimal (well-watered) conditions.

2. Materials and methods

2.1 Environments and stress management

The study was conducted in eastern and southern Africa, in Ethiopia, Kenya, Zambia and Zimbabwe from 2006 to 2008 (Table 1). Nine environments at Harare (HAOM), Rattray Arnold (RAOM), Mpongwe (MPOM), Bako (BKOM), Melkassa (MLOM), Pawe (PWOM), Awassa (AWOM), Jimma (JMOM) and Kiboko (KBOM) research stations comprised optimum management (optimal fertilization and supplemental irrigation as needed to avoid water stress). Fertilizer rates at each location were adjusted to reflect the agronomic recommendations for each location. The trials were conducted during the summer (main cropping) seasons of the respective countries. Two experiments were grown under water stress during the winter (dry) seasons at Chiredzi, Zimbabwe (CHDS) and Kiboko, Kenya (KBDS) research stations.

Both Chiredzi and Kiboko are largely rain free during the winter season, allowing the control of water stress intensity by withdrawing or delaying irrigation for varying lengths of time during flowering and grain filling stages (Edmeades et al., 1999). At Chiredzi, water stress was achieved by applying a total of 220 mm irrigation water in the first 50 days from planting. This regime caused severe water stress at flowering and grain filling time. The trials at Kiboko were irrigated from planting until 15 days before male flowering after which watering was withheld until 15 days after male flowering when additional irrigation was applied to prevent zero yield (Banziger et al., 2000). Care was taken so that irrigation, and hence stress, was as uniform as possible and the water stress blocks were not contaminated with irrigation water from neighbouring blocks or leaking pipes and wind drift. Sufficient fertilization and crop management practices were applied, except irrigation management to avoid confined effects from other factors.

Location	Country	Year	Latitude	Longitude	Altitude	Rainfall mm	Temperature °C		Type of environment	Code	Fertilization kg ha^{-1}			Plot size (m x m)	Density plants ha^{-1}
							Min	Max			N	P_2O_5	K‡		
Harare	Zimbabwe	2006/7	17°49'S	31°1'E	1489	890	14.2	26.8	Optimum	HAOM	166	56	24	4.0 x 1.50	53 333
RARS	Zimbabwe	2006/7	17°16'S	31°03'E	1341	865	14.2	27.0	Optimum	RAOM	208	35	21	4.0 x 0.75	53 333
Mpongwe	Zambia	2006/7	13°32'S	28°03'E	1300	1500	n/a†	n/a	Optimum	MPOM	208	35	21	4.0 x 0.75	53 333
Bako	Ethiopia	2007	9°06'N	37°09'E	1650	1245	14.0	28.1	Optimum	BKOM	100	100	-	4.8 x 1.50	44 444
Melkasa	Ethiopia	2007	8°24'N	39°21'E	1550	680	14.6	28.6	Optimum	MLOM	50	25	-	4.8 x 1.50	44 444
Pawe	Ethiopia	2007	11°09'N	36°03'E	1100	1577	16.6	33.4	Optimum	PWOM	64	46	-	4.8 x 1.50	44 444
Awassa	Ethiopia	2007	7°08'N	38°48'E	1700	1100	12.6	26.8	Optimum	AWOM	110	46	-	4.8 x 1.50	44 444
Jimma	Ethiopia	2007	7° 46' N	36°00'E	1753	1530	12.0	26.2	Optimum	JMOM	75	70	-	4.8 x 1.50	44 444
Chiredzi	Zimbabwe	2007	21°02' S	31°58' E	433	300	14.0	34.2	Stress	CHDS	148	56	24	4.0 x 1.50	53 333
Kiboko	Kenya	2007	2°10'S	37°40'E	975	561	14.0	33.0	Stress	KBDS	156	92	-	4.0 x 1.50	53 333
Kiboko	Kenya	2008	2°10'S	37°40'E	975	561	14.0	33.0	Optimum	KBOM	156	92	-	4.0 x 1.50	53 333

† n/a= not available; ‡K= potassiu m fertilizer was not used in Ethiopia and Kenya; RARS= Rattray Arnold Research Station

Table 1. Locations and environments used to evaluate F_1 hybrids, with their characteristics and codes

2.2 Germplasm

Fifteen inbred lines were selected based on diverse pedigree backgrounds. These lines showed better combining ability in top-cross evaluations and *per se* performance across a range of tropical and subtropical environments (data not shown). Most of the lines are resistant/tolerant to major foliar diseases of the tropics (CIMMYT, 2004). Diallel crosses were made among the 15 inbred lines in the winter of 2006 at Muzarabani, Zimbabwe. Seeds from reciprocal crosses were bulked to form a set of 105 F_1 hybrids. The F_1 hybrids were evaluated along with two QPM (SC527Q and CML144/CML159//CML176) and one normal maize (SC633) hybrid checks in all experiments conducted in Kenya, Zambia and Zimbabwe, and two normal maize (BH540 and BH541) and one QPM (BHQP542) hybrid checks in all experiments conducted in Ethiopia.

2.3 Experimental design and field measurements

All experiments were laid out as 9 x 12 alpha-lattice designs (Patterson and Williams, 1976) with two replications (Table 1). Measurements were recorded on well-bordered plants by excluding the plant nearest to the alley of each row. Days to anthesis and silking were calculated as the number of days from planting to 50% pollen shed and silk emergence. Anthesis silking interval was calculated as the difference between days to silking and anthesis (ASI = DS - DA). Two weeks after pollen shed, plant height and ear height were measured as the distance from ground level to the first tassel branch or to the node bearing the main ear. Number of ears per plant was obtained by dividing the number of ears by number of plants harvested. An ear was counted if it had at least one fully developed grain. Grain weight from all the ears of each experimental unit was measured and used to calculate grain yield (expressed in ton ha^{-1} and adjusted to 12.5% moisture content).

2.4 Statistical analysis

Before data analysis, anthesis-silking interval (ASI) was normalized using $\ln\sqrt{(ASI+10)}$ as suggested by Bolanos and Edmeades (1996). Analysis of variance per environment was conducted with the PROC MIXED procedure in SAS (SAS, 2003) considering genotypes as fixed effects and replications and blocks within replications as random. Entry means adjusted for block effects generated from individual location analyses according to a lattice design (Cochran and Cox, 1960) were used to perform across environments combined analyses using PROC GLM in SAS (SAS, 2003) and combining ability analysis using a modification of the DIALLEL-SAS program (Zhang and Kang, 1997).

GCA effects of the parents and SCA effects of the crosses were estimated following Griffing's Method IV (crosses only) and Model I (fixed) of diallel analysis (Griffing, 1956). Combined analyses of variance were conducted for each trait that showed significant entry mean squares in individual environment analysis. Combining ability was analyzed, and GCA and SCA effects were estimated accordingly. The mean squares for hybrids and environments were tested against the mean squares for hybrid x environment (E) as error term while hybrid x E interactions mean squares were tested against pooled error.

Since means (over replication) of each of the genotypes were used for combined analysis of variance, estimate of pooled error mean squares were calculated following the procedure of Dabholkar (1999) as: $\sum_{i=1}^{n} K_i S_i^2 \Big/ \sum_{i=1}^{n} K_i r$, where K_i and S_i^2 are error degrees of freedom and error mean square at i^{th} environment, respectively, n is the number of environments and r is the number of replications in each environment. The significance of GCA and SCA sources of variation was determined using the corresponding interactions with the environment as error terms. Error mean squares calculated above were used to test the significance of GCA and SCA interactions with environment; because the combining ability mean squares were calculated based on entry means of each genotype from each environment (Griffing, 1956; Singh, 1973; Dabholkar, 1992). For GCA effects of the inbred lines, the restriction $\sum gi = 0$ was imposed. Significance of GCA effects was determined by the t-test, using standard errors of GCA effects (Griffing, 1956; Singh and Chaudhary, 1985).

3. Results

Analysis of variance for each environment revealed the existence of significant differences among hybrids for most traits except anthesis-silking interval at Harare, Mpangwe and Pawe optimal (Table 2). Mean squares due to GCA were highly significant for all traits studied at all environments. SCA effects were also significant for most traits. Mean grain yields for the QPM hybrids (excluding the checks) ranged from 0.6 t ha^{-1} under severe water stress at Chiredze to 8.4 t ha^{-1} under optimum management at Mpongwe (Table 3). At Kiboko, average grain yield of the hybrids tested under water stress was 35.7% of grain yield under optimal conditions (KBOM).

Environment†	Source of variation	DF	GY	DA	DS	ASI	PH	EH	EPP
HAOM	Hybrids	104	4.6**	13.3**	15. 0**	2.0	425.4**	279.1**	0.04**
	GCA	14	6.6**	39.7**	42.4**	-	885.5**	537.9**	0.04**
	SCA	90	1.7**	1.5	2.1	-	108.1**	77.6*	0.02**
RAOM	Hybrids	104	5.5**	12.7**	15.9**	2.1**	341.5**	298.1**	0.05**
	GCA	14	6.9**	35.9**	42.6**	3.2**	467.5**	364.5**	0.07**
	SCA	90	2.1**	1.7**	2.6**	0.7	124.6**	115.5**	0.02**
MPOM	Hybrids	104	10.1**	3.6**	3.3**	0.8	986.7**	459.0**	0.03*
	GCA	14	15.8**	6.4**	5.1**	-	1278.1**	703.1**	0.02**
	SCA	90	3.4**	1.1	1.1	-	371.3**	155.8**	0.01
BKOM	Hybrids	104	5.0**	51.2**	51.0**	2.5**	589.4**	280.9**	0.13**
	GCA	14	8.3**	161.9**	157.8**	5.2**	850.5**	627.9**	0.30**
	SCA	90	1.6**	4.4**	4.9**	0.7	208.2**	64.6	0.03**
MLOM	Hybrids	104	3.9**	22.2**	22.2**	1.5**	570.6**	357.9**	0.11**
	GCA	14	6.2**	66.3**	62.5**	2.5**	633.0**	420.7**	0.26**
	SCA	90	1.3**	2.5*	3.1*	0.5*	231.2**	141.4*	0.02**
PWOM	Hybrids	104	3.6**	35.3**	36.6**	1.8	367.7**	212.8**	0.04**
	GCA	14	4.4**	77.5**	85.9**	-	335.9*	150.5**	0.03**
	SCA	90	1.4	8.4**	7.8**	-	160.2**	99.5**	0.02**
AWOM	Hybrids	104	2.6**	20.0**	18.3**	4.4**	571.9**	278.4**	0.12**
	GCA	14	2.9**	63.1**	55.7**	5.4**	757.9**	381.2**	0.16**
	SCA	90	1.0**	1.8**	1.9*	1.7*	212.5**	101.5**	0.04*
JMOM	Hybrids	104	3.2**	39.9**	35.4**	1.7**	598.3**	308.6**	0.07**
	GCA	14	4.8**	124.9**	110.2**	1.1**	924.7**	409.7**	0.13**
	SCA	90	1.1**	3.6**	3.3**	0.8**	201.9**	114.6	0.02**
CHDS	Hybrids	104	0.3**	71.2**	309.2**	108.4**	762.6**	293.3**	0.08**
	GCA	14	0.5**	207.6**	900.5**	271.4**	1598.7**	314.3**	0.19**
	SCA	90	0.1*	8.9**	38.6**	20.4*	191.9**	120.6**	0.02**
KBDS	Hybrids	104	3.8**	34.5**	90.9**	20.6**	-	-	0.09**
	GCA	14	10.2**	115.0**	254.7**	33.1**	-	-	0.22**
	SCA	90	0.6**	2.0**	12.9**	6.8**	-	-	0.02*
KBOM	Hybrids	104	9.4**	17.4**	18.4**	2.2**	427.0**	204.5**	0.03**
	GCA	14	14.7**	56.2**	56.7**	5.5**	807.6**	492.3**	0.05**
	SCA	90	3.2**	1.3**	1.8**	0.4	121.1**	41.5*	0.01**

HAOM=Harare optimal, RAOM=Rattray optimal, MPOM=Mpongwe optimal, BKOM=Bako optimal, MLOM=Melkasa optimal, PWOM=Pawe optimal, AWOM=Awassa optimal, JMOM=Jimma optimal, CHDS=Chiredzi stress, KBDS=Kiboko stress, KBOM=Kiboko optimal* P≤ 0.05 ; ** P≤ 0.01; DF= degrees of freedom; GY= grain yield; AD= days to anthesis; DS= days to silking; ASI= anthesis-silking interval; PH= plant height; EH= ear height; EPP= ears per plant

Table 2. Mean squares for hybrids, general (GCA) and specific (SCA) combining ability for grain yield and agronomic traits in stressed and optimal environments, 2006 - 2008

Combined analysis of variance across water stress environments revealed highly significant mean squares due to environments and hybrids for all traits analyzed (Table 4). Mean grain yield across water stress environments ranged from 0.3 to 3.7 t ha^{-1} with a mean of 1.8 t ha^{-1}. Higher grain yields were recorded for VL052 x VL05561 (3.7 t ha^{-1}), VL05561 x CML159 (3.5 t ha^{-1}), VL054178 x VL06375 (3.4 t ha^{-1}), VL05482 x VL05561 (3.3 t ha^{-1}) and VL054178 x VL05561 (3.0 t ha^{-1}). Mean grain yield across water stress environments (Table 4) was 27.4% of the mean grain yield across optimal environments (Table 5). Mean days to anthesis was 92.3 with a range of 82.8 – 103.5. Days to silking ranged from 83.7 to 120.0 d with a mean of 102.0. Anthesis-silking interval ranged from 0.4 to 21.4 with a mean of 9.7. Ears per plant ranged from 0.10 to 0.88 with a mean 0.50. Combining ability analysis revealed non-significant GCA mean squares for grain yield but significant GCA mean squares for days to anthesis and silking, anthesis-silking interval and ears per plant. SCA mean squares, however, were not significant for all traits. Hybrid x E, GCA x E and SCA x E interaction mean squares were significant for all traits tested.

Across optimal environments, the effects of environments, hybrids, GCA and SCA were highly significant for all the traits evaluated (Table 5). Grain yields ranged from 1.8 to 9.4 t ha^{-1} with a mean of 6.5 t ha^{-1}. The highest yielding hybrids were VL05483 x CML491 (9.4 t ha^{-1}), CML511 x CML491 (8.8 t ha^{-1}), VL05561 x CML491 (8.7 t ha^{-1}), CML159 x CML491 (8.5 t ha^{-1}) and VL054178 x CML491 (8.1 t ha^{-1}). Mean days to anthesis was 73.8 with a range of 66.9 – 80.4. Days to silking ranged from 68.9 to 82.8 with a mean of 75.1. Mean plant and ear height was 225.5 and 110.9 cm with ranges of 189.0 – 248.4 cm and 89.9 – 131.7 cm. Mean ears per plant was 1.14 with ranges of 0.79 – 1.48. Anthesis-silking interval ranged from -0.2 to 3.3 d with a mean of 1.6 d. Hybrid x E, GCA x E and SCA x E interactions were highly significant for all traits except SCA x E for ear height and anthesis-silking interval.

	HAOM	RAOM	MPOM	BKOM	MLOM	PWOM	AWOM	JMOM	CHDS	KBDS	KBOM
Grand mean	7.7	6.5	8.4	6.5	6.7	4.9	4.7	4.6	0.6	2.9	8.1
Hybrid mean	7.7	6.4	8.4	6.5	6.7	4.9	4.7	4.6	0.6	2.9	8.2
Best hybrid	12.9	10.6	13.8	9.7	10.6	8.8	7.2	7.9	2.1	5.7	12.1
Best QPM check	9.1	7.8	9.2	6.4	5.9	3.8	4.7	4.9	1.0	2.4	7.5
Best normal check	11.6	10.0	8.6	7.4	6.5	6.1	6.0	2.5	0.3	3.2	5.5
SE (m)	0.6	0.9	0.7	0.3	0.5	0.8	0.5	0.4	0.2	0.5	0.5
% high yielding hybs‡	1.0	1.0	41.9	31.4	61.0	14.3	10.5	41.0	13.3	45.7	73.3

HAOM=Harare optimal, RAOM=Rattray optimal, MPOM=Mpongwe optimal, BKOM=Bako optimal, MLOM=Melkasa optimal, PWOM=Pawe optimal, AWOM=Awassa optimal, JMOM=Jimma optimal, CHDS=Chiredzi stress, KBDS=Kiboko stress, KBOM=Kiboko optimal. ‡ proportion of QPM hybrid with higher grain yield than the best check (normal maize or QPM); SE(M)= standard error of the mean

Table 3. Means of QPM hybrids, and best normal and QPM checks for grain yield in stress and optimal environments, 2006 -2008

Sources of variation	DF	GY	DA	DS	ASI	EPP
Environment (E)	1	275.0**	65614.0**	116716.3**	7339.6**	14.24**
Hybrid	104	1.3**	46.7**	159.8**	42.4**	0.07**
GCA	14	7.3	308.8**	1030.3**	233.0**	0.39**
SCA	90	0.4	5.9	24.3	12.7	0.02
Hybrid x E	104	0.7**	6.1**	40.3**	22.1**	0.02**
GCA x E	14	3.3**	13.8**	124.9**	71.5**	0.02**
SCA x E	90	0.3**	5.0**	27.1**	14.4*	0.02**
Error	164	0.2	2.5	11.8	8.8	0.01
Mean		1.8	92.3	102.0	9.7	0.50
Minimum		0.3	82.8	83.7	0.4	0.10
Maximum		3.7	103.5	120.0	21.4	0.88
SE (m)		0.3	1.1	2.4	2.1	0.07
CV%		24.1	1.7	3.4	30.7	20.0

* $P \leq 0.05$; ** $P \leq 0.01$; ASI= Anthesis silking interval; CV= coefficient of variation; DA= days to anthesis; DF= degrees of freedom; DS= days to silking; EPP= ears per plant; GCA= general combining ability; GY= grain yield; SCA= specific combining ability; SE (m)= standard error of the mean

Table 4. Mean squares from combined analysis of variance and means for grain yield and agronomic traits of QPM hybrids across water stress environments at Chiredzi and Kiboko, 2007

Sources of variation	DF	GY	DA	DS	PH	EH	EPP	ASI
Environment (E)	8	231.4**	5263.9**	7362.0**	61527.2**	46311.2**	3.09**	509.0**
Hybrids	104	14.4**	69.8**	71.1**	1356.2**	643.5**	0.11**	2.8**
GCA	14	46.8**	477.1**	472.7**	5289.2**	2925.7**	0.58**	13.2**
SCA	90	9.4**	6.5**	8.6**	744.4**	288.5**	0.04**	1.2**
Hybrids x E	832	1.2**	4.7**	4.6**	135.4**	87.0**	0.02**	0.9**
GCA x E	112	2.6**	17.2**	16.2**	183.5**	129.1**	0.05**	1.6**
SCA x E	720	0.8**	2.2**	2.2**	110.5**	69.3	0.02**	0.6
Error	738	0.6	1.5	1.8	73.1	61.7	0.01	0.6
Mean		6.5	73.8	75.1	225.5	110.9	1.14	1.6
Minimum		1.8	66.9	68.9	189.0	89.9	0.79	-0.2
Maximum		9.4	80.4	82.8	248.4	131.7	1.48	3.3
SE (m)		0.3	0.4	0.5	2.9	2.6	0.03	0.3
CV%		11.6	1.7	1.8	3.8	7.1	8.8	47.8

* $P \leq 0.05$; ** $P \leq 0.01$; ASI= anthesis-silking interval; CV= coefficient of variation; DA= days to anthesis; DF= degrees of freedom; DS= days to silking; EH= ear height; EPP= ears per plant; GCA= general combining ability; GY= grain yield; PH= plant height; SCA= specific combining ability; SE (m)= standard error of the mean

Table 5. Mean squares from combined analysis of variance and means for grain yield and agronomic traits of QPM hybrids across nine optimal environments, 2006 – 2008

Estimates of GCA effects for grain yield showed that inbred lines VL05561, VL05483, CML511, CML159 and VL06375 combined well in most of the environments (Table 6). These inbred lines mostly showed positive and highly significant GCA effects in most environments. On the other hand, VL052, VL052887, VL0523 and CML144 showed negative and highly significant GCA effects in most of the environments. Inbred lines VL05561, VL05483 and CML511 showed high positive GCA effects across optimum and combined environments.

For days to anthesis, VL054178, VL05482, VL05561, VL05483, CML511 and VL06375 had negative and highly significant GCA effects in most environments (Table 7). On the other hand, inbred lines VL05200, VL054178, VL052887, VL0523, VL05561 and CML144 showed positive and highly significant GCA effects in most environments. VL054178, VL05482, VL05561, VL05483, CML511, CML159 and VL06375 had highly significant negative GCA effects for days to silking for both water stress and optimal environments.

Inbred lines VL054178, VL05482, VL05561, VL05483 and VL06375 had negative and highly significant GCA effects for days to silking (Table 8). On the other hand, VL05468, VL052887, VL0523, VL0524, CML144 and CML491 showed positive and highly significant GCA effects. VL054178, VL05482, VL05561, VL05483, CML159 and VL06375 had highly significant negative GCA effects for days to anthesis for both water stress and optimal environments.

The GCA effects for anthesis-silking interval were negative and highly significant for VL05561 but positive and highly significant for VL054178 in almost all environments (Table 9). Across water stress environments, inbred lines VL054178 and VL05482 showed lower GCA effects. VL052887, VL05561 and CML144 had negative and highly significant GCA effects across optimal environments. VL054178, VL05561, VL05483 and VL06375 showed lower GCA effects for anthesis-silking interval over all environments.

Inbred lines VL05200, VL054178, VL05482, CML144 and CML159 showed negative and significant GCA effects for plant and ear height in most environments (Tables 10 and 11). However, VL05483 and VL06375 had positive and significant GCA effects for plant height while VL053, VL0524 and VL5561 had positive and significant GCA effects for ear height in most environments.

For ears per plant, inbred lines VL05482, VL05483 and CML511 showed positive and significant and VL05200, VL05468, VL0523, VL0524 and CML159 showed negative and highly significant GCA effects in water stress and optimal environments (Table 12). At Chirezi under water stress, VL05482, CML511 and CML491 showed negative and significant GCA effects.

4. Discussion

The results observed in various environments (Table 2) showed that water stress significantly affected grain yield, as previously reported (Bolanos and Edmeades, 1993, 1996; Banziger et al., 1997; Banziger and Lafitte, 1997; Banziger et al., 1999a; Derera et al., 2008). High levels of variation observed among hybrids under water stress, and optimal environments indicate the possibility of selecting for improved grain yield and agronomic traits under stress and non-stress conditions. The existence of genetic variability in maize evaluated under stress conditions has been reported by several investigators (Bolanos et al., 1993; Bolanos and Edmeades, 1996; Banziger and Lafitte, 1997; Beck et al., 1997; Banziger et al., 1997; 1999b; Betran et al., 2003; Derera et al., 2008). Significant GCA and SCA mean squares for most traits in each environment indicate the importance of both additive and non-additive effects for the traits studied. This suggests that effective selection or systematic hybridization could be employed in improving these traits.

Environment†	P1	P2	P3	P4	P5	P6	P7	P8	P9	P10	P11	P12	P13	P14	P15	SE(gi)
HAOM	-0.78**	-0.85**	-0.22	0.87**	-0.42*	-0.43*	-0.29	-0.34	0.84**	1.21**	0.55**	-0.96**	-0.46*	0.57**	0.71**	0.20
RAOM	-0.97**	-0.89**	-0.08	0.30	-0.57*	-0.22	-0.41	-0.27	1.17**	0.89**	1.28**	-0.88**	-0.19	0.50	0.34	0.28
MPOM	-0.73**	-0.75**	-0.58*	0.51*	-0.40	-0.27	-0.92**	-1.23**	1.73**	1.46**	1.85**	-1.50**	-0.76**	0.58*	1.00**	0.23
BKOM	-0.66**	-1.17**	-0.92**	-0.22*	0.39**	-0.38**	-0.49**	-0.32**	1.58**	0.73**	0.47**	-0.94**	0.94**	0.26*	0.72**	0.11
MLOM	-0.34*	-0.36*	-0.79**	0.12	-0.24	1.20**	-1.09**	-0.87**	0.84**	0.78**	0.30*	-0.66**	0.57**	0.32*	0.22	0.14
PWOM	-0.76**	-0.10	-0.08	0.03	-0.71*	0.59*	-0.43	0.19	0.85**	-0.41	0.26	-0.89**	0.99**	0.56*	-0.08	0.27
AWOM	-0.49**	0.11	-0.50**	-0.02	-0.18	0.23	-0.56**	-0.38*	0.88**	0.29	0.61**	-0.54**	0.35	0.56**	-0.35	0.18
JMOM	-0.30*	-0.01	0.33*	-0.89**	1.31**	0.43**	0.04	0.25	-0.97**	0.51**	0.12	0.43**	-0.74**	-0.57**	0.07	0.14
CHDS	-0.04	-0.04	-0.20**	0.30**	-0.18**	0.21**	-0.26**	-0.20**	0.19**	0.06	0.03	-0.19**	0.22**	-0.11	0.22**	0.06
KBDS	-0.57**	-0.49**	-0.84**	1.11**	-0.97**	1.00**	-0.72**	-0.68**	1.63**	0.44**	0.39*	-1.19**	0.55**	-0.43**	0.78**	0.16
KBOM	-1.08**	-1.19**	-0.59**	0.20	-0.52**	-0.32	-0.82**	-0.66**	2.26**	1.19**	1.24**	-1.42**	0.86**	0.16	0.69**	0.18

Table 6. Estimates of general combining ability effects of 15 QPM inbred lines for grain yield per environment, 2006 -2008

Table 7. Estimates of general combining ability (GCA) effects of 15 QPM inbred lines for days to anthesis per environment and across environments, 2006 - 2008

Environment†	P1	P2	P3	P4	P5	P6	P7	P8	P9	P10	P11	P12	P13	P14	P15	SE(gi)
HAOM	0.42	1.48*	1.63**	-2.75**	1.89**	-3.84**	1.39**	1.05**	-0.84**	-1.19**	0.26	1.41**	0.19	0.88**	-1.97**	0.29
RAOM	0.46*	0.69**	1.87**	-2.32**	1.81**	-2.80**	2.19**	1.52**	-1.03**	-1.56**	-0.61*	1.62**	-0.22	0.39	-2.03**	0.20
MPOM	-0.68**	-0.29	0.12	0.12	-0.46	0.68**	0.17	-0.36	0.14	0.14	1.74**	-1.50**	-0.25	0.16	0.28	0.25
BKOM	0.90**	0.91**	2.66**	-5.11**	4.87**	-6.15**	3.07**	3.34**	-0.26**	-3.71**	-1.44**	3.36**	-2.77**	3.51**	-3.19**	0.27
MLOM	0.58	0.78*	1.81**	-2.83**	3.11**	-5.02**	2.28**	2.04**	-0.63	-2.22**	-0.73*	1.58**	0.02	1.38**	-2.15**	0.35
PWOM	1.37*	1.42*	1.79**	-3.48**	3.98**	-4.36**	1.33*	2.16**	-2.89**	-1.39*	1.05	1.87**	-1.97**	0.92	-1.81**	0.58
AWOM	1.00**	1.51**	1.55**	-2.88**	3.00**	-4.22**	1.51**	1.39**	-0.76**	-2.30**	-0.99**	2.16**	-1.07**	2.32**	-2.22**	0.25
JMOM	1.01**	-2.74**	-2.05**	1.99**	-1.65**	2.72**	-2.74**	-0.16	2.66**	-4.26**	4.21**	-6.00**	3.26**	3.20**	0.55	0.36
CHDS	2.36**	1.72**	1.35*	-5.31**	3.50**	-5.78**	4.88**	2.79**	-0.81	-3.25**	-3.89**	5.17**	-3.51**	4.98**	-4.20**	0.54
KBDS	1.88**	1.21**	1.83**	-4.03**	4.46**	-5.00**	2.05**	2.06**	-1.23**	-3.25**	-2.37**	3.66**	-1.77**	2.66**	-2.17**	0.26
KBOM	1.01**	0.58**	1.51**	-2.81**	2.66**	-4.16**	2.14**	1.61**	-1.17**	-1.86**	-1.09**	2.57**	-0.89**	1.38**	-1.46**	0.20
ACDRT‡	2.12**	1.46**	1.59**	-4.67**	3.98**	-5.39**	3.46**	2.42**	-1.02**	-3.25**	-3.13**	4.41**	-2.64**	3.82**	-3.19**	0.30
ACOPT#	0.67**	0.77**	1.73**	-2.92**	2.79**	-3.99**	1.93**	1.77**	-0.77**	-1.87**	-0.43**	1.67**	-0.96**	1.52**	-1.92**	0.11

HAOM=Harare optimal, RAOM=Rattray optimal, MPOM=Mpongwe optimal, BKOM=Bako optimal, MLOM=Melkasa optimal, PWOM=Pawe optimal, AWOM=Awassa optimal, JMOM=Jimma optimal, CHDS=Chiredzi stress, KBDS=Kiboko stress, KBOM=Kiboko optimal; * $P \leq 0.05$; ** $P \leq 0.01$; ‡ ACDRT= across water stress environments; # ACOPT= across optimum environments; P1= VL052; P2= VL05200; P3= VL05468; P4= VL054178; P5= VL052887; P6= VL05482; P7= VL0523; P8= VL0524; P9= VL05561; P10= VL05483; P11= CML511; P12= CML144; P13= CML159; P14= CML491; P15= VL06375; SE(gi)= standard error of GCA effects

Environment†	P1	P2	P3	P4	P5	P6	P7	P8	P9	P10	P11	P12	P13	P14	P15	SE(gi)
HAOM	0.57	0.52	1.93**	-3.03**	1.59**	-3.84**	1.57**	1.81**	-0.73	-1.36**	0.64	1.08**	0.35	0.93**	-2.05**	0.34
RAOM	0.51	0.75*	3.36**	-2.48**	1.55**	-2.36**	1.89**	1.28**	-1.78**	-1.68**	-0.55	1.44**	-0.16	0.61*	-2.36**	0.29
MPOM	-0.53*	-0.33	0.13	0.04	-0.52*	0.45	0.27	-0.04	0.05	0.22	1.56**	-1.35**	-0.35	0.21	0.19	0.24
BKOM	1.27**	0.52	3.35**	-4.69**	3.50**	-5.58**	3.31**	3.98**	-1.59**	-4.04**	-1.41**	3.11**	-2.52**	4.11**	-3.31**	0.36
MLOM	0.83*	0.97*	2.55**	-2.40**	2.62**	-4.75**	2.09**	1.87**	-1.57**	-2.22**	-0.56	0.99*	0.2	1.60**	-2.20**	0.39
PWOM	1.49*	1.47*	1.88**	-3.66**	4.16**	-4.59**	1.50*	2.14**	-3.17**	-1.68**	0.14	1.91**	-1.84**	1.87**	-1.60**	0.57
AWOM	0.61*	0.75*	2.53**	-2.25**	2.56**	-3.40**	2.25**	1.86**	-1.59**	-2.78**	-1.22**	1.24**	-0.83**	2.00**	-1.72**	0.30
JMOM	0.71	-2.73**	-2.16**	1.99**	-1.49**	2.53**	-2.28**	0.26	2.44**	-4.16**	3.71**	-5.51**	3.20**	3.27**	0.23	0.37
CHDS	4.28**	2.92*	7.35**	-13.41**	6.00**	-11.57**	11.49**	8.14**	-4.04**	-8.23**	-6.18**	9.08**	-5.59**	7.12**	-7.37**	1.14
KBDS	3.18**	1.43*	4.33**	-5.32**	6.65**	-6.50**	3.28**	3.24**	-4.01**	-4.48**	-3.72**	5.16**	-2.54**	3.11**	-3.82**	0.62
KBOM	0.93**	1.20**	2.56**	-2.78**	2.11**	-3.28**	1.66**	1.52**	-2.51**	-1.35**	-0.71**	2.45**	-1.35**	1.73**	-2.18**	0.23
ACDRT‡	3.73**	2.18**	5.84**	-9.36**	6.32**	-9.04**	7.39**	5.69**	-4.02**	-6.35**	-4.95**	7.12**	-4.06**	5.12**	-5.60**	0.65
ACOPT#	0.71**	0.68**	2.30**	-2.82**	2.36**	-3.65**	1.97**	1.97**	-1.41**	-1.96**	-0.47**	1.43**	-0.89**	1.73**	-1.95**	0.12

Table 8. Estimates of general combining ability effects of 15 QPM inbred lines for days to silking per environment and across environments, 2006 – 2008

Environment†	P1	P2	P3	P4	P5	P6	P7	P8	P9	P10	P11	P12	P13	P14	P15	SE(gi)
RAOM	0.05	0.09	1.47**	-0.14	-0.22	0.47**	-0.30	-0.22	-0.80**	-0.14	0.01	-0.18	0.05	0.20	-0.34	0.23
BAOM	0.34	-0.47*	0.76**	0.41*	-1.28**	0.64**	0.10	0.57**	-1.24**	-0.24	-0.01	-0.39*	0.22	0.61**	-0.05	0.18
MLOM	0.24	0.16	0.74**	0.39*	-0.45**	0.28	-0.30	-0.18	-1.03**	0.01	0.16	-0.53**	0.16	0.32*	-0.03	0.15
AWOM	-0.35	-0.78**	0.92**	0.57*	-0.51	0.84**	0.68*	0.61*	-0.78**	-0.39	-0.32	-0.93**	0.34	-0.35	0.45	0.28
JMOM	-0.31	0.02	-0.03	-0.05	0.06	-0.19	0.52**	0.37*	-0.27	0.12	-0.46*	0.45*	-0.01	0.10	-0.31	0.18
CHDS	1.85	1.18	6.01**	-8.08**	2.45*	-5.71**	6.62**	5.39**	-3.24**	-5.00**	-2.27*	3.89**	-2.12*	2.16*	-3.11**	1.01
KBDS	1.24*	0.23	2.47**	-1.23*	2.23**	-1.53**	1.17*	1.16*	-2.73**	-1.19*	-1.30**	1.47**	-0.84	0.50	-1.65**	0.49
KBOM	-0.10	0.69**	1.07**	-0.02	-0.55**	0.88**	-0.55**	-0.02	-1.29**	0.46**	0.35*	-0.14	-0.46**	0.39*	-0.69**	0.16
ACDRT‡	1.55**	0.71	4.24**	-4.65**	2.34**	-3.62**	3.89**	3.27**	-2.99**	-3.09**	-1.79**	2.68**	-1.48**	1.33*	-2.38**	0.56
ACOPT#	-0.02	0.01	0.78**	0.22**	-0.58**	0.59**	-0.10	0.14	-0.91**	-0.05	0.03	-0.37**	0.06	0.16*	-0.02	0.08

HAOM=Harare optimal, RAOM=Rattray optimal, MPOM=Mpongwe optimal, BKOM=Bako optimal, MLOM=Melkasa optimal, PWOM=Pawe optimal, AWOM=Awassa optimal, JMOM=Jimma optimal, CHDS=Chiredzi stress, KBDS=Kiboko stress, KBOM=Kiboko optimal; * $P \leq 0.05$; ** $P \leq 0.01$; ‡ ACDRT= across water stress environments; # ACOPT= across optimum environments; § P1= VL052; P2= VL05200; P3= VL05468; P4= VL054178; P5= VL052887; P6= VL05482; P7= VL0523; P8= VL0524; P9= VL05561; P10= VL05483; P11= CML511; P12= CML144; P13= CML159; P14= CML491; P15= VL06375; SE(gi)= standard error of GCA effects

Table 9. Estimates of general combining ability effects of 15 QPM inbred lines for anthesis-silking interval per environment and across environments, 2006 - 2008

Environment†	P1	P2	P3	P4	P5	P6	P7	P8	P9	P10	P11	P12	P13	P14	P15	SE(gi)
HAOM	-6.25**	-16.06**	0.03	-3.29	7.26**	-13.04**	3.44	5.13*	1.98	12.57**	9.99**	-9.01**	1.75	-0.72	6.21**	2.01
RAOM	-1.55	-12.35**	2.35	-0.05	4.05*	-10.13**	0.91	4.76*	-2.21	5.98**	9.25*	-6.85**	4.38*	-1.1	2.56	2.01
MPOM	-3.02	-15.17**	-12.29**	-0.36	8.10*	-4.4	8.10*	5.21	9.25**	7.17*	15.41**	-19.21**	-3.25	-2.48	6.94*	3.21
BKOM	-3.37	-18.01**	-2.22	-8.64**	3.58	-5.62**	5.34**	7.35**	6.78**	6.98**	5.16**	-10.82**	1.66	0.29	11.52**	1.89
MLOM	2.43	-12.27**	-1.40	-3.68	3.26	-10.33**	-0.60	-1.53	4.41	12.82**	11.71**	-3.66	0.04	-5.83*	4.63*	2.31
PWOM	-0.98	-7.42**	-0.23	-6.34*	-0.18	-3.61	5.24*	5.36*	-3.79	7.33**	-1.28	-7.63**	3.79	3.05	6.68*	2.60
AWOM	-2.37	-17.74**	-8.71**	-10.92**	7.69**	1.69	3.13	6.67**	-0.85	8.85**	9.82**	-1.18	1.26	0.17	2.47	2.26
JMOM	-0.40	6.40**	4.96*	-6.58**	6.03*	0.88	4.72*	-22.85**	-8.58**	-5.79*	10.04**	0.52	9.74**	1.41	-0.48	2.31
CHDS	-4.07	-13.24**	-5.92*	9.93**	-11.98**	15.15**	-3.53	-1.70	2.23	9.56**	9.05**	-17.05**	9.13**	-13.83**	16.25**	2.76
KBOM	-6.88**	-11.76**	-6.03**	-7.09**	1.95	-5.18**	2.21	3.08	6.29**	10.22**	13.77**	-11.29**	4.32*	-1.91	8.31**	1.66

Table 10. Estimates of general combining ability effects of 15 QPM inbred lines for plant height per environment and across environments, 2006 – 2008

Environment†	P1	P2	P3	P4	P5	P6	P7	P8	P9	P10	P11	P12	P13	P14	P15	SE(gi)
HAOM	-5.51**	-6.18**	1.40	-6.45**	6.34**	-12.83**	6.95**	9.42**	4.53*	2.34	2.25	-5.95**	-4.51*	5.62**	2.58	1.91
RAOM	-0.38	-8.54**	5.81**	-1.72	2.83	-8.24**	4.39*	6.25**	6.01**	-1.07	5.78**	-7.69**	-2.62	1.98	-2.79	2.13
MPOM	2.16	-8.57**	-6.48**	-13.21**	4.66	-6.88**	3.68	6.21*	12.01**	6.96**	7.58**	-8.82**	-2.28	2.93	0.04	2.38
BKOM	0.11	-9.96**	0.88	-6.39**	4.60*	-5.46**	6.11**	9.01**	14.76**	-3.31	0.31	-3.85	-9.00**	-3.31	5.49**	1.99
MLOM	-0.2	-12.66**	0.34	1.80	1.57	-3.3	5.07*	5.17*	11.35**	3.30	1.57	-7.27**	-5.17*	-0.30	-1.28	2.5
PWOM	-0.77	-4.67*	-0.82	-1.56	-0.64	-1.46	6.28**	6.68**	-0.70	1.18	-3.87	-4.14*	-0.35	3.87	0.97	1.98
AWOM	-0.72	-10.80**	-4.93*	-7.43**	5.07*	-0.29	4.08*	8.02**	2.36	5.20*	5.77**	0.63	-6.38**	-0.21	-0.36	1.99
JMOM	-0.3	-0.43	0.32	-1.51	-4.02	0.54	-1.50	-10.83**	-6.90**	-2.95	8.32**	-2.11	10.35**	7.03**	4.00	2.44
CHDS	0.83	-6.67**	-4.77*	3.86	-5.04*	1.33	5.64*	5.95**	9.31**	0.9	-0.21	-6.35**	-4.39	-2.84	2.46	2.24
KBOM	-1.92	-10.05**	-0.87	-3.90**	1.82	-8.61**	8.35**	9.46**	10.55**	-0.45	1.48	-3.61*	-6.18**	1.64	2.30	1.40

HAOM=Harare optimal, RAOM=Rattray optimal, MPOM=Mpongwe optimal, BKOM=Bako optimal, MLOM=Melkasa optimal, PWOM=Pawe optimal, AWOM=Awassa optimal, JMOM=Jimma optimal, CHDS=Chiredzi stress, KBDS=Kiboko stress, KBOM=Kiboko optimal; * $P\leq 0.05$; ** $P\leq 0.01$; § ACALL= across all environments; # ACOPT= across optimum environments; P1= VL052; P2= VL05200; P3= VL05468; P4= VL054178; P5= VL052887; P6= VL05482; P7= VL0523; P8= VL0524; P9= VL05561; P10= VL05483; P11= CML511; P12= CML144; P13= CML159; P14= CML491; P15= VL06375; SE(gi)= standard error of GCA effects

Table 11. Estimates of general combining ability (GCA) effects of 15 QPM inbred lines for ear height per environment and across environments, 2006 – 2008

Environment†	P1	P2	P3	P4	P5	P6	P7	P8	P9	P10	P11	P12	P13	P14	P15	SE(gi)
HAOM	-0.04*	-0.06*	-0.07**	-0.05*	0.01	0.07**	-0.02	0.00	0.04*	0.13**	0.03	0.00	-0.06*	0.02	0.00	0.02
RAOM	-0.08**	-0.08**	-0.01	-0.03	0.07**	-0.01	-0.05*	0.03	0.02	0.07*	0.19**	-0.09**	-0.07**	0.05*	0.00	0.02
MPOM	-0.01	0.00	-0.07**	-0.01	0.01	0.03	0.00	-0.07**	0.03	0.08**	0.05	0.02	-0.07*	0.00	0.00	0.03
BKOM	-0.01	-0.19**	-0.11**	-0.27**	0.34**	0.12**	-0.03	-0.01	-0.03	0.12**	0.06*	0.07**	-0.16**	0.15**	-0.04	0.03
MLOM	0.01	-0.04*	-0.05*	-0.18**	0.24**	0.20**	-0.18**	-0.09**	-0.09**	0.14**	0.10**	0.09**	-0.21**	0.12**	-0.07**	0.02
PWOM	-0.01	-0.03	-0.05*	-0.02	-0.05	0.07*	0.00	-0.06*	0.07**	-0.01	0.03	-0.02	0.01	0.09**	-0.01	0.03
AWOM	-0.01	-0.04	-0.08	-0.09*	0.17**	0.21**	-0.06	0.09	-0.05	0.05	0.00	0.03	-0.21**	0.09	-0.10*	0.05
JMOM	-0.06*	0.00	0.12**	-0.01	-0.04	0.09**	-0.07**	0.00	-0.18**	0.02	0.23**	0.10**	-0.07*	-0.06*	-0.07**	0.03
CHDS	-0.03	-0.03	-0.13**	0.19**	-0.11**	0.16**	-0.18**	-0.16**	0.05	0.06*	0.08**	-0.09**	0.15**	-0.04	0.10**	0.03
KBDS	-0.10**	-0.06*	-0.12**	0.13**	-0.14**	0.15**	-0.18**	-0.15**	0.19**	0.13**	0.07**	-0.14**	0.09**	0.02	0.12**	0.03
KBOM	-0.06**	-0.05**	-0.01	-0.07**	0.13**	0.04*	-0.08**	-0.05**	0.02	0.11**	0.05**	-0.03	-0.04*	0.02	0.03	0.02
ACDRT‡	-0.07**	-0.04*	-0.13**	0.16**	-0.13**	0.15**	-0.18**	-0.16**	0.12**	0.10**	0.07**	-0.12**	0.12**	-0.01	0.11**	0.02
ACOPT#	-0.03**	-0.06**	-0.07**	-0.08**	0.13**	0.09**	-0.05**	-0.02*	-0.01	0.08**	0.07**	0.01	-0.10**	0.07**	-0.03*	0.01

HAOM=Harare optimal, RAOM=Rattray optimal, MPOM=Mpongwe optimal, BKOM=Bako optimal, MLOM=Melkasa optimal, PWOM=Pawe optimal, AWOM=Awassa optimal, JMOM=Jimma optimal, CHDS=Chiredzi stress, KBDS=Kiboko stress, KBOM=Kiboko optimal; * $P \leq 0.05$; ** $P \leq 0.01$; ‡ ACDRT= across water stress environments; # ACOPT= across optimum environments; P1= VL052; P2= VL05200; P3= VL05468; P4= VL054178; P5= VL052887; P6= VL05482; P7= VL0523; P8= VL0524; P9= VL05561; P10= VL05483; P11= CML511; P12= CML144; P13= CML159; P14= CML491; P15= VL06375; SE(gi)= standard error of GCA effects

Table 12. Estimates of general combining ability effects of 15 QPM inbred lines for ears per plant per environment and across environments, 2006 - 2008

Combined analysis of variance across water stress (Table 4) and optimal (Table 5) environments indicated the existence of significant variation among hybrids and environments for all traits. Both additive and non-additive genetic effects were not important for grain yield across water stress environments while only additive effect was important for days to anthesis and silking, anthesis-silking interval and ears per plant. This finding is contrary to the reports of other researchers (Betran et al., 1999; 2003; Makumbi et al., 2004; Derera et al., 2008), who reported the importance of additive effects for grain yield of normal maize under water stress. When genetic variance for grain yield is not apparent, secondary traits of adaptive value whose genetic variability increases and whose heritability remains high under water stress can increase selection efficiency (Bolanos and Edmeades, 1996; Edmeades et al., 1997; Banziger and Lafitte, 1997; Banziger et al., 1999b).

Highly significant GCA and SCA mean squares for all traits under optimal environments indicate the importance of both additive and non-additive gene effects for the inheritance of these traits. Similar results have been reported in diallel studies of QPM inbred lines under optimal environments (Pixley and Bjarnason, 1993; Bhatnagar et al., 2004; Hadji, 2004; Fan et al., 2004). Derera et al. (2008) reported the importance of both additive and non-additive effects in conditioning grain yield, days to anthesis and silking, and anthesis-silking interval in Design-II crosses of normal maize inbred lines. Similarly, additive and non-additive effects were important for all traits evaluated across environments except anthesis silking interval which had non-significant SCA effects. Significant mean squares of Hybrid x E, GCA x E and SCA x E interactions for most traits across environments indicate that these effects were not consistent over environments. This implies that different genes are involved in controlling these traits under water stress and optimal conditions. Cooper and Byth (1996) explained that the larger the degree of genotype-by-environment interaction, the more dissimilar the genetic systems controlling the physiological processes conferring adaptation to different environments.

Even though significant cross-over interactions were observed for GCA effects of the inbred lines, some inbred lines were identified with consistent GCA effects across environments. This implies that the genetic systems controlling a given trait under different stress and non-stress conditions are at least partially similar. Hence, it is possible to identify QPM hybrids that perform well across stress levels in Africa. Similar conclusions have been drawn by Betran et al. (2003) who evaluated tropical normal maize inbred lines and their hybrids for grain yield under optimal and water stress conditions.

Inbred lines VL054178, VL05561, VL05483, CML511, CML159 and VL06375 were good general combiners for grain yield in both water stress and optimal environments indicating that these inbred lines contributed to increased grain yield in their crosses under all environmental conditions. Inbred lines VL054178, VL05482, VL05561, VL05483, CML159 and VL06375 contributed to earliness under most environments as inferred from the negative and highly significant GCA effects of days to anthesis and silking. VL05561 was the best general combiner for anthesis-silking interval. Inbred lines VL05200, VL054178, VL05482, CML144 and CML159 were good combiners for plant stature as they contributed to reduced plant and ear height in the crosses. VL05482, VL05483 and CML511 contributed to increased ears per plant in the crosses. Anthesis- silking interval and ears per plant are important secondary traits to be considered in increasing the efficiency of selection for grain yield under stress. The highest grain yielding genotypes under water stress tended to show lower anthesis-silking interval, delayed senescence, and a higher number of ears per plant (Bolanos and Edmeades, 1993; Banziger and Lafitte, 1997; Banziger et al., 1999c; Diallo et al., 2004).

Higher SCA variances than GCA variances for grain yield in most optimal environments indicate that additive variability was of greater importance in the inheritance of grain yield under optimal conditions. Under water stress conditions, however, additive variability was more important than non-additive variability. The predominance of additive effects under water stress conditions has been reported by several researchers (Betran et al., 2003; Diallo et al., 2003; Makumbi et al., 2004; Derera et al., 2008).

Additive effects were more important that non-additive effects in the inheritance of days to anthesis and silking in all cases. Similarly, additive effects were more important for anthesis-silking interval, plant and ear height, and ears per plant in most cases.

According to Baker (1978), when SCA mean squares are not significant, the hypothesis that the performance of a single-cross progeny can be adequately predicted on the basis of GCA would be accepted. On the other hand, if the SCA mean squares are significant, the relative importance of GCA and SCA should be assessed by estimating components of variance in determining progeny performance.

5. Conclusions

A large proportion of the maize crop in Africa is grown by small scale farmers under low input systems, without adequate fertilization and irrigation. Significant yield losses due to water stress were realized in this study. The results indicated the availability of considerable variation among QPM hybrids and the possibility of making selections for grain yield and agronomic traits under stress and non-stress conditions. Significant GCA and SCA mean squares, and hence the importance of both additive and non-additive effects was observed for most traits in most environments. Neither additive nor non-additive genetic effects were important for grain yield across water stress environments. In this case, secondary traits such as anthesis-silking interval and ears per plant with high genetic variability and heritability can be used to increase selection efficiency.

Estimates of GCA effects showed that inbred lines VL054178, VL05482, VL05561, VL05483, CML511, CML159, CML491 and VL06375 had good GCA effects for most traits under stress and non-stress conditions. These inbred lines can be used for the development of QPM hybrids and synthetics that perform well across stress and non-stress environments. In general, the inbred lines used in this study were found to be useful sources for genetic variability for the development of new genotypes for stress tolerance and the study confirmed the possibility of achieving good performances across stress and non-stress conditions in QPM germplasm.

6. References

Baker, R.J. 1978. Issues in diallel analysis. Crop Science 18: 533-536.

Banziger, M and A.O. Diallo. 2004. Progress in developing water stress and N stress tolerant maize cultivars for eastern and southern Africa. In: D.K. Friesen and A.F.E. Palmer (Eds.). Integrated Approaches to Higher Maize Productivity in the New Millennium. Proceedings of the 7th Eastern and Southern Africa Regional Maize Conference. 5-11 February 2002, CIMMYT/KARI, Nairobi, Kenya. pp. 189-194.

Banziger, M. and H.R. Lafitte. 1997. Efficiency of secondary traits for improving maize for low-nitrogen target environments. Crop Science 37: 1110-1117.

Banziger, M., F.J. Betrán and H.R. Lafitte. 1997. Efficiency of high nitrogen selection environments for improving maize for low nitrogen target environments. Crop Science 37: 1103-1109.

Banziger, M., N. Damu, M. Chisenga and F. Mugabe. 1999a. Evaluating the water stress tolerance of some popular maize hybrids grown in sub-Saharan Africa. In: CIMMYT and EARO (Eds.). Maize Production Technologies for the Future: Challenges and Opportunities. Proceedings of the 6th Eastern and Southern African Regional Maize Conference. 21-25 September 1998, CIMMYT and EARO, Addis Ababa, Ethiopia. pp. 61-63.

Banziger, M., G.O. Edmeades and H.R. Lafitte. 1999b. Selection for water stress tolerance increases maize yields across a range of nitrogen levels. Crop Science 39: 1035–1040.

Banziger, M., G.O. Edmeades, D. Beck and M. Bellon. 2000. Breeding for Water stress and N Stress Tolerance in Maize: From Theory to Practice. CIMMYT, Mexico, D.F., Mexico.

Banziger, M., P.S. Setimela, D. Hodson and B. Vivek. 2006. Breeding for improved abiotic stress tolerance in maize adapted to southern Africa. Agricultural Water Management 80: 212-224.

Banziger, M., S. Mugo and G.O. Edmeades. 1999c. Breeding for water stress tolerance in tropical maize - conventional approaches and challenges to molecular approaches. In: J.-M. Ribaut and D. Poland (Eds.). Molecular Approaches for the Genetic Improvement of Cereals for Stable Production in Water-Limited Environments. A Srategic Planning Wokshop held at CIMMYT, El Batan, Mexico. 21-25 June 1999, CIMMYT, Mexico, D.F., Mexico. pp. 69-72.

Beck, D.L., F.J. Betran, M. Banziger and M. Willcox. 1997. From landrace to hybrid: Strategies for the use of source populations and lines in the development of water stress tolerant cultivars. In: G.O. Edmeades, M. Banziger, H.R. Milckekson and C.B. Pena-Valdiva (Eds.). Developing Water stress and Low N-Tolerant Maize. Proceedings of a Symposium. 25-29 March 1996, CIMMYT, Mexico, D.F, Mexico. pp. 369-382.

Betran, F.J., D. Beck, G.O. Edmeades, J.M. Ribaut, M.Banziger and C. Sanchez. 1999. Genetic analysis of abiotic stress tolerance in tropical maize hybrids. In: CIMMYT and EARO (Eds.). Maize Production Technology for the Future: Challenges and Opportunities. Proceedings of the 6th Eastern and Southern African Regional Maize Conference. 21-25 September, CIMMYT and EARO, Addis Ababa, Ethiopia. pp. 69-71.

Betran, J.F., J.M. Ribaut, D.L. Beck and D. Gonzalez de Leon. 2003. Genetic analysis of inbred and hybrid grain yield under stress and non stress environments. Crop Science 43: 807-817.

Bhatnagar, S., F.J. Betran and L.W. Rooney. 2004. Combining ability of quality protein maize inbreds. Crop Science 44: 1997-2005.

Blum, A. 1988. Plant Breeding for Stress Environments. CRC Press, Baco Raton, Florida, USA.

Bolanos, J. and G.O. Edmeades. 1993. Eight cycles of selection for water stress tolerance in lowland tropical maize. I. Responses in grain yield, biomass and radiation utilization. Field Crops Research 31: 233-252.

Bolanos, J. and G.O. Edmeades. 1996. The importance of the anthesis-silking interval in breeding for water stress tolerance in tropical maize. Field Crops Research 48: 65-80.

Bolanos, J., G.O. Edmeades and L. Martinez. 1993. Eight cycles of selection for water stress tolerance in lowland tropical maize. III. Responses in water stress-adaptive physiological and morphological traits. Field Crops Research 31: 269-286.

CIMMYT. 2004. Maize inbred lines release by CIMMYT: A compilation of 497 CIMMYT maize lines (CMLs), CML1 - CML497. April 2004, CIMMYT, Mexico D.F., Mexico.

Cochran, W.G and G.M Cox. 1960. Experimental designs. John Wiley and Sons, New York, USA.

Cooper, M and D.E. Byth. 1996. Understanding plant adaptation to achieve systematic applied crop improvement: A fundamental challenge. In: M. Cooper and G.L. Hammer (Eds.). Plant Adaptation and Crop Improvement. CAB International and IRRI, UK. pp. 5-23.

Dabholkar, A.R. 1992. Elements of Biometrical Genetics. Ashok Kumar Mittal Concept Publishing Company, New Delhi, India.

Derera, J., P. Tongoona, B.S. Vivek and M.D. Laing. 2008. Gene action controlling grain yield and secondary traits in southern African maize hybrids under water stress and non-water stress environments. Euphytica 162: 411-422.

Dhillon, B.S. and W.J. Pollmer. 1978. Combining ability analysis of an experiment conducted in two contrasting environments. EDV in Medizin und Biologie 9: 109-111.

Diallo, A.O., J. Kikafunda, L. Welde, O. Odongo, Z.O. Mduruma, W.S. Chivatsi, D.K. Friesen, S. Mugo and M. Banziger. 2004. Water stress and low nitrogen tolerant hybridsfor the moist mid altitude ecology of eastern Africa. In: D.K. Friesen and A.F.E. Palmer (Eds.). Integrated Approaches to Higher Maize Productivity in the New Millennium. Proceedings of the 7th Eastern and Southern Africa Regional Maize Conference. 5-11 February 2002, CIMMYT/KARI, Nairobi, Kenya. pp. 206-212

Edmeades, G.O., J. Bolanos, S.C. Chapman, M. Banziger and H.R. Lafitte. 1999. Selection improves water stress tolerance to mid/late season water stress in tropical maize populations. I. Gains in biomass, grain yield, and harvest index. Crop Science 39: 1306-1315.

Fan, X.M., J. Tan, J.Y. Yang and H.M. Chen. 2004. Combining ability and heterotic grouping of ten temperate, subtropical and tropical quality protein maize inbreds. Maydica 49: 267-272.

Griffing, B. 1956. Concept of general and specific combining ability in relation to diallel crossing system. Australian Journal of Biological Sciences 9: 463-493

Hadji, T.H. 2004. Combining ability analysis for yield and yield-related traits in quality protein maize (QPM) inbred lines. M. Sc. Thesis. School of graduate studies, Alemaya University, Ethiopia.

Kamara, A.Y., J.G. Kling, S.O. Ajala and A. Menkir. 2004. Vertical root-pulling resistance in maize is related to nitrogen uptake and yield. In: D.K. Friesen and A.F.E. Palmer (Eds.). Integrated Approaches to Higher Maize Productivity in the New

Millennium. Proceedings of the 7th Eastern and Southern Africa Regional Maize Conference. 5-11 February 2002, CIMMYT/KARI, Nairobi, Kenya. pp. 228-232.

Lafitte, H.R. and G.O. Edmeades. 1994. Improvement for tolerance to low soil nitrogen in tropical maize. II. Grain yield, biomass production, and N accumulation. Field Crops Research 39: 15-25.

Makumbi, D., M. Banziger, J.-M. Ribaut and F.J. Betran. 2004. Diallel analysis of tropical maize inbreds under stress and optimal conditions. In: M. Polland, J.Sawkins, J.-M. Ribaut and D. Hoisington (Eds.). Resilent crops for water limited ennvirments: Proceedings of a workshop Held at Cuernavaca, Mexico. 24-28 May 2004, CIMMYT, Mexico D.F., Mexico. pp. 112-113.

Moll, R.H., E.J. Kamprath and W.A. Jackson. 1982. Analysis and interpretation of factors which contribute to efficiency of N utilization. Agronomy Journal 74: 502-564.

Patterson, H.D. and E.R. Williams. 1976. A new class of resolvable incomplete block designs. Biometrika 63: 83-89.

Pixley, K.V. and M.S. Bjarnason. 1993. Combining ability for yield and protein quality among modified endosperm *opaque-2* tropical maize inbreds. Crop Science 33: 1229-1234.

Singh, R.K. and B.D. Chaudhary. 1985. Biometrical Methods in Quantitative Genetics Analysis. 2nd ed. Kalyani Publishers, New Delhi, India.

SAS Institute, Inc. 2003. SAS proprietary Software. SAS Institute, Inc, CARY, NC, Canada.

Singh, D. 1973. Diallel analysis for combining ability over several environments-II. Indian Journal of Genetics and Plant Breeding 33: 469-483.

Zhang, Y. and M.S. Kang. 1997. DIALLEL-SAS: A SAS program for Griffing's Diallel analyses. Agronomy Journal 89: 176-182.

11

Application of Molecular Breeding for Development of the Drought-Tolerant Genotypes in Wheat

Mohamed Najeb Barakat* and Abdullah Abdullaziz Al-Doss
Plant Genetic Manipulation and Genomic Breeding Group,
Center of Excellence in Biotechnology Research,
College of Food and Agricultural,
King Saud University, Riyadh
Saudi Arabia

1. Introduction

Drought is one of the most important environmental challenges growers have to face around the world. Drought is the cause for large grain losses every year, especially in developing countries, and the current trend in global climate change will likely lead to further losses. The worldwide water shortage and uneven distribution of rainfall makes the improvement of drought tolerance especially important (Lou and Zhang, 2001). Breeding for drought tolerance is a major objective in arid and semiarid regions of the world due to inadequate precipitation, shortage of irrigation water and high water demand for crop evapotranspiration in such climates. Little progress has been made in characterizing the genetic determinants of drought tolerance, because it is a complex phenomenon (Tripathy *et al*, 2000). Breeding for water stress tolerance by traditional methods is a time consuming and considered inefficient procedure. Improving the drought tolerance of a crop is difficult for a breeder because yield usually has a relatively low heritability even under ideal condition and an unpredictably variable water supply reduces heritability (Blum, 1988).

Drought tolerance is now considered by both breeders and molecular biologists to be a valid breeding target. In the past, breeding efforts to improve drought tolerance have been hindered by its quantitative genetic basis and our poor understanding of the physiological basis of yield in water-limited conditions (Passioura, 2002). Recently, Tuberosa and Saliva (2007) reported that genomics based approaches provide access to agronomically desirable alleles present at quantitative trait loci (QTLs) that affect such responses, thus enabling us to improve the drought tolerance and yield of crops under water limited conditions more effectively.

Compared to conventional approaches, genomics offers unprecedented opportunities for dissecting quantitative traits into their single genetic determinants (QTLs), thus paving the way to marker-assisted selection (MAS) (Ribaut *et al*,2002; Morgante and Salamini, 2003) and eventually, cloning of genes at target QTLs (Salivi and Tuberosa, 2005).Recently, Tuberosa

*Corresponding Author

and Salivi (2007) demonstrated that how the information on QTLs governing the response to drought and candidate genes responsible for QTL effects can be used to elucidate the physiological basis of drought tolerance and to select genotypes with an improved yield under water-limited condition.

Molecular markers improve the efficiency of breeding by allowing manipulation of the genome through marker-assisted selection. Leaf senescence is the sequence of biochemical and physiological events comprising the final stage of leaf development from the mature fully extended state, until death. It is induced either by internal hormonal factors related to ageing, or, prematurely, by external environmental factors such as high temperature and drought (Chandler, 2001). In wheat (*Triticum aestivum* L.), flag leaf senescence (FLS) relates to the period of reallocating resources from the source to the sink during grain filling. Since flag leaf photosynthesis in wheat contributes about 30–50% of the assimilates for grain filling (Sylvester-Bradley *et al.*, 1990), the onset and rate of senescence are important factors for determining yield potential (Evans, 1993). In order to identify molecular markers for flag leaf senescence, it is first necessary to construct a genetic map as a tool for discovering the genetic factors as quantitative trait loci (QTL). Though QTLs influencing senescence have been identified both in sorghum (Tuinstra *et al.*, 1997; Crasta *et al.*, 1999; Xu *et al.*, 2000; Kebede *et al.*, 2001; Harris *et al*, 2007) and maize (Beavis *et al.*, 1994). Mapping quantitative trait loci for flag leaf senescence as a yield determinant in winter wheat under optimal and drought-stressed environments have been reported (Verma *et al*, 2004).

In the near future, molecular markers can provide simultaneous and sequential selection of agronomically important genes in wheat breeding programs allowing screening for several agronomically important traits at early stages and efficiently replace time consuming bioassays in early generation screens (Patnuk and Khurana, 2001). They also reported that application of biotechnology will thus contribute greatly to improving yield stability by generating plants with improved resistance to biotic and stresses rather than raising the overall yield. The coming years will undoubtedly witness an increasing application of genomics-assisted breeding for the genetic improvement of wheat. The goal of wheat breeding is to combine desirable genes from different lines into new varieties. However, it is often difficult to monitor for the presence of multiple desirable genes during the selection process. Genomics has revolutionized plant breeding by providing tools for high-throughput marker evaluations, which can be used in Marker-Assisted Selection (MAS) strategies for variety improvement.

The application of molecular markers to plant breeding can be divided into three main categories:(I) the characterization of germplasm, known as fingerprinting; (II) the genetic dissection of the target trait- actually the identification and characterization of genomic regions involved in the expression of the target trait; and (III) following the identification of the genomic regions of interest, crop improvement through marker-assisted selection (MAS). The first two applications have proved themselves by generating knowledge about the genetic diversity of germplasm, thereby allowing placement into heterotic groups and a better understanding of the genetic basis of agronomic traits of interest. For simply inherited traits - those that have high heritability and are regulated by only a few genes–the use of molecular markers to accelerate germplasm improvement has been well documented (e.g. Johnson and Mumm, 1996; Mohan *et al.*, 1997; Young, 1999).Such work has proved successful in: (i) tracing favorable alleles in the genomic background of genotypes of interest; and (ii) identifying individual plants in large segregating populations that carry the favorable alleles. Moreover, with the recent development of PCR-based markers, for example, simple sequence repeats (SSRs) (Chin*etal.*, 1996; Powell*etal.*,1996) and single nucleotide polymorphisms(SNPs) (Gilles *et al.*, 1999), a substantial improvement in the capacity to efficiently screen large populations has been achieved, thereby increasing the efficiency of MAS experiments.

2. Development of drought tolerance-associated DNA markers

Twelve wheat genotypes were used in this study. These included the two recommended cultivars (Yecora Rojo and West Bread) as well as ten advanced lines (F8) selected from the wheat breeding program at the Plant Production Department, College of Food and Agriculture Sciences, King Saud University, Saudi Arabia. They evaluated phenotypically for drought tolerance and were planted under four irrigation treatments over two seasons to expose genotypes to different level of drought stress during filling period. The four irrigation treatments were formed by irrigation scheduled at cumulative pan evaporation (CPE) of T1:50, T2:100, T3:150 and T4:200 mm during the entire irrigation interval. The CPE was calculated as sum of daily recorded evaporation from USWB open pan. The pan was located at the Meteorological Station adjacent the experimental site. Two types of molecular markers, (RAPD) and (ISSR), were assayed to determine the genetic diversity of 12 wheat genotypes and to develop of drought tolerance-associated DNA markers.

The potential of using markers generated in the current study to develop drought tolerance - associated DNA markers is presented in Table 1. For the RAPD analysis presented here some wheat genotypes reported to be drought tolerant/sensitive (on the basis of field performance) were used. Figure 1 and Table 1 indicated that a DNA band at about 310 bp that are present in Ksu103 as a drought tolerance, but not in Yecora Rojo as drought sensitive, when primer OPE20 is used .On the other hand, specific DNA bands at 1400bp and 1200bp are present in Yecora Rojo as drought sensitive, but not in the Ksu103 and Ksu105 as a drought tolerance, when primer OPE20 was used. Moreover, specific DNA bands generated from RAPD primers (Table 1) could be used to characterize between Ksu103 and Ksu105 (drought tolerance) and Yecora Rojo (drought sensitive). For the ISSR analysis, a polymorphic DNA fragments of 950bp and 740bp were identified in Ksu103 as well as Ksu105 and were absent in the Yecora Rojo, when primer ISSR-811was used (Fig.1; Table 1). These fragments appear to be linked to drought tolerance genes .On the other hand, specific DNA bands at 1200bp and 1040bp are present in Yecora Rojo as drought sensitive, but not in the Ksu103 and Ksu105 as a drought tolerance. In addition, specific DNA bands generated from ISSR primers (Table 1) could be used to characterize between Ksu103 as well as Ksu105 (drought tolerance) and Yecora Rojo (drought sensitive). The reproducibility of these variety specific markers was confirmed in RAPD and ISSR analyses for which DNA isolation, PCR amplification, and gel electrophoresis were carried out separately. Molecular marker technology has allowed the identification and genetic characterization of QTLs with significant effects on stress tolerance during different stages of plant development and facilitated determination of genetic relationships among tolerance to different stresses (Foolad 2005). Comparatively, however, limited research has been conducted to identify genetic markers associated with drought tolerance in different plant species.

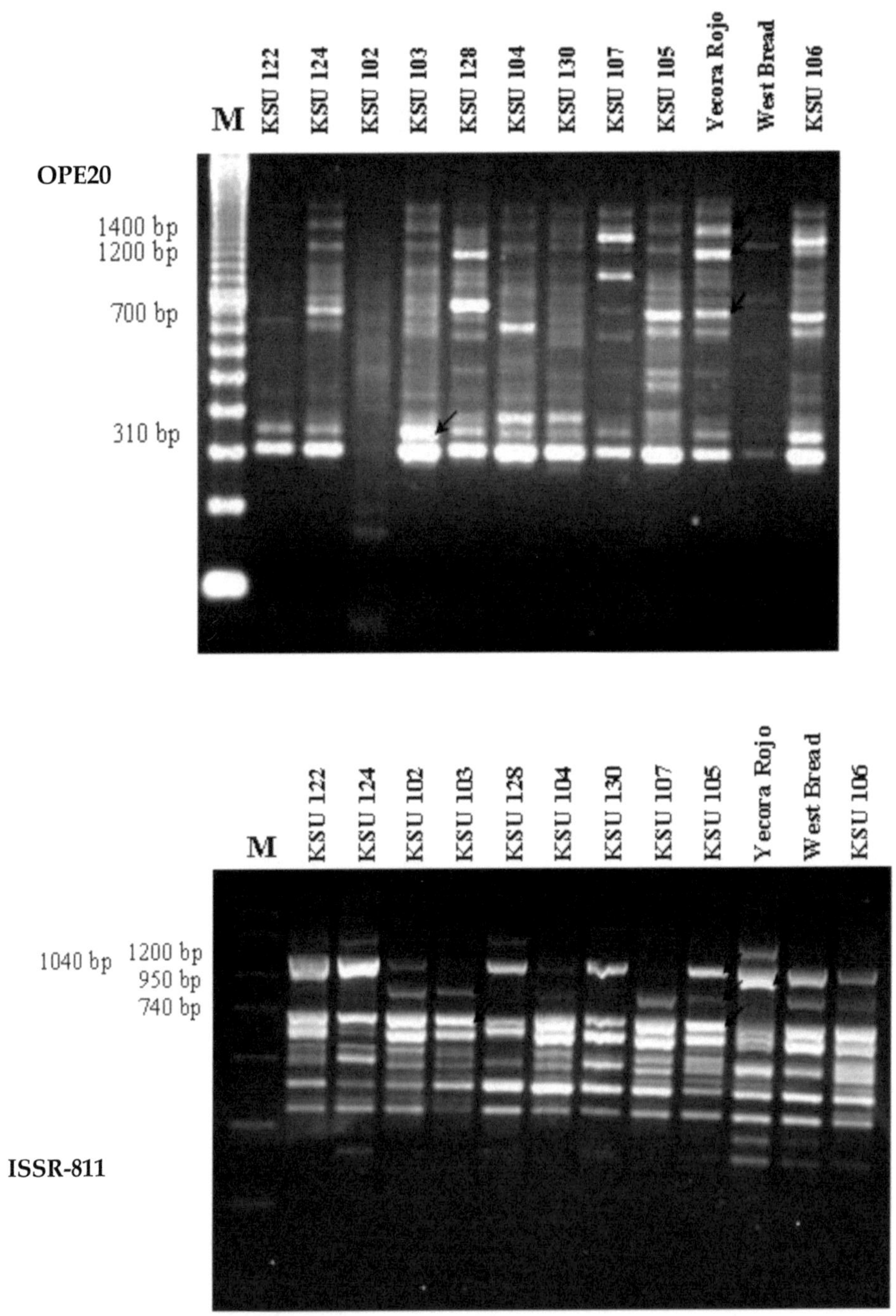

Fig. 1. Polymorphic DNA fragments linked to drought tolerance genes, generated by RAPD primer OPE20 (5` AACGGTGACC 3') and generated by ISSR primer ISSR-811 (5` G (AG) 7AC 3') M: Molecular weight, followed by wheat genotypes.

RAPD					ISSR				
Primer	Marker	Ksu105	Yocora Rojo	Ksu103	Primer	Marker	Ksu105	Yocora Rojo	Ksu103
Pr10	680	+	-	+	ISSR-811	1200	-	+	-
Pr16	750	+	-	-		1040	-	+	-
	500	+	-	+		950	+	-	+
	250	+	-	+		740	+	-	+
OPC04	300	+	-	-	ISSR-816	500	+	-	+
OPC15	630	-	+	+	ISSR-817	540	-	+	-
OPE20	1400	-	+	-		520	+	-	+
	1200	-	+	-	ISSR-11	280	-	+	+
	700	+	+	-	ISSR-821	900	-	+	-
	310	-	-	+		850	+	-	-
OPE15	620	-	-	+		320	-	+	-
	410	+	-	+					
	330	+	-	+					
OPU06	420	-	+	+					
	320	-	+	-					
OPZ03	1400	-	+	+					

Table 1. Specific DNA fragments generated from RAPD and ISSR analysis to develop drought tolerance-associated DNA markers between Yecora Rojo (drought-sensitive) and Ksu 105& Ksu 103 (drought-tolerance).

3. Drought QTL identification

Genetic mapping with dense marker maps can be used to identify the number and genetic positions of Quantitative Trait Loci (QTL) associated with a specific phenotype under drought stress. In addition, this process can be used to estimate effects of the segregating QTL and their contributions to trait variation (individually and in combined QTL models), and obtain estimates of their stability across environments (QTL x E interactions) and across genetic back- grounds (QTL x QTL interactions).

The timing of flag leaf senescence is an important determinant of yield under stress and optimal environments. A segregating populations from the two crosses; the first cross between drought sensitive genotype (Variant-2 which was derived from Gemeza 1 cultivar; using somaclonal variation (Barakat *et al.*, 2005)) and drought tolerant genotype (Cham-6) and the second cross between drought sensitive genotype (Variant-1 which was derived from Sakha 69 cultivar, using somaclonal variation tool (Barakat *et al.*, 2005)) and drought tolerant genotype (Veery), were made to identify molecular markers linked to flag leaf senescence in wheat under water-stressed conditions. This trait was utilized as an indicator for drought tolerance genes and mapping QTL for flag leaf senescence in F_2 populations using bulked segregant analysis. Thirty-eight RAPD primers, twenty-five ISSR primers and fourty -six SSR primers were tested for polymorphism among parental genotypes and F_2 population.

Drought tolerance evaluation have been made to identify molecular markers linked to flag leaf senescence: One hundred F_2 plants and their parents were planted in polyethylene bags (13 cm diameter, 15 cm height) under green house conditions in the winter season of 2008 to evaluate drought tolerance. All plants were grown at 20/15°C (day/night), with 50/70 % relative humidity and16-h photoperiod. The bags were filled with sandy soil (3.5 kg) and were given the total amount of daily irrigation until reaching booting stage. Drought tests were carried using 50% of the amount of daily irrigation. Daily irrigation water requirements were calculated by CROPWAT software (Smith 1991) from agro-meteorological data of the studied area and Kc of wheat as follows.

$$ETo = \frac{0.408\Delta(Rn - G) + \gamma \frac{37}{Thr + 273} u2\ (e^{\circ}\ (\mathrm{Thr}) - ea)}{\Delta\ +\ \gamma(1 + 0.34\mathrm{u2})}$$

$$ETc = \mathrm{Eto} * \mathrm{Kc}$$

Where, ETc = Evapotranspiration for crop, Kc = Crop coefficient, ETo =Reference evapo-transpiration (mm h^{-1}), Rn = net radiation at the grass surface, (MJ m-2h^{-1}), G = soil heat flux density (MJ m-2h^{-1}), Thr = mean hourly air temperature (°C), Δ =Saturation slope vapor pressure curve at Thr (kPa °C^{-1}), γ = psychrometric constant (kPa °C^{-1}), eo (Thr) = saturation vapor pressure at air temperature Thr, ea = average hourly actual vapor pressure, u2 = average hourly actual wind speed (ms^{-1}).

Calculated ETc, (crop evapotranspiration), which is equal 100% of daily water consumption use for the wheat was used to calculate irrigation requirements with the following equation:

Daily irrigation requirements (IR) = ETc +15% (leaching requirements).

The data of daily IR was adjusted to the volume of Polyethylene bags used and the Table 2 shows the volume of daily IR in cubic cm till the stage of flag leaf appearance (35 days old) then drought tests were carried for 21 days. After 21 days from the stress condition, the flag leaf of the main tiller of each plant was obtained during morning hours when leaves were fully turgid. The percentage of flag leaf area remaining green (% GFLA) was measured by using the leaf area meter (Portable Living Leaf Area Meter, Model: YMJ, Zhejiang Top Instrument Co., Ltd). These assessments were carried out by the same operator in the population to avoid any bias between operator's influencing results.

The present study in the first population (Cham-6 x Variant-2) indicated that one RAPD marker (Pr_9 primer (5` GGGTAACGCC 3')), four ISSR markers,(Pr8, AD5; AD2 and AD3 primers (5' (GTG)5 3'; 5' (CA)10C 3'; 5' (AGC)6G3' and 5' (ACC)6G3', respectively)) ,and one SSR *XGWM 382-2D* (Right; CTACGTGCACCACCATTTTC and Left; GTCAGATAACGCCGTCCAAT) linked to the flag leaf senescence in wheat, were identified

(Fig. 2). QTL for flag leaf senescence was associated with RAPD marker (Pr9$_{270bp}$), ISSR markers (Pr8$_{380bp.}$, AD5$_{900bp}$, AD2$_{600bp}$ and AD3$_{700\ bp.}$), and SSR marker (*XGWM 382-2D*$_{108bp}$) and explained 44.0%, 50.0%, 35.0%, 31.0%, 22.0%and 73% of the phenotypic variation, respectively. The genetic distance (Fig.3) between RAPD marker (Pr9$_{270bp}$) and flag leaf senescence gene was determined to be 10 cM, with LOD score of 22.9. The ISSR markers (Pr8, AD5, AD2 and AD3) have genetic distance of 10.5, 14.6, 15.6 and 18.1 cM, respectively, from flag leaf senescence gene. In addition, the genetic distance between SSR marker XGWM 382-2D$_{110bp}$ and flag leaf senescence gene was determined to be 3.9 cM, with LOD score of 33.8. Therefore, the RAPD, ISSR and SSR markers were linked to the QTL for the flag leaf senescence as indicator for drought tolerance gene in wheat. Once these markers are identified, they can be used in wheat breeding programs, as a selection tool in early generations.

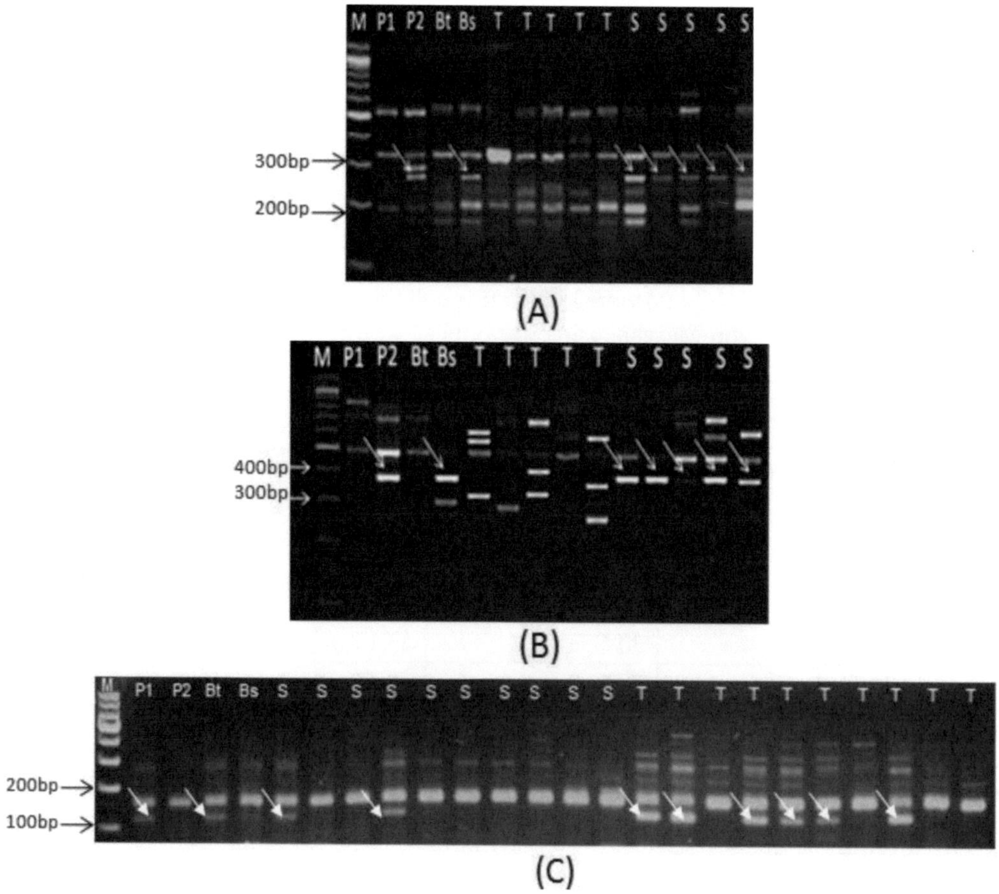

Fig. 2. RAPD fragments (A), produced by primer 9 (5` **GGGTAACGCC** 3'), ISSR fragments (B), produced by Pr.8 (5` (ACA CAC) 2ACA CG 3') and SSR fragments (C), produced by *XGWM 382-2D*. M: Molecular weight followed by P_I and P_2 parents Cham-6 and Variant-2, respectively. B_t, bulk tolerant; Bs, bulk sensitive, F_2 individuals in the cross, Cham-6 X Variant-2 (T: tolerant; S: sensitive)

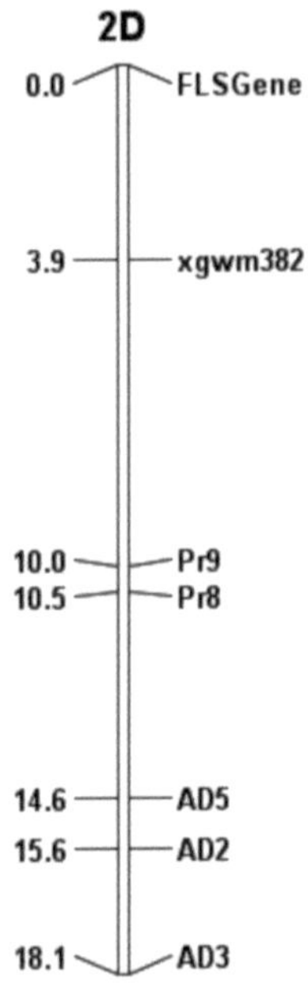

Fig. 3. RAPD marker (**Pr9** $_{\mathbf{270bp}}$), ISSR markers (Pr8$_{380bp}$, AD5$_{900bp}$; AD2$_{600bp}$ and AD3$_{700bp}$) and SSR marker (*XGWM 382-2D*$_{110bp}$) were located through the MAPMAKER-QTL analysis. All distances are given in centi-Morgan, using Kosambi's mapping function.

Days	**daily IR cm^3**	**50% of the amount of daily irrigation cm^3**
1-10 March	31.9	15.95
11-20 March	28.2	14.1
21-31 March	38.9	19.45
1-10 April	38.1	19.05
11-20 April	30.3	15.15
21-30 April	20.9	10.45

Table 2. Daily irrigation requirements (cm^3) from 1st March to 30 April. Drought tests were started with the stage of flag leaf appearance.

The present study in the second population (Veery x Variant-1) indicated that out of 38 RAPD arbitrary primers , screened for polymorphisms between the two tested parents (Veery and Variant-1), 24 RAPD primers (63.2 %), that gave polymorphic bands suitable to differentiate between the two parents, were identified. Of these 24 RAPD primers, Pr$_{11}$ primer (5' CAATCGCCGT 3'), produced one strong polymorphic band at 230 bp, that was present only in the tolerant parent (Veery), as shown in Figure (4). The Pr11 primer was selected for screening DNA bulks and their parental DNA .The Pr$_{11}$ primer , generated one polymorphic fragments at 230 bp, which was present only in tolerant bulk and Veery (tolerant parent) and were missing in sensitive bulk and Variant-1 (sensitive parent), as shown in Figure (4). In addition, primer Pr.19 (5` CAAACGTCGG 3'), produced a strong polymorphic band at 240 bp that was present only in the sensitive bulked DNA, but not in the tolerant bulked DNA and primer OPU06 (5` ACCTTTGCGG 3'), produced a strong polymorphic band at 340 bp that was present only in the sensitive bulked DNA, but not in the tolerant bulked DNA. Also, primer OPH13 (5` GACGCCACAC 3'), produced a strong polymorphic band at 450 bp that was present only in the sensitive bulked DNA, but not in the tolerant bulked DNA. These RAPD markers ($Pr11_{230bp}$, $Pr19_{240bp}$, $OPU06_{340bp}$ and $OPH13_{450bp}$) were regarded as candidate markers, linked to the flag leaf senescence gene as indicator for drought tolerance.These polymorphic markers; viz, $Pr11_{230bp}$, $Pr19_{240bp}$, $OPU06_{340bp}$ and $OPH13_{450bp}$, were further used to check their linkage to the flag leaf senescence gene, using a segregating F_2 population, derived from the cross between the tolerant parent (Veery) and the sensitive parent (Variant-1). When analyzing the individual plants of F_2 population, the $Pr11_{230bp}$ and $OPU06_{340bp}$ fragments were amplified in the DNA, obtained only in F_2 tolerant ones. In addition $Pr19_{240bp}$ and $OPH13_{450bp}$ fragments were amplified in the DNA, obtained only in F_2 sensitive ones. The RAPD markers, $Pr11_{230bp}$, $Pr19_{240bp}$, $OPU06_{340bp}$ and $OPH13_{450bp}$ 80; 33; 67 and 36 of 100 individuals, respectively, in the F_2 population, exhibited the amplified polymorphic fragments (230, 240, 340 and 450bp), while, the remaining did not. The ratio fitted the expected Mendalian ratio, 3:1 for all markers except for $OPH13_{450bp}$ which did not fitted the ratio (Table 3).To check for potential co-segregantion of DNA fragments and drought tolerant phenotypes, correlation and simple regression analysis were carried out in order to confirm an association between the $Pr11_{230bp}$, $Pr19_{240bp}$, $OPU06_{340bp}$ and $OPH13_{450bp}$ markers and the flag leaf senescence gene as indicator for tolerant to drought in all 100 F_2 progenies. The results showed that the correlation and the coefficient of determination (R^2) for the relationship between the four markers ($Pr11_{230bp}$, $Pr19_{240bp}$, $OPU06_{340bp}$ and $OPH13_{450bp}$) and the phenotypes of F_2 individuals were significant and they recorded r = 0.26, -0.71, 0.49 and -0.36, respectively, R^2 = 0.07, 0.50, 0.24 and 0.13, respectively (Table 3). This indicates that the four markers were associated with the flag leaf senescence gene as indicator for drought tolerance.

	Markers	Tested of plant			Expected Ratio	χ^2	r	R^2	QTL cM	LOD
		Total	T	S						
RAPD	$Pr11_{230bp}$	100	80	20	3 : 1	1.33^{ns}	0.26	0.07**	18.6	16.4
	$Pr19_{240bp}$	100	67	33	3 : 1	3.41^{ns}	-0.71	0.50**	15.0	19.0
	$OPU06_{340bp}$	100	67	33	3 : 1	3.41^{ns}	0.49	0.24**	13.2	20.7
	$OPH13_{450bp}$	100	64	36	3 : 1	6.4**	-0.36	0.13**	17.4	16.0
ISSR	$M1_{1100bp}$	100	67	33	3 : 1	3.41^{ns}	0.50	0.25**	12.5	21.6
	$AD2_{300bp}$	100	76	24	3 : 1	0.05^{ns}	-0.58	0.34**	10.2	26.0

T=Tolerant plants, S= sensitive plants
Ns, **= non significant and significant at .01 level of probability

Table 3. QTL analysis and significant association between drought tolerance and markers (RAPD and ISSR) in the 100 F_2 plants population of Veery x Variant-1, using Chi - square (χ^2), correlation (r) and coefficient of determination (R^2) analysis.

Out of 33 ISSR primers, screened for polymorphisms between the two tested parents, thirteen ISSR primers (34.2 %) that gave polymorphic bands suitable to differentiate between the two parents, were identified. Of these thirteen ISSR primers, M1 and AD2 primers (5'(AC) 8CG3' and 5'(AGC) 6G3', respectively), which produced two strong polymorphic bands at 1100 and 300 bp, respectively. The M1 primer produced one strong polymorphic band at 1100 bp that was present only in the tolerant parent (Veery), as shown in Figure (5). The M1 primer was selected for screening DNA bulks and their parental DNA .The M1 primer , generated one polymorphic fragments at 1100 bp, which was present only in tolerant bulk and Veery (tolerant parent) and were missing in sensitive bulk and Variant-1 (sensitive parent), as shown in Figure (5). In addition, AD2 primer was generated one polymorphic fragment at 300 bp that were present only in the sensitive parent (Variant-1). AD2 ISSR primer was selected for screening DNA bulks and their parental DNA. The primer AD2 generated the polymorphic fragments 300 bp, which were present only in the sensitive bulk and Variant-1 (sensitive parent) and were missing in tolerant bulk and Veery (tolerant parent). These ISSR markers ($M1_{1100bp}$ and $AD2_{300bp}$) were regarded as candidate markers linked to the flag leaf senescence gene as indicator for drought tolerance in wheat. These polymorphic markers, $M1_{1100bp}$ and $AD2_{300bp}$, were further used to check their linkage to the flag leaf senescence gene, using a segregating F_2 population, derived from the cross between the tolerant parent, Veery, and the sensitive one, Variant-1. When analyzing the individual plants of F_2 population, the $M1_{1100bp}$ fragments were amplified in the DNA, obtained only in F_2 tolerant ones and the $AD2_{300bp}$ fragments were amplified in the DNA, obtained only in F_2 sensitive ones. The ISSR markers, $M1_{1100bp}$ and $AD2_{300bp}$, 76 and 24 of 100 individuals respectively, in the F_2 population, exhibited the amplified polymorphic fragments (1100 and 300 bp), while, the remaining did not. The ratio fitted the expected Mendalian ratio, 3:1 (χ^2= 3.41 and 0.05, $P < 0.1$ respectively) (Table 3).To check for potential co-segregantion of DNA fragments and drought tolerant phenotypes, correlation and simple regression analysis were carried out in order to confirm an association between the each $M1_{1100bp}$ and $AD2_{300bp}$ markers and the flag leaf senescence gene as indicator for the tolerance to drought in all 100 F_2 progenies. The results showed that the correlation and the coefficient of determination (R^2) for the relationship between presence of the two markers, $M1_{1100bp}$ and $AD2_{300bp}$, and the phenotypes of F_2 individuals were significant and they recorded $r = 0.5$ and -0.58, respectively, $R^2 = 0.25$ and 0.34, respectively (Table 3). This indicates that the two markers were linked to the flag leaf senescence gene as indicator for drought tolerance.

The linkage relationship between the RAPD markers ($Pr11_{230bp}$, $Pr19_{240bp}$, $OPU06_{340bp}$ and $OPH13_{450bp}$) and the flag leaf senescence gene as indicator for drought tolerance was estimated, using F_2 population, derived from the cross, Veery X Variant-1. The genetic distance between RAPD markers ($Pr11_{230bp}$, $Pr19_{240bp}$, $OPU06_{340bp}$ and $OPH13_{450bp}$) and the flag leaf senescence gene were determined to be 15.6, 15.0, 13.2 and 17.4 cM, respectively, with LOD scores of 16.4, 19.0, 20.7 and 16.0, respectively (Table 3 and Fig.6). Therefore, RAPD markers ($Pr11_{230bp}$, $Pr19_{240bp}$, $OPU06_{340bp}$ and $OPH13_{450bp}$ primers) were linked to the quantitative trait loci (QTL) for the flag leaf senescence gene as indicator for drought tolerance.

After mapmaker linkage analysis on the F_2 population, the genetic distance between ISSR markers ($M1_{1100bp}$ and $AD2_{300bp}$) and the flag leaf senescence gene was determined to be 12.5 and 10.2 cM, respectively, with LOD scores of 21.6 and 26.0, respectively (Table 3 and Fig.6).

Therefore, ISSR markers ($M1_{1100bp}$ and $AD2_{300bp}$) were linked to the quantitative trait loci (QTL) for the flag leaf senescence gene as indicator for drought tolerance.
One-way ANOVA was carried out using marker genotypes as groups. The ANOVA on RAPD markers ($Pr11_{230bp}$, $Pr19_{240bp}$, $OPU06_{340bp}$ and $OPH13_{450bp}$) and ISSR markers ($M1_{1100bp}$ and $AD2_{300bp}$) genotypes as groups for flag leaf senescence established significant association between marker (RAPD markers and ISSR markers) and phenotype (flag leaf senescence) (Tables 3 and 4). The single marker ANOVA analysis revealed that RAPD markers-linked QTL ($Pr11_{230bp}$, $Pr19_{240bp}$, $OPU06_{340bp}$ and $OPH13_{450bp}$) accounted for 7.0%, 50.0%, 24.0% and 13.0% of the total variation, respectively, in flag leaf senescence in F2 population. While, ISSR markers-linked QTL ($M1_{1100bp}$ and $AD2_{300bp}$) accounted for 25% and 34% of the phenotypic variation, respectively, in flag leaf senescence in F2 population, in wheat under water-stressed conditions.

Marker	Source	DF	SS	MS	F	P
$Pr11_{230bp}$	Genotypes	1	520.75	520.75	4.578	0.034
	Error	98	11146.5	113.73		
	Total	99	11667.25			
$Pr19_{240bp}$	Genotypes	1	4057.45	4057.45	52.25	0.0001
	Error	98	7609.81	77.65		
	Total	99	11667.25			
$OPU06_{340bp}$	Genotypes	1	3029.35	3029.35	34.37	0.0001
	Error	98	8637.91	88.14		
	Total	99	11667.25			
$OPH13_{450bp}$	Genotypes	1	3208.85	3208.85	37.18	0.0001
	Error	98	8458.41	86.31		
	Total	99	11667.25			
$M1_{1100bp}$	Genotypes	1	3089.19	3089.19	35.29	0.0001
	Error	98	8578.07	87.53		
	Total	99	11667.25			
$AD2_{300bp}$	Genotypes	1	4073.52	4073.52	52.57	0.0001
	Error	98	7593.74	77.49		
	Total	99	11667.25			

Table 4. Analysis of variance of the difference for flag leaf senescence as determinant for drought tolerance in F_2 population.

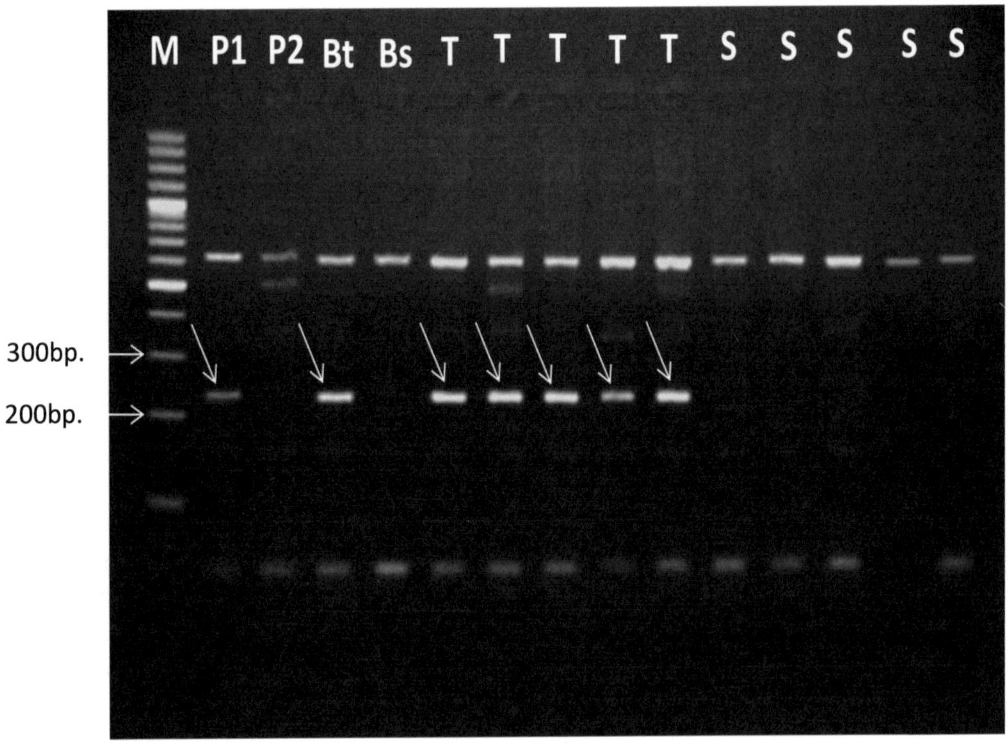

Fig. 4. RAPD fragments, produced by primer11 (5` CAATCGCCGT 3'), M: Molecular weight followed by P_I and P_2 parents Veery and Variant-1, respectively. Bt, bulk tolerant; Bs, bulk sensitive, F_2 individuals in the cross (T: Tolerant; S: sensitive). Arrow points to polymorphic bands of the $Pr11_{230bp}$ marker.

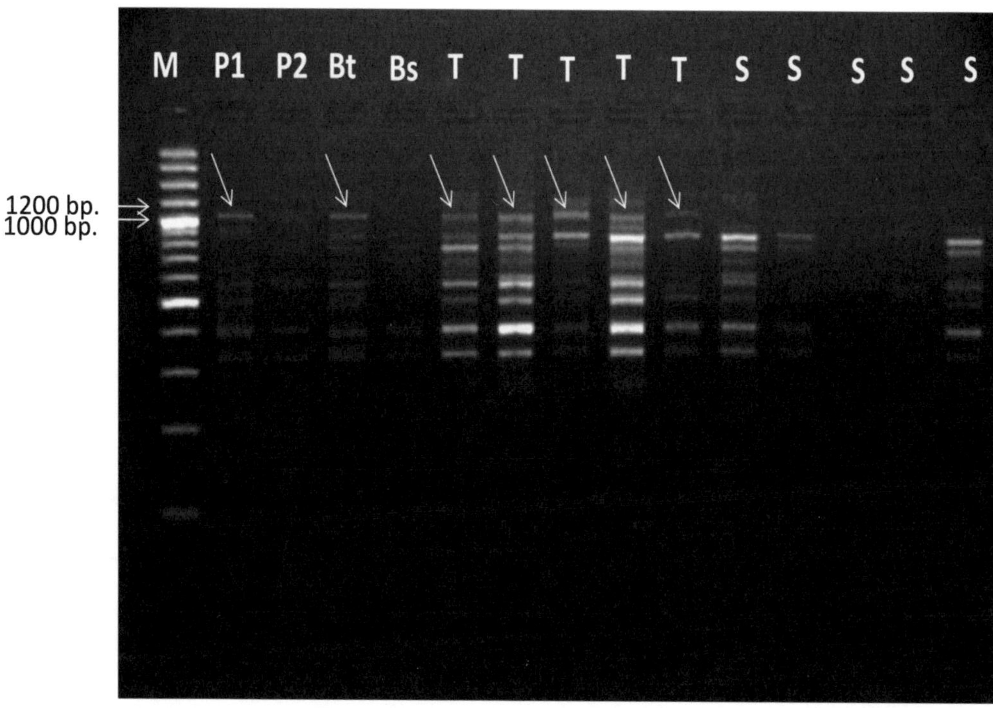

Fig. 5. ISSR fragments, produced by M1 (5` (AC)8CG 3'), M: Molecular weight, followed by P_I and P_2 parents Veery and Variant-1, respectively. Bt, bulk tolerant; Bs, bulk sensitive, F_2 individuals in the cross (T: Tolerant; S: sensitive). Arrow points to polymorphic bands of the $M1_{1100bp}$ marker.

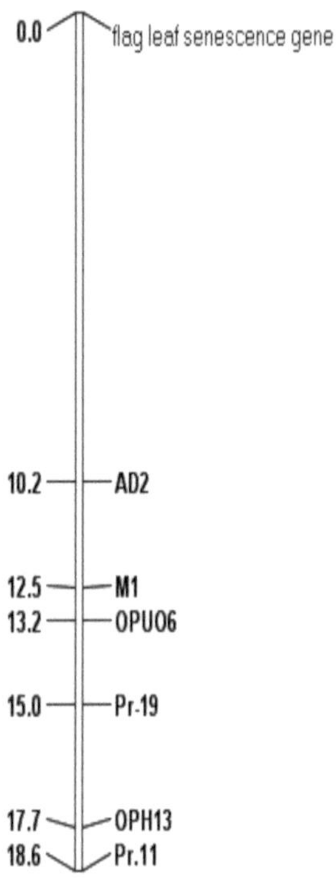

Fig. 6. RAPD markers ($Pr11_{230bp}$, $Pr19_{240bp}$, $OPU06_{340bp}$ and $OPH13_{450bp}$), ISSR markers ($M1_{1100bp}$ and $AD2_{300bp}$) and the flag leaf senescence gene were located through the MAPMAKER-QTL analysis. All distances are given in centi-Morgan, using Kosambi's mapping function.

Water-stress tolerance in wheat is a quantitatively inherited trait controlled by several genetic loci, and several of its genetic components are difficult to measure (Forster et al., 2000). Identification of associated molecular markers at a major locus contributing to water-stress tolerance would be useful for the indirect selection of wheat plants for water-stress tolerance (Visser, 1994). However, identifying molecular markers associated with important genes or traits in most instances requires screening of a relatively large number of individuals in the population (Lawson et al., 1994). Bulked segregant analysis (BSA) was originally developed to overcome this difficulty, because comparing bulk samples is easier than evaluating many individuals in different populations (Sweeney and Dannebeger, 1994).

The use of molecular markers can increase the efficiency of conventional plant breeding by identifying markers linked to the trait of interest, which are difficult to evaluate and/or are largely affected by the environment. Hence, there is a need to develop a rapid screening method to select for drought tolerance. Tight linkage between molecular markers and gene for flag leaf senescence can be of great benefit to drought tolerance breeding programs by allowing the investigator to follow the DNA markers (PCR-based markers) through early generation rather than waiting for phenotypic expression of the tolerance genes. Molecular markers that are closely linked with target alleles present a useful tool in plant breeding since they can help to detect the tolerant genes of interest without the need of carrying out field evaluation. Also, it allows for screening big number of breeding materials at early growth stages and in short time.

The present study indicated that RAPD, ISSR and SSR markers, combined with bulked segregant analysis, could be used to identify molecular markers linked to the flag leaf senescence gene as indicator for drought tolerance in wheat. Once these markers are identified, they can be used in wheat breeding programs as a selection tool in early generations.

4. References

Barakat, M. N., S.I. Milad & I.A. Imbaby.2005. Field evaluation for rust diseases and RAPD analysis for somaclonal variant lines in wheat. Alex. J. Agric. Res. 50: (3) 11-24, 2005.

Beavis, W.D., O.S. Smith, D. Grant & R. Fincher. 1994. Identification of quantitative trait loci using a small sample of top-crossed and F4 progeny from maize. Crop Sci 34: 882-896.

Blum, A. 1988. Plant breeding for stress environments. CRC Press. Boca Raton, Florida, US

Chandler, J.M. 2001. Current molecular understanding of the genetically programmed process of leaf senescence. Physiologia Plantarum 113: 1-8.

Chin, E.C.L., Senior, M.L., Shu, H. and Smith,J.S.C. (1996) Maize simple repetitive DNA sequences: abundance and allele variation. Genome 39, 866–873.

Crasta, O.R., W.W. Xu, D.T. Rosenow, J. Mullet and H.T. Nguyen.1999.Mapping of post-flowering drought resistance traits in grain sorghum: association between QTLs influencing premature senescence and maturity. Mol & General Genet 262: 579–588.

Evans, L.T., 1993. Crop Evolution, Adaptation and Yield. Cambridge University Press, Cambridge, UK.

Foolad M (2005) Breeding for abiotic stress tolerances in tomato. In: Ashraf, M., Harris, P.J.C. (Eds.), Abiotic Stresses: Plant Resistance Through Breeding and Molecular Approaches. The Haworth Press Inc., New York, USA, pp. 613–684.

Forster, B.P., Ellis, R.P., Thomas, W.T.: The development and application of molecular markers for abiotic stress tolerance in barley. J. Exp. Bot. 51: 19-27, 2000

Gilles, P.N., Wu, D.J., Foster, C.B., Dillon, P.J.and Chanock, S.J. (1999) Single nucleotide polymorphic discrimination by an electronic dot blot assay on semiconductor microchips. Nature Biotechnology 17, 365–370.

Johnson, G.R. and Mumm, R.H. (1996) Marker assisted maize breeding. In: Proceedings of the 51st Annual Corn and Sorghum Research Conference, Chicago, Illinois, 11–12 December 1996. American Seed Trade Association, Washington, DC, pp. 75–84.

Harris K., P. K. Subudhi , A. Borrel , D. Jordan , D. Rosenow , H. Nguyen, P. Klein, R. Klein and J. Mullet . 2007. Sorghum stay-green QTL individually reduce post-flowering drought-induced leaf senescence. J. of Exp. Botany. 58: 327-338.

Kebede, H., P.K. Subudhi, D.T. Rosenow and H.T. Nguyen. 2001. Quantitative trait loci influencing drought tolerance in grain sorghum (Sorghum bicolor L. Moench). Theor App Genet 103:266–276.

Lawson, W.R., Henry, R.J., Kochman, J.K., Kong, G.A.: Genetic diversity in sunflower (*Helianthus annus L.*) as revealed by random amplified polymorphic DNA analysis. Aust. J. Agric. Res. 45: 1319-1327, 1994.

Luo, L. J., and Q. F. Zhang. 2001.The status and strategy on drought resistance of rice (Oryza sativa L.). Chinese J. Rice Sci. 15: 209–214.

Morgante M. and F. Salamini.2003.From plant genomics to breeding practice. Curr.Opin.Biotechnol.14:214-219.

Mohan, M., Nair, S., Bhagwat, A., Krishna, T.G.,Yano, M., Bhatia, C.R. and Sasaki, T. (1997) Genome mapping, molecular markers and marker-assisted selection in crop plants. Molecular Breeding 3, 87–103.

Passioura J.B.2002. Environmental biology and crop improvement. Funct.Plant Biol.29:537-546.

Powell, W., Machray, G.C. and Provan, J. (1996) Polymorphism revealed by simple sequence repeats. Trends in Plant Science 1, 215–222.

Ribaut,j.m. et al.2002. Use of molecular markers in plant breeding drought tolerance improvement in tropical maize .In Quantitative Genetics, Genomics, and Plant Breeding(Kang,M.S.,ed),pp.85-99,CABI Publishing

Salvi S. and R.Tuberosa .2005.To clone or not to clone plant QTLs: present and future challenges. Trends Plant Sci.10:297-304.

Smith, M.: "CROPWAT: Manual and Guidelines.", FAO of UN, Rome, 1991

Sweeney, P.M., Danneberger, T.K.: Random amplified polymorphic DNA in perennial ryegrass: a comparison of bulk samples vs. individuals. Hort Science 29: 624-626, 1994.

Sylvester-Bradley, R., R.K. Scott and C.E. Wright. 1990. Physiology in the production and improvement of cereals. Home-grown Cereals Authority Research Review 18. HGCA, London.

Tripathy, J. N., J. Zhang, S. Robin and H. T. Nguyen. 2000 .QTLs for cell-membrane stability mapped in rice oryza sativa L.) under drought stress. Theor. Appl. Genet. 100: 1197–1202.

Tuberosa R. and S. Salvi, 2007.Genomics-based approaches to improve drought tolerance of crops.Trends in Plant Science 11:405-412.

Tuinstra, M.R., E.M. Grote, P.B. Goldsbrough & G. Ejeta, 1997.Genetic analysis of post-flowering drought tolerance and componentsof grain development in Sorghum bicolor (L.) Moench.Mol Breed 3: 439–448

Verma, V., Foulkes, M.J., Caligari, P., Sylvester-Bradley, R., Snape, J.: Mapping QTLs for flag leaf senescence as a yield determinant in winter wheat under optimal and droughted environments. Euphytica 135: 255-263, 2004.

Visser, B.: Technical aspects of drought tolerance. Biotechnology Dev. Monitor 18: 5, 1994.

Xu, W.W., P.K. Subudhi, O.R. Crasta, D.T. Rosenow, J.E. Mullet& H.T. Nguyen, 2000. Molecular mapping of QTLs conferringstay-green in grain sorghum (Sorghum bicolor L. Moench).Genome 43: 461–469.

Young, ND 1999 QTL mapping and quantitative disease resistance in plants.Annu.Rev. Phytopathol.34:479-501.

12

Integrated Agronomic Crop Managements to Improve Tef Productivity Under Terminal Drought

Dejene K. Mengistu[1,*] and Lemlem S. Mekonnen[2]
[1]*Department of Dryland Crop and Horticultural Sciences, Mekelle University, Mekelle,*
[2]*Department of Plant Science, Aksum University, Axum,*
Ethiopia

1. Introduction

1.1 Origin, distribution and botany of tef

Tef (*Eragrostis tef* [Zucc.] Trotter) is an allotetraploid ($2n = 4x = 40$) cereal crop grown primarily in Ethiopia. Ethiopia is the center of origin and diversity of tef. It is entirely cultivated only in Ethiopia as food crop and distributed to several other countries in the 19th century, and it is now cultivated as a forage grass in Australia, India, Kenya and South Africa (Costanza *et al.*, 1979). Intensive studies carried out on tef in USA universities initiated its cultivation for both grain and forage has also begun in USA. Studies so far carried out on morphological, cytological and biochemical characters of wild and cultivated species of tef revealed that Ethiopia is the origin and center of diversity of tef even though the wild relative, *Eragrostis pilosa,* a weedy species, occurs throughout the world in tropical and temperate regions (e.g. Vavilov, 1951). This wild relative is the closest relative of the cultivated tef, *E. tef. E. pilosa* is also an allotetraploid and has a karyotype similar to *E. tef* (Tavassoli, 1986). These two species are similar morphologically. The only known consistent morphological distinction between *E. pilosa* and *E. tef* is spikelet shattering of *E. pilosa.*

The multi-floreted spikelets of *E. pilosa* readily break apart at maturity as a means of natural seed dispersal, whereas the lemmas, paleas, and caryopses of *E. tef* remain attached to the rachis at maturity and thereby facilitate harvesting (Phillips, 1995). It is speculated that the transition from shattering to non-shattering is one of the most common traits altered during the domestication process as it allows farmers to control seed dispersal. The current tef breeding program makes interspecific crosses between *E. pilosa* and *E. tef* with fully fertile resultant progenies. Hence, it is highly likely that Ethiopian farmers domesticated tef from *E. pilosa* and altered key agronomic features such as seed mass and spikelet shattering through generations of selections. Furthermore, Endeshaw *et al.* (1995) reported as there is anthropological evidence that *E. pilosa* is harvested and used as a food source in much the same fashion as *E. tef* during times of food scarcity.

Tef is a C4, self-pollinated annual grass, 40 - 80cm tall. It has a shallow fibrous root system with mostly erect stems, although some cultivars are bending or elbowing types (Plate 1). Its sheaths are smooth, glabrous, open and distinctly shorter than the internodes. It has a panicle type of inflorescence showing different forms - from loose to compact, the latter appearing like a spike. The flowers of tef are hermaphroditic with both the stamens and pistils being found in the same floret (Hailu *et al.*, 1990). Florets in each spikelet consist of three anthers, two stigmas and two lodicules that assist in flower opening. Its grain is tiny with 0.9 - 1.7mm long and 0.7 - 1mm wide and its colour varies from white to dark brown (Tadesse, 1975).

A) B) C)

Plate 1. Morphological structure of tef crop: A) the whole plant; B) root and C) panicle

Plants with C4 pathway have a 'Kranz' type of leaf anatomy referring to the bundle sheath, a vascular tissue containing large and thick cell wall with prominent chloroplasts. Possessing such leaf structure helps C4 plants to increase the concentration of CO_2 available to the Calvin cycle even under stress conditions by inhibiting photorespiration. Previous studies such as Hirut *et al.* (1989) and Etagegnehu (1994) showed that tef possesses typical C4 leaf structure. It has two layers of bundle sheath and a single layer of mesophyll cell. Granal chloroplasts are present in both tissues with higher concentration in the bundle sheath cells.

1.2 Grain chemical composition and use

The grain of tef is used to make a variety of food products, including *injera,* a spongy fermented flatbread that serves as the staple food for the majority of Ethiopians. Chemical composition analysis showed that tef has comparable nutritional content with the major cereal crops: maize, barley, wheat and sorghum, cultivated in Ethiopia. Tables 1 and 2 present nutritional and amino acid contents of tef in comparison of other major Ethiopian cereal crops adopted from Agren and Gibson (1968), Alemayehu (1990) and Jansen *et al.* (1962), respectively with minor modification.

Content item	Tef			Barley (whole	Maize (whole)	Wheat (whole)	Sorg-hum (whole)
	Nech (white)	Key (Brown)	Sergegna (mixed)				
Food energy (cal.)	339	336	336	334	356	339	338
Moisture (%)	10.4	11.1	10.7	11.3	12.4	10.8	12.1
Protein (g)	11.1	10.5	7.2	9.3	8.3	10.3	7.1
Fat (g)	2.4	2.7	2.9	1.9	4.6	1.9	2.8
Carbo-hydrate (g)	73.6	73.1	75.2	75.4	73.4	71.9	76.5
Fibre (g)	3.0	3.1	3.6	3.7	2.2	3.0	2.3
Ash (g)	2.5	3.1	3.0	2.0	1.3	1.6	1.5
Calcium (mg)	156.0	157.0	140.0	47.0	6.0	49.0	30.0
Phosphorus (mg)	366.0	348.0	368.0	325.0	276.0	276.0	282.0
Iron (mg)	18.9	58.9	59.0	10.2	4.2	7.5	7.8

Source: Agren and Gibson (1967)

Table 1. Nutrient content of major Ethiopian Cereals per 100gm

Amino acid	Tef	Barley	Maize	Rice	Sorg-hum	Wheat	Pearl millet	FAO pattern	Whole egg
Lysine	3.68	3.48	2.67	3.79	2.02	2.08	2.89	4.20	6.60
Laoleucine	4.00	3.58	3.68	3.81	3.92	3.68	3.09	4.20	7.50
Leucine	8.53	6.67	12.5	8.22	13.3	7.04	7.29	4.80	9.40
Valine	5.46	5.04	4.45	5.50	5.01	4.13	4.49	4.20	7.20
Phe-alanine	5.69	5.14	4.88	5.15	4.90	4.86	3.46	2.80	5.80
Trysosine	3.84	3.10	3.82	3.49	2.67	2.32	1.41	2.80	4.40
Trypto-phan	1.30	1.54	0.70	1.25	1.22	1.07	1.62	1.40	1.40
Threonine	4.32	3.31	3.60	3.90	3.02	2.69	2.50	2.80	4.20
Histidine	3.21	2.11	2.72	2.50	2.14	2.08	2.08	-	2.10
Arginine	5.15	4.72	4.19	8.26	3.07	3.54	3.48	-	6.90
Methionine	4.06	1.66	1.92	2.32	1.39	1.46	31.35	2.20	3.80
Cystine	2.50						3.19	2.00	2.40

Source: Alemayehu (1990); Jansen et al. (1962)

Table 2. Amino acid contents of tef compared with other cereals, the FAO pattern and whole egg (g per 16 grams of nitrogen)

The three tef types are especially rich in mineral nutrients calcium, phosphorus and iron compared to the other major cereal crops maize, barley, wheat and sorghum. There is also slight difference in nutrient content among the three tef types. The carbohydrate content of tef is comparable with other cereal crops ranging from 73.1% (brown) to 75. 2% (mixed tef) with average value of 73.9%. Since tef is the major component of Ethiopian recipe, it provides the major requirement of energy. *Nech* (white seed color) tef has 11.1% protein content which exceeds its content for other major cereal crops during the time of analysis. All the three tef types contain significantly higher mineral nutrients (Calcium, potassium and iron) compared to maize, sorghum, wheat and barley and also have reasonably high fiber (3.2%) and ash (2.9%) contents averaged over the three tef types. Amino acid composition of the three tef types was reported to be the same (Tadesse, 1975; Endeshaw, 1989) regardless of their seed color.

Tef is used in various forms by Ethiopians. The dominant form of usage is *injera,* unleavened pan cake made of tef flour, which is the mainstay of Ethiopian diet. It is also consumed in the form of porridge and bread. Its straw is a nutritious and highly preferred feed for livestock compared to the straw of other cereals particularly during dry season. Besides its local use, it is the major cash earning crop for the farming community as market price for both its grain and straw is higher compared to other cereal crops. It is also among export commodity at national level.

2. Tef productivity under changing climate

In tropics, the most important climatic factors that influence growth, development and yield of crops are rainfall, temperature and solar radiation even though relative humidity and wind velocity can also influence crop growth to some extent. Yielding potential of any crop is mainly depends on climate and more than 50% of variation in yield of a crop is due to climatic variability (Reddy & Reddi, 1992). Rainfall is the most dominating factor that influences crops productivity in tropical environment. Precipitation is reaching of atmospheric humidity either as rain to the ground in regions characterized by high temperature. In Ethiopia, the entire precipitation occurs as rainfall.

The amount of precipitation above the basic minimum required to enable the crop to achieve maturity determines its yield. This requirement varies for various crops and different developmental stages of the same crop. Intensity and distribution of the rainfall are very crucial for satisfactory growth and development of crops. If the intensity of rainfall much exceeds the rate of infiltration of the soil, the consequences are runoff and development of anaerobic conditions in the root zone of the crop. These conditions affect crop performance through nutrient deprivation and oxygen deficiency. Similarly, if its intensity is less to satisfy infiltration and evaporative demands, the crop is subjected to water deficiency which greatly affects its productivity. The amount of rainfall received at periodic interval also determines the final productivity of crops as crops response to moisture varies from stage to stage. Temporal and spatial distribution of rainfall also greatly varies. In tropics the spatial and temporal variability of rainfall is greater compared to the temperate. Records from meteorological stations show much spatial and temporal variability of rainfall in Ethiopia and as a result the country is characterized by many agro ecologies. More than 70% of the rainfall is received in the months of July and August in most parts of the country despite the fact that the cropping period extends to mid October. There is usually water deficiency during the later developmental stages of crops in arid and semiarid parts of the country, this uneven distribution of rainfall within the same cropping season has been blamed for causing significant yield losses. The variability across years is presented in figure 1, based on rainfall data obtained from satellite.

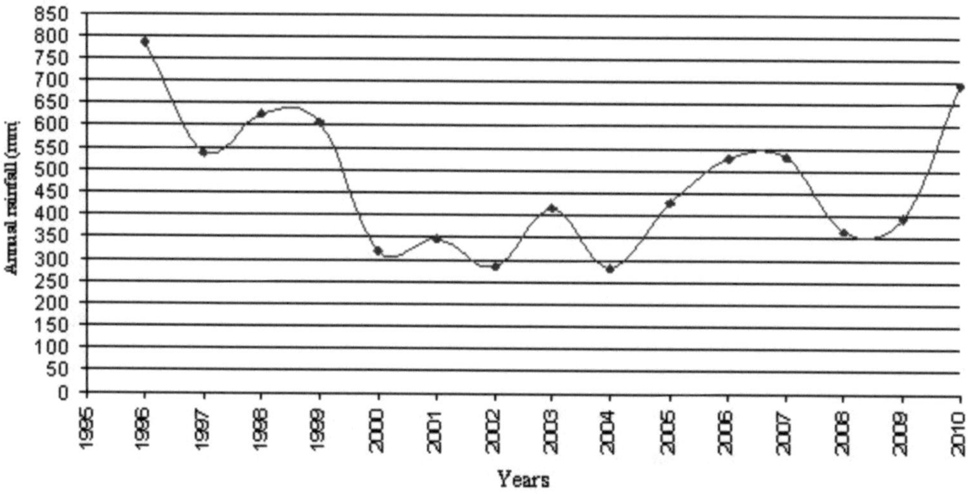

Fig. 1. Trend of annual rainfall distribution of the last 15 years at *Adiha*, northern Ethiopia

More than 50% of the cropping seasons received annual rainfall of less than 500mm which indicates every other year is dry. Early and late season droughts are commonly affecting the agricultural sector in Ethiopia. Early season drought could delay sowing and /or causes poor germination of sown crops as soil moisture content is the major environmental factor affecting crop germination and its establishment. Most of the crop seeds germinate well within the moisture regime of field capacity to 50 percent available soil moisture (Reddy & Reddi, 1992). It is stated that 'drought is an insidious natural hazard characterized by lower than expected or lower than normal precipitation that, when extended over a season or longer period of time, is insufficient to meet the demands of human activities and the environment'. The cropping seasons of 2000, 2001, 2002 and 2004 can be regarded as droughty as the annual rainfall received is by far less than the requirement of most crops. Terminal drought, which usually occurs due to early cessation of rainfall, affects the productivity of crops as it coincides with the most water deficit critical development stages. Water deficit at these critical stages leads to irreversible yield loss. These stages are known as critical period or moisture sensitive stages. In tef, the most important moisture deficit sensitive later developmental stages include flag leaf initiation, flowering, panicle initiation and early grain filling (Dejene, 2009).

Water stress affects various plant growth phases, which starts with activation of the embryo and ends with maturation of the seed, depending on period of its occurrence. Early season stress affects the germination, establishment and crop stands while late season stress affects flowering, fruit or seed setting and fruit or seed quality. Investigations carried out so far on this crop indicated that both its agronomic and physiological traits are affected if the crop is subjected to different levels of water stress.

2.1 Physiological responses

Photosynthesis, which links the inorganic and organic worlds, is an important metabolic process that link water (H_2O) and carbon dioxide (CO_2) in the presence of light to form organic compounds, sugar. So, continuous supply of the raw materials, H_2O and CO_2 and harvesting as much solar energy as possible, is vital for maximum photosynthesis and thereby maximum dry matter accumulation. The main principle in agronomy and physiology is aimed at ensuring appropriate supply of crop with sufficient water and nutrients to keep its health for maximum light interception and carbon dioxide fixation. Hence, the understanding of critical water stress sensitive developmental stages is vital for management of this scarce resource. Water deficit imposed during the reproductive stages of tef can cause reduction in net assimilation rate depending on its severity. For instance, in study conducted to investigate the influence of various soil moisture regimes on physiological processes of tef, it was found that severe water stress (75% of water withhold) has caused 92.8% and 60% reduction in net assimilation and respiration, respectively (Dejene, 2009).

Upon impose of water stress during vegetative developmental stage, tef respiration increased for sometimes and gradually declined below its value obtained under normal growth conditions (Fig. 2a). Similarly the rate of photosynthesis fallen below the control treatment upon exposure to water deficit during vegetative stage and gradually increased to maintain its maximum photosynthesizing potential (Fig. 2b). Stress imposed during flag leaf emergence had significant impact on tef metabolism even though the rate of recovery to normal state is fast after the plant relieved from the stress. Both net assimilation and respiration rates are severely affected by water stress imposed during grain filling stage which shows clearly that the crop possess differential response to water stress during different developmental stages. Reduction of net photosynthesis by moisture stress could be due to reduction in photosynthesis rate, chlorophyll content, leaf area, increase in assimilate saturation in the leaves and closure of stomata.

The stomatal conductance of tef gradually decreased as severity of water deficit increased even though the effect greatly varies for various developmental stages. Highest (97.5%) reduction in stomatal conductance, measured on flag leaf, of tef was reported so far up on exposure of the crop to severe water deficit (75% water withhold), compared to the control treatment, during grain filling stage. The sensitivity of tef physiological processes to water stress imposed during the later developmental stages dictates the need of judicious agronomic crop management in areas where terminal drought is prevalent. Understanding of the physiological processes affected by moisture stress is necessary to ameliorate the stress effects either by management practices or by plant improvement.

2.2 Grain yield

Moisture regimes during flowering and grain development stages determine the number of grains and size of individual grain weight. The effect of water stress depends largely on what proportion of the total dry matter produced is considered as useful material to be harvested (Reddy & Reddi, 1992). For cereals moisture stress during anthesis phase is detrimental. Naturally occurring terminal drought usually coincides with reproductive stages of the crop and consequently the associated yield loss is immense. The magnitude of yield loss actually varies depending on various factors such as crop type, variety, crop developmental stage at which the stress develops and other environmental conditions. Water stress affects yield attributes and final yield. In tef, water stress imposed during reproductive stages significantly affects its yield attributes and the final yield (Tables 3 and 4).

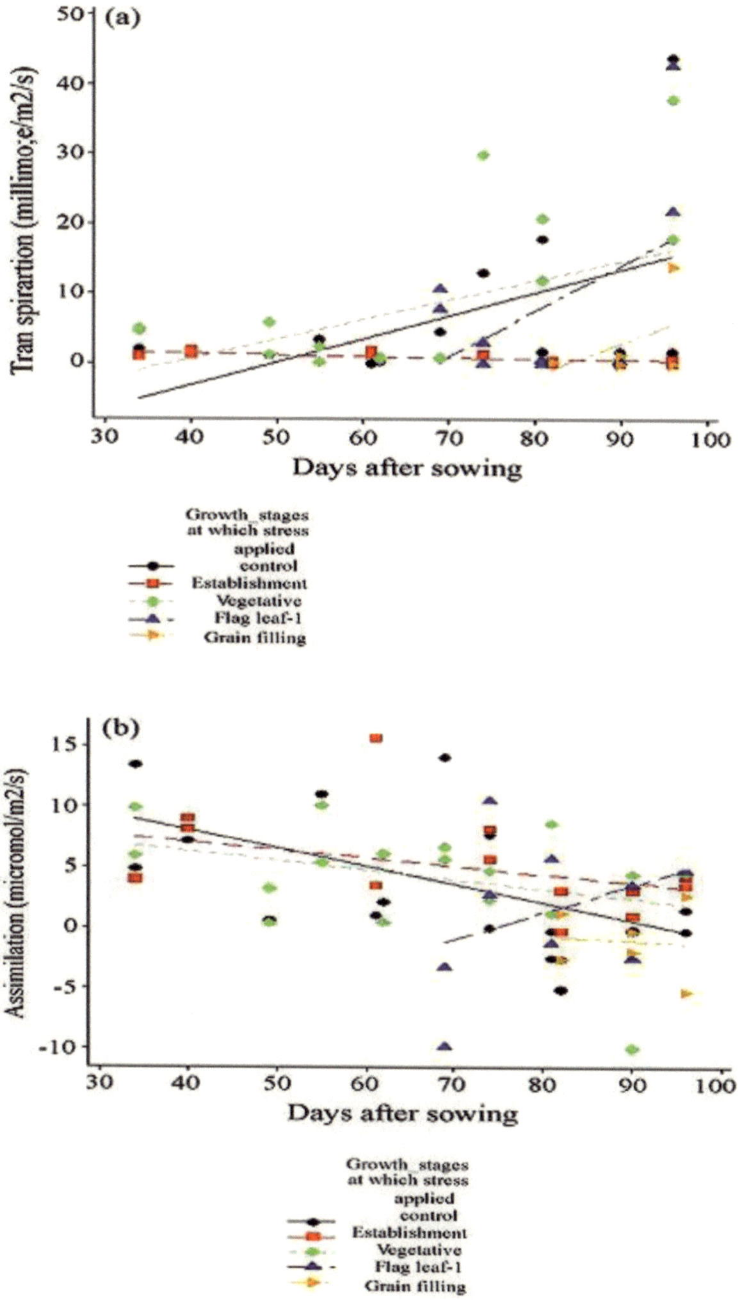

Adopted from Dejene (2009)

Fig. 2. (a) The trend of tef transpiration rate during (start of each line) and after relieved from the stress (b) The trend of net CO_2 assimilation, averaged over stress levels, during (starting point) and after relieving from the stress during different developmental phases compared with the control.

Source of variation	D.f	Mean square values		
		DM	PL	GY
Rep	1	19.636	8.439	0.86970
Variety	10	56.645**	107.218**	1.07495**
Water stress	2	92.924**	64.753**	8.61455**
Variety × water stress	20	24.191**	14.551**	0.14631**
Residual	32	5.574	5.123	0.06765
Total	65			

** *Significant at significance level of 1%*

Table 3. Mean square values of ANOVA results for days to maturity (DM), panicle length (PL) and grain yield (GY) (t ha^{-1}) of tef varieties grown under three water regimes

Stress level	GY	DM	PL
0%	2.30	92.23	31.4
35%	1.49	95.45	29.7
65%	1.09	96.06	28.5
LSD (5%)	**0.16**	**1.45**	**1.07**

Table 4. Effect of water deficit on mean maturity time (DM) mean panicle length (PL) and mean grain yield (GY) (t ha^{-1}) of tef, averaged over varieties under greenhouse conditions

Table 3 shows main effect of both water stress and varieties and their interaction effects on all measured tef attributes is highly significant. The yield loss, due to moderate (35%) and severe (65%) water stresses imposed from booting to grain filling stages, of tef was 35.3% and 52.3% respectively compared with fully irrigated plots (Table 4). For cereals, water deficit at panicle initiation is critical. As the panicle is the organic that growing most rapidly, it is most affected by stress due to reduction in cell expansion (Reddy & Reddi, 1992). The delay in maturity due to water deficit imposed during reproductive stage might imply that tef does not use developmental plasticity as escaping mechanism of the stress period. Even though significant yield reduction recorded under this particular case, the overall performance of the crop under water deficit condition shows that the crop has good level of tolerance to water deficit compared to other cereal crops such as wheat and maize (*data not shown*). The significant variation in response of the imposed water stress among tested tef varieties implied an opportunity of selecting tolerant varieties in drought prone areas (see section 3.2).

This calls for adaptation strategies that ensure minimization of this irreversible yield loss as drought occurrence is inevitable. Drought, because of climate change, is becoming a vital challenge to agricultural sector in the world. Various studies show that without the knowledge of adaptation, climate change is generally detrimental to the agriculture sector; but with adaptation, vulnerability can largely be reduced (e.g. Elizabet *et al.*, 2009; Maddison, 2006). According to Gbetibouo (2009), adaptation is widely recognized as a vital component of any policy response to climate change. The degree to which an agricultural system is affected by climate change depends on its adaptive capacity, which refers to the ability of a system to adjust to climate change (including climate variability and extremes) to moderate potential damage, to take advantage of opportunities, or to cope with the consequences (IPCC, 2001). Thus, the adaptive capacity of a system or society describes its ability to modify its characteristics or behavior so as to cope better with changes in external conditions. Adaptation to climate change requires that farmers' first notice that the climate has been changing and then identify potential adaptation strategies to be implemented (Maddison 2006).

3. Crop management strategies to avert the effect of terminal drought

There are various agricultural management practices in place for adaptation to water stress including supplementary irrigation, diversification of crop varieties, adjustment of cropping calendar and diversification of different enterprises.

3.1 Spate irrigation

Irrigation is needed to be scheduled whenever soil moisture is depleted to critical soil moisture level to avoid irreversible yield loss. Under limited water supply conditions, irrigation should be scheduled targeting moisture deficit sensitive stages of crops and skipped at non - sensitive stages. In drylands, water is the scarcest natural resource competed for by many sectors. Moreover, many drylands are deprived of running water bodies to use for irrigation. Where the mean annual potential Evapotranspiration exceeds the mean annual rainfall, rainfed crop production is uneconomical, unless supported by irrigation. Runoff flood diversion or spate irrigation, which can be defined as *flood harvesting and management system, involving the diversion of flowing flood using some deflecting technologies (using simple deflectors of bunds constructed from earth, sand, stones, brushwood and recently gabions, masonry or concrete structures) on the beds of normally dry creeks or river channels in to a farmland*, was proved to be an alternative water management system to improve agriculture productivity in drylands. It is believed that spate irrigation was started in the present day Yemen and has been practice there for around five thousand years (Lawrence *et al.*, 2005). This type of traditional irrigation is most commonly practice in arid and semi arid parts of the Middle East, East Africa (e.g. Ethiopia, Eritrea, Somalia and Sudan), North Africa (e.g. Morocco, Algeria and Tunisia) and West Asia (e.g. Pakistan, Iran and Afghanistan). Communities in these areas have developed this irrigation practice to cope with the unpredictable erratic rainfall in the regions (Lawrence *et al.*, 2005).

Spate irrigation is characterized by a great variation in the size and frequency of floods from year to year and season to season, which directly influence the availability of water for agriculture. According to Abraham (2007) spate irrigation is practiced in lowland areas where there is surrounding mountainous with better rainfall pattern that can serve as source of flood and deep soils that are capable of storing ample water to support crops during period of low precipitation. The use of spate irrigation systems varies based on hydro - geological (catchments characteristics, rainfall pattern), geographical and sociological (land tenure, social structure) situations. It is also distinct from other irrigation systems such as river diversions that use water from perennial rivers. In spate irrigation systems there is high uncertainty. Lawrence *et al.* (2005) related this uncertainty to the unpredictability in timing, volume and sequence of floodwater. This system of irrigation is mainly managed by farmers.

Two types of spate irrigation are known in Ethiopia: highland and lowland systems (Wallingford *et al.*, 2007; Catterson *et al.*, 1999). The highland spate system is usually referred as run-off system diverts flashy floods received from the same catchments to the relatively small irrigable land. On the other hand, the lowland spate irrigation system is practiced in the foothills of mountainous water shades and covers larger command areas. Flood that comes from the neighboring mountains becomes steady and lasts for longer time. Spate irrigation in Ethiopia differ form those in the Middle East and South East Asia where farming is more unpredictable and entirely dependent on one or two flood events and rainfall events. In contrast farming in Ethiopia relies more on rainfall and spate irrigation can serve as supplementary to rainfall. More than 140,000 hectare of land in Ethiopia is estimated to be under spate irrigation.

Spate irrigation enriches the moisture content of the soil and assists crops water uptake even after cessation of precipitation. Many crops complete their growth on this residual moisture. The productivity of many cereal and vegetable crops such as sorghum, millets, pepper and cabbage has been increasing in marginal areas due to spate irrigation. For instance, Berhanu (2001) shows a yield increment of 400% and 100% for pepper and sorghum, respectively in Ayub and Jarota lowlands of northern Ethiopia because of spate irrigation (fig.3).

The author reported that there is no any yield advantage for tef due to supplementary spate irrigation despite various studies such as Hailu *et al.* (2000) and Dejene (2009) came up with opposite result. Both studies concluded that tef is susceptible to particularly terminal drought which coincides with its reproductive stages and respond to supplementary irrigation. Strong evidence also came out from 2009 trail conducted to test the effectiveness of spate irrigation for tef based farming system in central Tigray, Ethiopia on fragile soil where the seasonal rainy period last only for two months. Diversion of flood water into tef farm particularly during the later developmental stages significantly increased grain and straw yields over the control plot (Fig. 4).

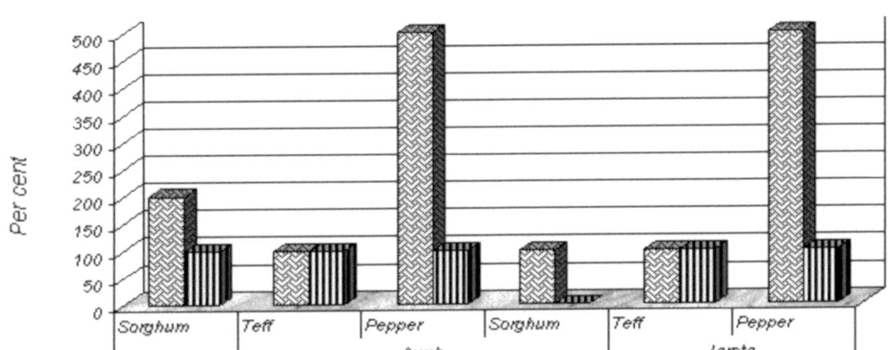

	Sorghum	Teff	Pepper	Sorghum	Teff	Pepper
	Ayub			Jarota		
Yield with flood (%)	200	100	500	100	100	500
Yield without flood (%)	100	100	100	0	100	100

Source: Berhanu, 2001

Fig. 3. Yield differences of various crops due to spate irrigation

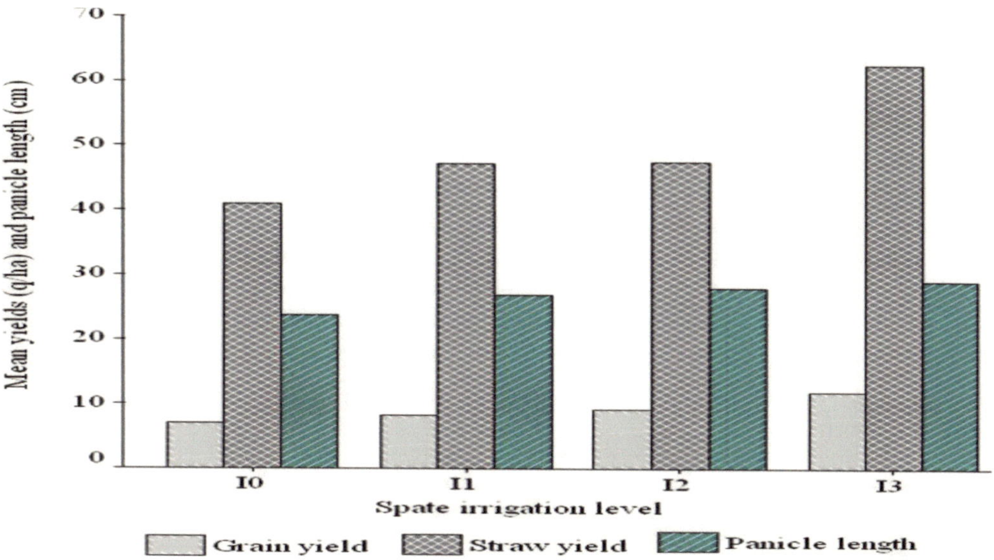

I0 = rainfed, no flood irrigation, I1 = supplementary flood irrigation at tillering stage, I2 = supplementary spate irrigation at tillering and flag leaf stages, I3 = supplementary spate irrigation at I2 plus grain filling stages of the crop.
Source Dejene et al. (2010)

Fig. 4. Tef mean grain yield, straw yield and panicle length as affected by spate irrigation at *Adiha*, Tigray, northern Ethiopia

Supplementing tef during reproductive stage with flood irrigation could fetch more than 50% yield advantage over non supplemented plots. The increment for straw yield was also appreciable (Fig. 4). The increment in yield is the result of improvement of soil moisture content in crops root zone. This implies that supplementing tef farm with any form of irrigation could potentially increase its productivity and hence the economical return of involving farmers.

3.2 Varietal selection

There are several varieties of tef cultivated in wider agroecologies of Ethiopia which could not have similar performance elsewhere. These varieties are classified as early, intermediate and late based on their maturity period. Some are engineered for highland areas, others to mid - altitude and still others to lowland areas. Lowland areas are characterized by high temperature and low and erratic rainfall compared to the highlands. Genotype × environment analysis made on common tef varieties by Tiruneh (2000) using bi - plot principal component analysis approach clearly showed differential adaptation of these varieties to different localities. Intermediate maturing varieties are preferred to both early and late maturing varieties in areas characterized by low moisture. Hence, varieties that perform best in highlands might not perform well in low lands and vise versa. Similarly the performance of farmers' landraces and modern improved varieties also differ greatly.
Testing for their performance and selecting best performing ones should be done as varietal selection is suggested as one of the management approaches to adapt to changing climate. Crop varieties have differential response to water deficit developed at various developmental

stages. Because of this, several genotype × environment interaction studies have been conducted on different crops to come up with recommendations that consider the differential response of varieties in various environments to avoid blanket recommendation. The yield of tef varieties under water deficit is presented below (table 5) from experimental finding conducted with the objective of evaluating and screening of tef varieties under three water regimes. Better performance of local varieties (*Abat Key, Abat Nech, Kobo and Wofey*) than improved varieties should be attributed to their adaptation to naturally occurring terminal drought. Even though there is significant reduction of grain yield and other yield components, the level of reduction for local varieties is small as the stress level increases compared to improved varieties. Variability in response among the varieties might be related to difference in their morphological, physiological and biochemical reactions against the stress.

Stress level	Varieties											
	Abat Keyi	Abat Nech	Berkay	DZ- cr-358	DZ-01-1281	DZ-01-1285	DZ-01-1681	DZ-01-99	DZ-Cr-37	Kobo	Wofey	LSD2 (5%)
No stress (0%)	3.19	2.23	1.77	2.68	3.32	2.04	1.96	1.74	2.44	1.96	2.00	
Mode-rate stress (35%)	2.15	1.73	0.70	1.39	1.80	1.56	1.18	1.06	1.97	1.62	1.28	**0.15**
Severe stress (65%)	1.43	1.33	0.32	0.93	1.33	1.32	0.60	0.31	1.48	1.42	1.27	
LSD1 (5%)	**0.53**											

LSD1 (5%) = Least significance difference value for comparing the stress levels effect on particular variety
LSD2 (5%) = Least significance difference value for comparing the differential response of varieties at a particular stress level

Table 5. Mean grain yield (t ha^{-1}) as affected by the interaction of tef varieties (six improved and five local) and three water regimes

Tef adapts drought by using various strategies such as leaf rolling as morphological adaptation (Dejene, 2009); osmotic adjustment by solute accumulation (Mulu, 1999); control of stomatal aperture (Belay & Baker, 1996 and lower excised leaf water loss (ELWL) (Mulu, 1993). These authors, except Dejene (2009), evaluated various tef genotypes for these strategies and their result indicated that there exist different degrees of response to drought among the genotypes. Genotypes with high osmotic adjustment, lower stomatal conductance and gradual water loss (lower ELWL) under drought are high yielder than genotypes having the opposite trait.

3.3 Changing of sowing date

Farm level decision making is vital to adapt to changing climate. Farmers' knowledge and perception of changing climate is vital to adjust their farming practices as response.

Currently, only few farmers adjust their sowing time in response to perceived climate change as they have no access to information based on long term data (Elizabet *et al.,* 2009). Under such situations, the cropping calendar of farmers remains as it is despite change in timing of rainfall. Tef production activity calendar varies from location to location, and its production takes place mainly during the long (Meher) rainy season. Tef sowing starts from end of June and extends to early September (Kenea *et al.,* 2000) depending on growing length of particular location and beginning of rainfall. However, in most places, tef may be sown between mid - July and early August (Yilma & Cajuste, 1980). This cropping calendar was adopted when appreciable amount of rainfall received in the months of September and the first two - weeks of October. And still practiced despite of great departure of rainfall from normal distribution due to global climate change.

In recent days there is change in trend of rainfall across the country where its starts in May and ceases early - to - mid September. This early cease of rainfall matches crop's reproductive stages, stages most sensitive to water deficit, with periods of water shortage. Coincidence of water deficit sensitive developmental stages of crop with soil moisture deficit causes both quantity and quality loss to the final yield as indicated in section 2.2. Date of sowing has a profound influence on crop performance because it determines the kind of environmental conditions to which the various phenological stages of the crop will be exposed. The sensitivity to some diseases and insect pest damage as well as length of growing period is also related to the effect of date of sowing. Table 6 presents the effect of sowing date on performance of tef at *Adiha,* site severely affected by terminal drought.

A trial involved three sowing dates: July 12, July 22 and Aug., 02 was conducted in 2009 to test the effect of sowing date on tef performance at *Adiha,* a site characterized by severe terminal drought due to early cessation of rainfall and sandiness of the soil. Tef is commonly sown starting from the end of July and lasts up to the mid- August in the study area. Late sowing is preferred despite early cessation of rainfall due to the danger of shoot fly (*Delia arambourgi* (Seguy)) infestation on early sown plots. Similarly during the experimental season early sown plots were seriously infested with shoot fly as a result of which the performance of early sown plots was significantly inferior to the other two treatments. Both grain and straw yields of the early sown plots were significantly (*p<0.01*) lower than the other two sowing dates due to infestation by shoot fly (*data not shown*). Had it been no infestation of shoot fly, the early sown plots should have given higher grain and straw yields as the crop takes longer period for maturity which allows it to intercept more light.

Sowing date	Ph	TN	DM	PL	GY	BM
12/7/09	55.0	1.83	88.17	22.4	0.374	2.46
22/7/09	**55.4**	**2.87**	**81.33**	**23.3**	**0.523**	**3.212**
02/8/09	64.4	3.00	75.83	24.3	0.465	3.01
LSD (0.05)	5.5	1.12	5.1	1.74	0.065	0.35
P value	<0.01	0.08	<0.001	0.09	<0.01	<0.01

Table 6. The effect of sowing date on mean plant height (cm), tiller number, days to maturity, panicle length (cm) and grain and straw yields (t ha^{-1}) of tef at Adiha

The second sowing time, 22/7/09, advanced tef sowing calendar of the area by one week and increased grain and straw yields by 12.5% and 6.7% respectively over the third sowing period (Table 6), which is the sowing period of local farmers in the study area. This implies that adjusting cropping calendar is vital to adapt to the changing climate of the area.

3.4 Soil fertility management

Soil factors influencing its water holding capacity are its texture, structure and chemical composition. In dryland areas where the soil is dominantly sandy or sandy loam, the effect of terminal drought on crops performance is reported very significant. Maintenance of soil fertility using various sources of fertilizers has several advantages apart nutrient supply to the plant. The fact on the ground is that soil nutrient status of most farming systems is widely constrained by the limited use of inorganic and organic fertilizers and by nutrient loss mainly due to erosion and leaching (Balesh *et al.*, 2007). Many smallholder farmers do not have access to synthetic fertilizer because of its high price, lack of credit facilities, poor distribution, and other socio-economic factors. Consequently, crop yields are low, in fact decreasing in many areas, and the sustainability of the current farming system is at risk (Stangel, 1995). This declining soil fertility (Fekadu & Skjelvag, 2002) coupled with terminal drought (Edmeades *et al.*, 1989; Hailu *et al.*, 2000; Dejene, 2009) is posing serious threat to crop production and consequently food security in Ethiopia as elsewhere in Sub – Saharan Africa.
To combat the challenges of climate change, various adaptation strategies have been studied and tried for their effectiveness. Among the different strategies for dryland areas, use of organic fertilizers such as manures and compost applied either in sole or combined with inorganic fertilizers has been attracted special attention. Organic fertilizers have several advantages such as supplying plants with nutrients including micronutrients and improvement of soil structure and water holding capacity (Reddy & Reddi, 1992). Farm generated resources such as crop residues; farmyard manure and compost are regarded as best solutions (CIAT, 2007; Devi *et al.*, 2007) to sustain crop production in dryland areas. Study by Edwards (2007) indicates that organic fertilizers improve crops yield even under arid and semi-arid conditions comparable to chemical fertilizers. Nevertheless sole dependence on organic fertilizers is less feasible due to unavailability of large plant biomass to produce organic fertilizers and competing demands from short-term needs such as for fuel. There should be an alternative approach which reduces heavy reliance on both organic and inorganic fertilizers. Combined application of organic and inorganic fertilizers increase agricultural productivity, improve soil fertility and decrease environmental pollution (Erkossa & Teklewold, 2009; Mugwe *et al.*, 2007; Wakene *et al.*, 2007; Blaise *et al.*, 2003; Corbeels *et al.*, 2000). However, this approach has not been yet widely tested in arid and semiarid parts of Ethiopia. Thus, this work presents the effect of combined application of organic [FYM and compost] and inorganic (NP) fertilizers, from two years experiments, on yield and other yield attributes of tef at *Adiha*, a site prone to frequent terminal drought in northern Ethiopia.

3.4.1 Crop improvement

The effects of organic, inorganic and their combined application on yield and yield components of tef were tested on sandy loam soil. Crop response to one factor is influenced by the availability of the other as to the law of the minimum. Yield of crops is the function of many interacting soil and climatic factors. In drylands, water is the most limiting factor of crops productivity even though other factors meet the demand of the crop. Exploring agronomic practices that improve soil water content in the root zones potentially ensures attainment of optimum crop yield. Application of the right type and dose of fertilizer is one of the agronomic practices to improve crops performance. The effect of combined application of organic - inorganic fertilizers, half dose of the recommended rate for both, on crop phenology, grain and straw yields of tef presented in sections 3.4.1.1 and 3.4.1.2.

3.4.1.1 Crop phenology

Organic and inorganic fertilizers affect differently and significantly tef morphological traits such as number of effective tillers per plant, panicle length and plant height (Table 7).

Fertilizer types	Application rate (kg ha^{-1})	Morphological traits of tef		
		Plant height (cm)	Tiller number	Panicle length (cm)
Compost	7000	43.9	4.3	17.2
FYM	7000	44.5	4.9	18.9
NP	23N +30P	68.0	6.6	24.9
½[NPComp]	11.5N+15P+3500Comp	56.7	8.8	22.7
½[NPFYM]	11.5N+15P+3500FYM	64.1	8.0	24.0
Unfertilized	0	40.1	2.5	16.8
LSD (5%)		**5.8**	**1.0**	**2.2**

Values are average of four measurements

Table 7. Effect of organic, inorganic and combined organic - inorganic fertilizers on mean performance of morphological traits of tef at *Adiha*, northern Ethiopia

Combined application of, half dose of the sole application, organic - inorganic fertilizers resulted in taller plant height and panicle length and many tillers over sole application of both farmyard manure and compost. The tallest plant and panicle lengths were recorded from inorganically treated plants though the difference is not statistically significant.

3.4.1.2 Straw and grain yields

The importance of tef straw has been becoming as equal as its grain yield as it is preferred as animal feed during dry period and also sold at reasonable price. Farmers prefer varieties having larger biomass and give quite good yield. In dryland areas where soil moisture is deficient for pulverization of and quick release of nutrients from organic fertilizers, mixed application of organic - inorganic fertilizers proved to have more positive effect than sole application of both. Figure 5 presented tef straw (BM) and grain yield (GY) obtained from two years experiment.

Combined application of organic - inorganic fertilizers improves straw yield of tef. The figure shows that the highest (5.2t ha^{-1}) straw yield was obtained from ½[NPFYM2] followed by ½[NPComp2]. ½[NPFYM2] and ½[NPComp2] implies half of the recommended dose for both organic (compost and FYM) and inorganic (NP) fertilizers in the study area. Sole application of both compost and farmyard manure at different rates did not show any improvement over inorganically treated plots for both straw and grain yields. Considering this result as general truth might not be possible as this fact derived from experiment conducted at single location over years. Tef has all cultural, nutritional and economical importance for Ethiopian farmers. The grain is used to make a variety of food products, including *injera,* a spongy fermented flatbread that serves as the staple food for most Ethiopians, and porridge. *Injera* is a cultural food consumed almost by every Ethiopian on daily basis. Ethiopian farmers cultivate tef both for home consumption and for sale. There are three classes of tef based on their seed color: white, red and mixture of both (*sergegna*).

The white seeded tef treated as commercial crop as it fetches them higher market price and not produced for home consumption unless 1) they do not have the red or sergegna tef or 2) if they have surplus white seeded tef. The crop being consumed in Europe and United States of America by Ethiopians and other accustomed nations, its export to these parts of the world has been escalating. On the other hand, as tef is proved to be gluten free, its preference by diabetic people has also been increasing.

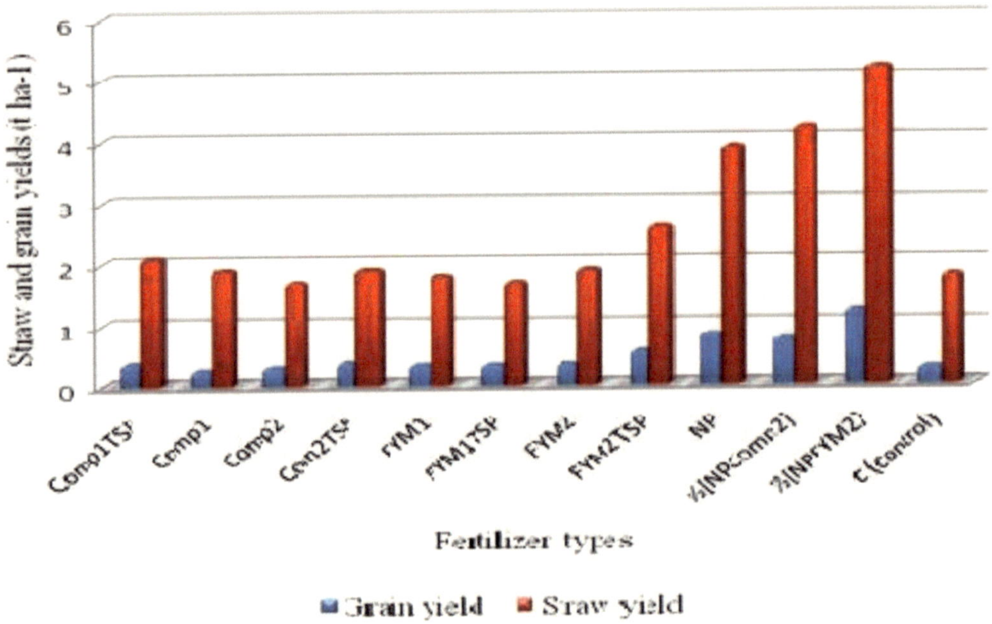

Comp1TSP = compost @3.5t ha^{-1} + TSP, Comp1= compost (3.5t ha^{-1}), Comp2= compost (7t ha^{-1}), Comp2TSP = compost (7t ha^{-1}) + TSP, FYM1= FYM (7t ha^{-1}), FYM1TSP = FYM (3.5t ha^{-1}) + TSP, FYM2=FYM@7t ha^{-1}, FYM2TSP = FYM (7t ha^{-1}) +TSP, NP = P@20kg ha^{-1} and N@23kg ha^{-1}, ½[NPComp2) = combined half dose NP and comp2, ½[NPFYM2) = combined half dose NP and FYM2 and C(control) = no fertilizer application
Adopted from Dejene et al. (2011)

Fig. 5. Effect of organic, inorganic and combined organic - inorganic fertilizers, averaged over years, and year on mean Grain yield (GY) and straw yield (DM) of tef

It could be imagined that the productivity of tef is comparable to the productivity of major cereal crops such as wheat and maize having such great cultural, nutritional and economic values. However, due to several production constraints, its yield is far below the average annual yield of other major cereal crops. Although its genetic make - up is to be blamed, suboptimal crop managements contribute the lion share for its low productivity. Recent studies confirm this fact. Recent trials on supplementary application of micronutrients, row planting and transplanting of tef clued that grain yield of tef could increase from 1.2t ha^{-1}, actual current national average productivity, to up to 6t ha^{-1}(*data not presented*).

The response of tef to different fertilizer types significantly vary under various environmental conditions. In drylands, where rainfall both in amount and distribution is erratic, crops response to fertilizers is not the same as when they grow under conducive growing conditions. For instance, result from trial conducted to evaluate the response of tef for compost, farmyard manure, inorganic NP and combination of organic - inorganic fertilizers in area characterized by terminal drought demonstrated that combining FYM with half dose of the recommended rate of NP fertilizers gave the highest grain yield than sole application of either organic or inorganic fertilizers. Grain yield increment of 45% was obtained due to combined application of FYM - inorganic NP over inorganically treated plots. Similarly, application of combined compost and NP fertilizers also increased the grain yield over sole compost and NP even though the increment was not statistically significant for this particular event. Difference in nutrient content (Chong, 2005) could be the reasons for differential impacts of FYM and compost either in sole or combined application. This higher yield from combined application of organic and inorganic fertilizers presumably attributed to continuous supply of nutrient throughout the developmental stages of the crop. Inorganic fertilizers release nutrient during early growth stages and that of organic fertilizers do release during the later developmental stages as they are slow nutrient releasers (Lorry *et al.*, 2006). Furthermore, organic fertilizers potentially increase moisture retention capacity of the soil, which enables the crop to access water even during the dry period. Edwards (2007) indicated that crops treated with organic fertilizers resist wilting for about two weeks longer than inorganically treated plots when encountered terminal drought. Similar reports were came out from Balesh *et al.* (2007) on tef and Mugwe *et al.* (2009) on maize which described the role of organic fertilizers in improving soil water retention capacity which in turn improves water access of the crop.

4. Conclusion

Erraticiness of rainfall and increase of temperature is inevitable due to global climate change. Consensus has been reached among international community that change in climate will have various degrees of negative impacts mainly on agricultural sector. Consequently the vulnerability of agrarian communities to risks related to climate change hazards is very significant. A number of drought mitigation and adaptation strategies were tested in different countries to minimize the effect of climate change on various sectors. The effectiveness of each of the adaptation strategies varies from country to country, region to region and sector to sector. This calls for intensive studies to sort out appropriate and adoptable adaptation strategies. The effectiveness of four agronomic practices: sowing date, varietal test, spate irrigation and fertilizer application was tested for improving tef performance in northern Ethiopia, region characterized by terminal drought. Advancing the sowing day of tef by one week has increased grain yield of tef by 12.5% over sowing date currently practiced by local farmers. Tef varieties have shown significant differential response to water deficit that develop during the later developmental stages.

The significance of interaction effect between tef varieties and stress level shows the possibility of selecting tolerant tef variety to terminal drought prone areas. Local varieties such as *Abat Nech, Abat Keyi* and *Kobo* are more tolerant to terminal drought than improved varieties. This urges the need of paying important attention in tef breeding program to incorporate drought tolerance gene in our improved tef varieties to improve their adaptation to the wider drought prone areas of the country. Tef is highly responsive to water. Supplementing it with any form of irrigation during reproductive developmental stages significantly increase its productivity. On the other hand combined application of organic - inorganic fertilizers revealed to increase both straw and grain yields of tef significantly over inorganic fertilizer. This implies that combining organic - inorganic fertilizers has had positive and synergistic effect on tef productivity in arid and semi-arid areas where terminal drought is acute. Integrating sowing date, selection of tolerant varieties, spate irrigation and maintaining soil fertility will undoubtedly ensures sustainable production of tef even under terminal drought and reduces the vulnerability of communities settled in terminal drought prone areas of Ethiopia.

5. Acknowledgment

We have not written a review of the extensive literature, which records the conclusions of earlier works: references are given mostly to attribute ideas, to back up assertions and to supply elaborations. But we build very largely on other scholars work, and hence take credit which is mostly due to others. We, therefore, are thankful for the contribution of all authors cited in this work.

6. References

Abraham, M. 2007. A Tradition in Transition: Water Management Reforms and Indigenous Spate Irrigation Systems in Eritrea. *Ph.D. thesis*, Wageningen University, the Netherlands.

Agren, G. and Gibson, R. 1968. *Food composition table for use in Ethiopia*. I. ENI, Addis Ababa, Ethiopia.

Alemayehu, A. 1990. Studies on the nutritional composition of tef [*Eragrostis tef* (Zucc) Trotter] and the interactive influence of Environmental and genotype. *PhD thesis*, University of London, London, UK.

Belay, S. and Baker, D. 1996. An evaluation of drought screening techniques for Eragrostis *tef. Tropical Science 36*: 74 – 85.

Balesh, T, Bernt Aune, J. and Breland, T. 2007. Availability of organic nutrient sources and their effects on yield and nutrient recovery of tef [*Eragrostis tef* (Zucc.) Trotter] and on soil Properties. *Journal of Plant Nutrition and Soil Science. 170:* 543–550.

Berhanu, F. 2001. *Draft PRA Report on Evaluation of RWH Technologies in Ayub and Jarota KAs, Kobo Wereda*. Ethiopian Rainwater Harvesting Association (ERHA), Kobo, Ethiopia.

Blaise, D, Singh, J, Venugopalan, M. and Mayee, C. 2003. Effect of continuous application of manures and fertilizers on productivity of cotton - sorghum rotation. *Acta Agronomica Hungarica, 51*(1): 61–67.

Catterson, T., Worku. M., Endalew, M., Abate, C., Brockman, F., Woldeamaneul, A., Mamusha, K. 1999. Programmatic Environmental Assessment of Small-Scale Irrigation in Ethiopia, Catholic Relief Services, U.S. Catholic Conference Baltimore, Maryland.

Chong, R. 2005. *Using organic fertilizer*. Food and Fertilizer technology center for the Asia and Pacific regions, China.

CIAT. 2007. Formulating soil fertility decision guides with Ethiopian farmers. *CIAT in Africa no.40*, Kampala, Uganda.

Corbeels, M, Abebe Shiferaw and Mitiku Haile. 2000. Farmers knowledge of soil fertility and local management strategies in Tigray, Ethiopia. *Managing Africa's soils no.10*. IIED, London.

Costanza, S., DeWet, J. and Harlan, J. 1979. Literature review and taxonomy of Eragrostis tef (tef). *Econ. Bot.*, 33:413 – 424.

Dejene, K., Dereje, A. and Daniel, G. 2011. Synergistic effects of combined application of organic and inorganic fertilizers on the yield and yield components of tef (*Eragrostis tef* (Zucc.) Trotter) under terminal drought at *Adiha*, Northern Ethiopia. *Journal of the Drylands 3(1)*: In press.

Dejene, K., Dereje, A. and Daniel, G. 2010. Teff (*Eragrostis tef*) agronomic research in Adiha, Central Tigray, Ethiopia. Oxfam America-Horn of Africa Risk Transfer for Adaptation (HARITA)-MU Progress Report no. 02. Mekelle, Ethiopia.

Dejene, K. 2009. The influence of soil water deficit imposed during various developmental phases on physiological processes of tef (*Eragrostis tef*). *Agriculture, Ecosystems and Environment 132*, 283–289.

Devi, R., Kumar, A. and Bishaw, D. 2007. Organic farming and sustainable development in Ethiopia. *Scientific Research and Essay* 2(6): 199-203.

Edmeades, G., Bolaños, J., Lafitte, R., Rajaram, S., Pfeiffer, W. and Fischer, R. 1989. Traditional approaches to breeding for drought resistance in cereals. *In*: Baker, F.(ed.). *Drought Resistance in Cereals.* ICSU and CABI, Wallingford, UK. pp. 27-52.

Edwards, S. 2007. Greening Ethiopia: Ecological Agriculture with Smallholder Farmers in Ethiopia. *www.indsp.org* [Accessed on 24/02/11].

Elizabet, B., Temesgen, D., Gbetibouo, G. and Ringler, C. 2009. Adaptation to climate change in Ethiopia and South Africa: Options and Constraints. *Environmental Science and Policy 12*: 413 - 426.

Endeshaw., Fido, R., Tatham, A. and Shewry, P. 1995. Heterogeneity and polymorphism of seed proteins in tef (*Eragrostis tef*). *Hereditas* 122: 67-72.

Endeshaw, B. 1989. Lysine and other essential amino acids in various fractions of major tef seed proteins. In: Cereals of the semi arid tropics. IFS, Cameroon. Pp230 - 240.

Erkossa, T. and Teklewold, H. 2009. Agronomic and Economic Efficiency of Manure and Urea Fertilizers Use on Vertisols in Ethiopian Highlands. *Agricultural Science in India 8*(3): 352 - 360.

Etagegnehu, G. 1994. A comparative study of the leaf ultrastructure of wheat (*Triticum aestivum* L.) (C3), maize (*Zea mays*) (C4) and tef (*Eragrostis tef* (Zucc.) Trotter). *M.Sc thesis*. Wye college, University of London, UK.

Fekadu, W. and Skjelvag, A. 2002. Sowing date effects on growth rate and straw yield of local barley cultivars. *NLH PhD thesis 36*, 1–20.

Gbetibuou, G. 2009. Understanding farmers' perceptions and adaptations to climate change and variability: The case of the Limpopo Basin, South Africa. IFPRI Discussion Paper 00849. http://www.ifpri.org [Accessed on 12/07/2010].

Hailu, T., Simane, B., Tuinstra, M., 2000. The influence of drought stress on yield of tef (*Eragrostis tef*). *Tropical Science 40*, 40–45.

Hailu, T., Seyfu, K. and Tesfaye T. 1990. Variability, heritability and genetic advance in tef (Ergarostis tef (Zucc.) Trotter) cultivars. *Tropical Agriculture 67*: 317 - 320.

Hirut, K., Johnson, R. and Ferris, 1989. Physiological responses of Eragrostis tef to temperature. *Physiologia Plantarum 77*: 262 - 266.

Intergovernmental Panel on Climate Change (IPCC). 2001. *Climate change 2001: Impacts, adaptation, and vulnerability. Intergovernmental panel on climate change.* Cambridge University Press, Cambridge, UK.

Jansen, G., Dimaio, L. and Hause, N. 1962. Amino acid composition and lysine supplementation of tef. *J. Agric.Food Chem.* 10: 62 - 64.

Kenea, Y., Gezahegn, A. and Workneh, N. 2000. Farming Systems Research on Tef: Smallholders' production practices. Hailu, T., Getachew, B. and Sorrells, M. (eds.).

Narrowing the Rift: Tef Research and Development. *Proceeding of the international workshop on Tef genetics and improvement*, 16 – 19 October 2000, Addis Ababa, Ethiopia

Lawrence, S. 2005. Improving community spate irrigation, HR Wallingford limited. http://www.spate-irrigation.org [Accessed on Feb 13, 2010]

Lory, J., Massey, R. and Joern, B. 2006. Using Manure as a Fertilizer for Crop production. http://www.epa.gov [Accessed on 20/09/2010].

Maddison, D. 2006. The perception of and adaptation to climate change in Africa. *CEEP Discussion paper No.10.* Center of Environmental Economics and Policy in Africa, Pretoria, South Africa: University of Pretoria.

Mugwe, J., Mugendi, D., Kungul, J. and Mucheru-munal, M. 2007. Effect of plant biomass, manure and inorganic fertiliser on maize yield in the central highlands of Kenya. *African Crop Science Journal 15* (3): 111 – 126.

Mulu, A. 1999. Genetic diversity in tef [*Eragrostis tef* (Zucc). Trotter] for osmotic adjustment, root traits and amplifies fragmented length polymorphism. *PhD thesis*, Texas Tech. University, Texas, USA.

Mulu, A. 1993. Use of excised leaf water content in breeding tef [*Eragrostis tef* (Zucc). Trotter] for moisture stress area. *Acta Agronomica Hungarica 42*: 261 – 265.

Phillips, S. 1995. Flora of Ethiopia and Eritrea, vol. 7. Poaceae (Gramineae). National Herbarium, Addis Ababa University, Addis Ababa, Ethiopia and Department of Systematic Botany, Uppsala University, Uppsala, Sweden

Reddy, T. and Reddi, S. 1992. *Principles of Agronomy*. Kalyani publisher, New Delhi, India.

Stangel, P. 1995. Nutrient cycling and its importance in sustaining crop-livestock systems in sub-Saharan Africa: An overview. *In*: Powell, J., Fernandez-Rivera, S., Williams, T. and Renard, C. (eds.). *Livestock and sustainable nutrient cycling in mixed farming systems of sub-Saharan Africa.* Volume II: Technical papers. Proceeding of an International Conference, 22–26 November 1993. ILCA, Addis Ababa, Ethiopia. pp. 37–59.

Tadesse, E. 1975. Tef (*Eragrostis tef*) cultivars. Morphological and classification. Part II. *Agricultural Experimental Station Bulletien*, 66, Addis Ababa University, College of Agriculture, Dire Dawa, Ethiopia.

Tavassoli, A. 1986. The cytology of *Eragrostis* with special reference to *E. tef* and its relatives. Ph.D. dissertation, London University, London, UK

Tiruneh, K. 2000. Genotype × Environment Interaction in Tef. In: Hailu, T., Getachew, B. and Sorrells, M. (eds.). Narrowing the Rift: Tef Research and Development. *Proceeding of the international workshop on Tef genetics and improvement*, 16 – 19 October 2000, Addis Ababa, Ethiopia

Vavilov, N. 1951. *The origin, variation, Immunity and Breeding of cultivated plants.* Ronald Press, New York. [Translated from Russian by Sarrchester, K]

Wakene, N., Fite, G., Abdena, D. and Berhanu, D. 2007. Integrated Use of Organic and Inorganic Fertilizers for Maize Production. Paper presented on workshop of *Utilization of diversity in land use systems: Sustainable and organic approaches to meet human needs*, October 9 - 11, 2007, Witzenhausen.

Wallingford, R. and Meta, M. 2007. Spate Irrigation in Ethiopia. http://www.spate-irrigation.org [Accessed on Feb 17, 2010].

Yilma, S. and Cajuste, J. 1980. *Tef production guideline.* Published by IAR Extension and Liaison Department, Addis Ababa, Ethiopia.

13

Sugarcane Responses at Water Deficit Conditions

Sonia Marli Zingaretti[1], Fabiana Aparecida Rodrigues[2], José Perez da Graça[2], Livia de Matos Pereira[2] and Mirian Vergínia Lourenço[1]
[1]*Universidade de Ribeirão Preto*
[2]*Universidade Estadual Paulista*
Brazil

1. Introduction

Sugarcane (*Saccharum* spp.), a plant of the Poaceae family, is an important source of sucrose and ethanol in many tropical and sub-tropical countries. It was introduced in Brazil during the colonization period, and today represents one of the main cultures of the economy. Brazil is currently the world's largest sugarcane producer. The 08/09 crop yielded approximately 569 millions of tons, totaling 27.5 million liters of ethanol and 31.0 million tons of sugar (ÚNICA, 2011). This makes Brazil responsible for providing 25% of the sugar consumed all over the globe (ÚNICA, 2011), sharing with the United States the leadership in ethanol production. The search for alternative, renewable sources of energy, placed the spotlight on Brazilian ethanol, causing an increase in the production of bi-fuel automobiles, which estimates foresee will represent 85% of the market in a few years.

Today, in Brazil, more than 9 million ha are currently used to cultivate sugarcane, but the demand for sugar and ethanol will have it increased to 64 million ha by the years 2018/2019. The growing demand for ethanol leads to an increase in production, which subsequently demands expansion of cultivated areas as well as the search for new technologies which will enable higher agricultural and industrial productivity. The current occupied agricultural area is reportedly, only 2.4%, and expansion has occurred in regions of the country where this culture will need adjustments due to soil and weather characteristics.

Plants are exposed to adverse environmental conditions, and drought is the major abiotic factor that can damage its growth and development. Drought also limits the areas suitable to agriculture. It is known that, as for any crop, during vegetative growth water is essentially required to obtain maximum yield and drought events in this stage can significantly decrease productivity.

Sugarcane is among the crops which produce a higher amount of biomass per unit of cultivated area and water requirement varies throughout the developing stages, thus for higher tillering and development of culms, there is a higher water requirement than during the maturation stage, when this need is diminished.

In some countries as Australia, Sudan and South Africa about 60% of sugar produced is grown in irrigated areas (Inman-Bamber & Smith, 2005), a practice which always results in production cost increase.

Plants under water deficit suffer a disturbance on cellular homeostasis and must change their metabolism in order to protect themselves against stress. Some plants can undergo an acclimation process and delay effects of stress by using the water available in the soil to slow the process. The level of drought tolerance exhibited by plants can be evaluated by growth analysis and plant productivity under water stress conditions. Drought tolerance is a complex process that varies according to the severity of stress, age of the plant and the water use efficiency. Moreover, other abiotic or biotic stresses that might occur simultaneously can become more stressful to plants. Some cultivars can tolerate the stress more effectively than others. As well as the empirical observations performed in field during growing seasons, the first investigation includes the evaluation of physiological parameters that can indicate more tolerant or sensitive plants. Once the tolerance level is known, cultivars with different drought tolerance levels can be used in gene expression studies to investigate the molecular basis of tolerance.

2. Water deficit

The growing food production demand has led to an agricultural area increase involving areas never before cultivated and frequently poor in nutrients and water deficient. Moreover, climate changes have been directly affecting water availability globally, due to variations in precipitation levels which estimate serious consequences for the water balance in plants (Ryan, 2011). Correspondingly, drought is among the greatest limits to productivity and geographic distribution of the cultures (Fisher & Turner, 1978; Le Rudulier et al., 1984; Delauney & Verma, 1990).
It is well known that water is a very important solvent, essential for all types of living cells because it is directly related to various growth, development and reproductive metabolic functions. Plants are constituted of up to 85% - 95% of water and any losses in water content produce changes in physiological and biochemical cellular reactions (Sawhney & Sing, 2002). Water deficit may be considered as a deviation in the optimal condition to life, causing changes and responses in all functional levels of the organism, which are primarily reversible, but may become permanent and is frequently observed in many crops, with effects varying according to species, duration of stress, season and severity. Water deficit can trigger a negative impact upon growth and development of the crop, compromising plant productivity especially if the drought duration exceeds the capacity of drought-tolerance of the plant species (Inman-Bamber, 2004; Smit & Singles, 2006). Regardless of the temporary nature of the stress, the set of properties and vital functions of the plant gradually decrease according to its duration (Larcher, 2004).
While moderate water deficit causes significant morphological and physiological changes in plants (Creelman et al., 1990), severe deficit may lead to plant death (Cheng et al., 1993). Among the physiological effects are changes in a variety of processes such as radiation capture, foliar temperature, stomatal conductance, transpiration, photosynthesis and respiration, which determine culture productivity (Qing et al., 2001, Silva et al., 2007). On a cellular level, water deficit occurs when transpiration rates exceed water absorption, being the result of different types of stress and potentially leading to changes in concentration of solutes, cell volume and shape of membrane, changes in water potential gradients, loss of cell turgor, compromising of membrane integrity, protein denaturation, total loss of free water leading to dryness and dehydration (Bray, 1997).
Hence, a decrease in water potential is directly related to the reduction in plant photosynthesis rates, where during water stress less carbon assimilation occurs, resulting in

insufficient dissipation of electrons generated by the electron transfer chain and subsequently an overproduction of reaction oxidative species (ROS) (Edreva, 2005).

Other abiotic stresses may be related to water stress acting as limiting factors in production, such as salinity and temperature extremes (Bray, 2004). Therefore, these stressing factors prevent plants from performing to their full genetic potential (Zhu, 2002). Albert et al., (2011), studied the interaction of adverse environmental effects, namely CO_2 concentration increase, high nocturnal temperature and water stress, as to photosynthesis capacity. They observed that estimated future weather conditions will essentially affect physiological processes of plants in unforeseeable ways regarding isolated factors; though dependant on the interaction of growth strategy of the species and their capacity to adjust to the conditions of water availability in the soil.

Other studies indicate that the effects of drought, salinity, temperature extremes and oxidative stress are interconnected and may induce similar damage on a cellular level, such as osmotic changes resulting in homeostasis interruption and ion distribution in the cell (Zhu, 2001; Wang et al., 2007). Accordingly, it is clear that abiotic stresses occur normally as a pool of factors which interact and that plants react to these environmental adversities in a complex manner (Shao et al., 2007).

The interaction of water with other environmental factors can be explained by the interconnectivity of plant, soil and the atmosphere. Transpiration occurs as a response to certain atmospheric conditions that determine a gradient of water under continuous tension, or total potential difference, from its highest point to the surface of the root interacting with the soil, which promotes ascension and soil water absorption. However, low soil humidity affects absorption, so when atmospheric demand increases and the soil is not able to produce enough water, the plant loses more water, there is a delay in absorption and water stress is established (Pires et al., 2010).

Generally, quick changes in water availability occur in superficial layers of the soil, and many studies report that seedling mortality is higher when compared to plant mortality under similar conditions (Schlesinger et al., 1982; Wellington, 1984), showing water absorption by plants occurs in extracts on soil profile according to the distribution of the radicular system and its water content (Pires et al., 2010). Therefore, we may conclude that adult plants with a well developed radicular system and deeply rooted are more drought tolerant. In this context, it was observed that plants can present mechanisms or strategies of tolerance and susceptibility that are expressed during their developmental stages (Osmond et al, 1097; Lichtenthaler, 1996; Chaves et al., 2003).

Different mechanisms and processes are involved in order to avoid the effects of drought and promote tolerance, among which phenology is considered one of the essential factors because water stress is highly variable in duration and severity (Witcombe et al., 2008). In regard to plant evolution, water availability is classified as the most relevant factor among elements which collaborate to abiotic stresses. Since signal transduction induces plant responses, one can infer that signals induced by stress are highly responsible for plant adjustment (Khu, 2002). Because water stress can lead to serious damage in plants, they developed complex transduction pathways to overcome injuries caused by stress (Fujita et al., 2005)

Plant changes aiming tolerance are normally based on manipulation of genes related to protection and maintenance of the structure and function of cells. In contrast to monogenetic characters of resistance to pests and herbicides, abiotic stresses involve more complex genetic responses, and therefore are more difficult to control and manipulate. Genetic engineering involves the transfer of one or more genes involved in signal transduction, regulation, or those that codify functional and/or structural genes.

Eventhough efforts to improve plant tolerance to stress through genetic engineering have resulted in significant findings, the nature of the genetic mechanisms in answer to stress makes these improvements extremely hard (Wang et al., 2003). Many studies have been conducted for the understanding of the mechanisms of plant response to stress, in order to identify gene products that play a role on water stress tolerance (Bray, 2004). Thus, elucidation of the effects of water stress on plants in a broader, interdisciplinary context is increasingly relevant.

3. Physiological parameters in sugarcane

According to Taiz & Zeiger, (2006) current crop productivity in many regions is only partly due to the genetic potential of the plants. Water stress is the major factor that can limit the potential yield of crop plants by up to 70% (Gosal et al., 2009; Morison et al., 2008), and sugarcane is especially affected by drought. Sugarcane presents four developmental stages (Gascho & Shih, 1983), described as: 1) Germination: the development of buds and roots, taking from 30 to 35 days; 2) Tillering: issuance of secondary and tertiary tillers, beginning approximately on the fortieth day after planting and lasting up to 120 days; 3) Grand growth: tillers growth and development with height gain and basal sugar accumulation taking up to 9 months after planting; 4) Maturity: accumulation of photoassimilates and fast sugar synthesis, lasting until the harvesting period. Tillering and grand growth phases are known as critical stages of drought-sensitivity due to the high need of water for sugarcane growth (Ramesh, 2000). During these phases the relationship between water content and the respective physiological responses can be used to identify and distinguish sugarcane genotypes tolerant to drought (Endres et al., 2010).

To uptake CO_2, plants need to maintain stomata open (Taiz & Zeiger, 2006), but this process requires a high transpiration rate, which can be a limiting factor in some areas due to the low water availability (Molina, 2002). However, plants present morphological, biochemical and physiological mechanisms that allow the use of available water more efficiently (Azevedo et al., 2011; Ghannoum, 2009). The C4 metabolism of plants such as sugarcane can probably facilitate their development in hot and dry areas by reducing the photorespiration rate and water loss.

In plants with C4 metabolism (C4 plants) the first stable molecule produced from fixation of carbon dioxide (CO_2) is the 4-carbon dicarboxylic acid oxaloacetate (OAA), which is converted into malate or aspartate. Besides the enzyme ribulose-1.5-bisphosphate carboxylase/oxygenase (Rubisco; EC 4.1.1.39) present in bundle sheath cells, plants C4 also have the phophoenolpyruvate carboxylase (PEPc, EC 4.1.1.31) in mesophyll cells. PEPc enzyme can increase CO_2 fixation due to its high affinity for CO_2 molecules. The spatial compartmentalization of these enzymes causes the fixed CO_2 to be translocated as malato and aspartate to the bundle sheath cells, where decarboxylation occurs with the input of carbon into the Clavin-Benson cycle (Lopes et al., 2011; Taiz & Zeiger, 2006).

In addition to the efficiency of CO_2 fixation, C4plants may have three forms of decarboxylation, and sugarcane shows evidences suggesting the presence of two of them, NADP-malic (NADP-ME, EC 1.1.1.40) and the phosphoenolpyruvate carboxykinase (PCK, EC 4.1.1.49), in a process where one enzyme activity probably supplements the other

(Christin et al., 2007; Ghannoum, 2009). Despite mechanisms that facilitate reactions of CO_2 fixation, sugarcane as any given plant species is subject to water deficit. Longer periods of drought on sugarcane crops can significantly decrease growth, productivity and quality of product (Wiedenfeld, 2000). In C4plants some evidences demonstrate that photosynthesis is highly sensitive to water deficit (Ghannoum, 2009). Moreover, these plants present low recovery capacity mainly when water deficit exceeds the plant recovery capacity limiting the photosynthesis metabolic pathways (Ripley et al., 2010). In Brazilian sugarcane cultivars the capacity to recover the physiological responses was compromised from the tenth day of stress and damaged the photosynthetic apparatus, as can be observed through the low photosynthetic rate, stomatal conductance and transpiration rate (Graça et al., 2010).

Plants under water deficit conditions show modifications in their metabolism to tolerate water loss. Concerning these changes, the root system is the first part to detect the stress and signal to other tissues. The hydraulic perturbation (Buckley, 2005) stimulates plants to send chemical signals through roots to trigger changes on stomata during the water deficit. Therefore, the abscisic acid (ABA) (Kholová et al., 2010), the pH (Schachtman & Goodger, 2008) and the ionic distribution (Bahrun et al., 2002) seem to play an important role on signaling throughout the plant under water stress. Molecular studies have identified a wide range of genes expressed by sugarcane plants under water stress conditions (Iskandar et al., 2011; Prabu et al., 2010; Rodrigues et al., 2011). Regarding the responses to stress, signaling pathways regulated by hormones are highly drought-responsive, mainly those associated with the increased ABA synthesis (Pinheiro & Chaves, 2011). Under drought stress condition, the level of endogenous ABA was increased and its function upon the stomatal closure can protect plants against immediate desiccation (Yoshida et al., 2006). In sugarcane cultivars some genes showed similarity to ABA-regulated proteins and genes directly or indirectly involved in its biosynthesis in plants submitted to water deficit, as well as a reduction in stomatal conductance (Rodrigues et al., 2011). Soil water content seems to be more influent in stomatal conductance than plant water content (Davies et al., 2002; Taiz & Zeiger, 2006). Sugarcane plants also presented a decreased soil water content under moderate (42%) and severe stress (22%) which produced changes in all photosynthetic apparatus, such as stomatal closure, reduction of transpiration and photosynthetic rate, as well as in RWC, photochemical efficiency of photosystem II (PS II), and increase in leaf temperature in plants submitted to water deficit (Rodrigues et al., 2009, 2011).

Different methods have been applied in plant genetic breeding programs, and considering the drought tolerance, programs have focused mainly on the characterization of genotypes under water stress conditions (Condon et al., 2004). Thus, the analyses of physiological parameters have allowed the selection and classification of cultivars through comparative tests using genotypes with known potential to drought tolerance or to drought sensitivity. Among the physiological parameters, the relative water content (RWC) is considered a fast and cheap tool to perform this type of physiological research in breeding programs (Matin et al., 1989; Silva et al., 2007). RWC represents an indicator of plant water balance because it expresses the absolute water amount the plant requires to reach artificial full saturation (González & González-Vilar, 2001). In fact, RWC indicates the level of cellular and tissues hydration which is important for the physiological plant metabolism (Silva et al, 2007). The control of physiological functions is related to plant water content and changes in RWC seem to directly affect all photosynthetic apparatus in sugarcane plants (Graça et al., 2010). When plants under water deficit start to lose water, RWC decreases and triggers a significant reduction in the CO_2 uptake rate due to the stomatal closure (Buckley, 2005). In sugarcane a decrease of 10 to 20% in RWC caused reduction in all photosynthetic apparatus of tolerant and sensitive sugarcane plants submitted to water deficit (Graça et al 2010). RWC applied to distinguish sugarcane genotypes demonstrates that tolerant cultivars show a higher percentage than sensitive plants, and the probable hydration of the protoplast in this cultivars can ensure it productivity in areas with low water availability (Silva et al., 2007).

In sugarcane the increase in the metabolites level (free proline, soluble sugars, glycinebetaine, soluble phenolic compounds, carotenoids and anthocyanins) was essential to improvement of net assimilation and heat tolerance (Wahid, 2007). Osmoprotectants are molecules that play roles in cell osmotic adjustment and due to their importance in stress events they are applied in breeding programs (Cha-um et al., 2008). However, biochemical and physiological evaluations of transgenic sugarcane improved with *Δ^1- pyrroline-5-carboxylate synthetase* demonstrate that the increased proline biosynthesis was more related to reactive oxygen species (ROS) scavengers than to osmoprotector (Molinari et al., 2007). In sugarcane the accumulation of proline and the photosynthetic activity were used as effective indicators to select drought tolerant cultivars (Cha-um et al., 2008). Under salt stress and water deficit sugarcane plants seem to up the osmoprotectant proline synthesis in response to both stresses. In the same study, the photochemical efficiency of photosystem II, stomatal conductance and transpiration rate were also reduced as consequence of the stress (Cha-um et al., 2008). Photosynthetic rate measured in plants under drought stress can present variation according to species and severity of stress. *Eragrostris curvula* submitted to water stress showed higher photosynthetic rate and RWC than the ones observed in tolerant cultivar (Colom & Vazzana, 2003). In sugarcane under dehydration conditions low photosynthetic rates at moderate (8 days) and severe stress (10 days) were observed (Rodrigues et al., 2009, 2011). Submitting the same sugarcane cultivars to other water-limited regime where irrigation was completely suppressed, produced low photosynthetic rates under a moderate stress. Both sensitive and tolerant plants exhibited decreased photosynthesis, although under normal water irrigation tolerant sugarcane plants showed higher photosynthetic activity compared to sensitive plants (Graça et al., 2010).

In addition to damages caused by water deficit, the stressed plant can also suffer the effects of a secondary stress, such as oxidative stress, as a consequence of the first stressful situation. Accumulation of reactive oxygen species [singlet dioxygen ($^1O_2$) and superoxide ($O_2^{\bullet -}$)] occurs naturally during the electron transport in photosynthesis reactions (Miller et al., 2010). As plants close the stomata under water deficit and reduce the internal CO_2 concentration, the generation of reactive oxygen species seems to stimulate mechanisms that reduce oxidative stress and so it may play an important role in drought tolerance (Arora et al., 2002).

In tylakoid membranes, photosystems I and II capture photons from sunlight and convert light energy into chemical energy by using water as base in this biochemical process. Reduction in water availability can produce low efficiency on photosystem II, consequently few molecules of ATP and NADPH are produced, reducing the CO2 fixation (Souza et al., 2004; Taiz & Zeiger, 2006). Under normal water resources, sensitive and tolerant genotypes presented few differences in photochemical efficiency (PSII). Nonetheless, the efficiency on photochemical efficiency (PS II) determined by using drought-stressed sugarcane, showed variations between the sensitive and tolerant cultivars, where the tolerant plants exhibited better use of the photosynthetic apparatus. Based on this result, it is suggested that tolerant plants can maintain the oxidative process at normal levels in photochemical efficiency (PS

II), differently than the sensitive cultivar (Graça et al., 2010). In general water deficit decreases the photochemical efficiency (PS II), and the ability of the cultivar to maintain a high level of F_v/F_m can be an indicative of the radiation use efficiency and carbon assimilation, which has become a promising tool to select cultivars more tolerant to drought (Silva et al., 2007).

Maintenance of leaf temperature requires large amounts of water transpirated throughout the plant in order to keep it under the ambient temperature and the appropriate functioning of photosynthetic apparatus (Machado & Paulsen, 2001). Under high temperatures (above 32°C) sugarcane cultivars limit the internode growth and reduce the sucrose content (Bonnett et al., 2006). In addition, the acclimation process during the cooling also includes small size of plant, leaf orientation, leaf rolling that minimize the area exposed to the environment (Taiz & Zeiger, 2006). Leaf rolling in sugarcane plants is described as a sensitivity characteristic, however it can be understood as part of the acclimation process used by plants to limit the leaf surface area and then avoid water deficit rather than endure it (Inman-Bamber & Smith, 2005). Under high temperatures, photosynthesis and respiration are inhibited mainly by the reduction of cell membranes stability. Damages in photosynthesis are more closely related to changes in membrane properties and with the decoupling of the mechanisms of energy transfer in chloroplasts than to a protein denaturation (Prasad et al., 2008; Sage & Kubien, 2007). As a C4 plant, sugarcane exhibits its highest productivity under temperatures of 30 to 34°C. Since the plant hydric status is not compromised, high temperature seems not to be a limiting factor either to the photosynthetic capacity or to wheat and sorghum development (Machado & Paulsen, 2001). Nevertheless, in drought-stressed sugarcane cultivars the increase of leaf temperature occurred due to the reduction in transpiration, which was triggered by stomatal closure (Graça et al., 2010). In tolerant plants, a higher water status seems to support the stomatal aperture and to maintain the leaf cooling (Silva et al., 2007). Sugarcane cultivars respond differently to water deficit. The tolerant cultivar CTC15 showed a reduction of 4% in TRA, which probably produced the stomatal closure that consequently increased leaf temperature. In the tolerant cultivar SP83-2847, the increase in leaf temperature was significant only when RWC was reduced to 20% in stressed plants (Graça et al., 2010). In sugarcane, signaling between roots and leaves, that leads to stomatal closure, seems to be more closely related to water availability in soil than to leaf water potential (Inman-Bamber & Smith 2005; Smit & Singels, 2006).

Thus, RWC, photochemical efficiency (PS II), stomatal conductance and the photosynthetic rate are some physiological parameters that have been useful in characterizing genotypes tolerant to drought (Buckley 2005; Vinocur & Altman, 2005; Shao et al., 2008; Tezara et al., 2008). Physiological parameters and the identification of genes can be applied as a base for research and development of new sugarcane cultivars (Hotta et al., 2010). Some physiological and biochemical methods used to select cultivars sensitive and tolerant to water deficit in breeding programs have showed promising results. They have also shown a wide applicability based on the low cost of some tools such as RWC as well as the availability of data (Azevedo et al., 2011; Silva et al., 2007).

4. Genetic analysis

Long periods of drought can be tolerated by plants, but the ability to maintain growth and development under limited water resource is considered a characteristic of tolerance to water deficit. Molecular biology, associated with classic genetic breeding programs, became an important tool to detect genetic variability by reducing the time and maximizing the efficiency. The identification and characterization of genes involved in drought tolerance brings knowledge about the perception of the stress and how plants respond to this adverse condition. Concerning this, studies have been performed aiming to identify new sources of variability in different crops (Cramer et al., 2007; Micheletto et al., 2007; Poroyko et al., 2007; Zhuang et al., 2007).

Drought is firstly detected by root tissues. Once roots detect a decrease in soil water content it emits a signal to leaves triggering the stomatal closure (Taiz & Zeiger, 2006). Synthesis of endogenous ABA might be related to signaling between plant tissues as other chemical signals (Schachtman & Goodger, 2008). Growth of the primary root is stimulated; probably regulated by the increase in abscisic acid (ABA) content. Plants subjected to water deficit lose the integrity of membranes and integral proteins (Larcher, 2003). As the severity of water stress increases, the photosynthetic rate decreases and the cell metabolism homeostasis becomes unbalanced. The plant hormone ABA appears to play an important role in protecting cell and signaling the expression of some stress-responsive genes (Shinozaki & Yamaguchi-Shinozaki, 2007).

The early perception of water stress as well as the signal transduction is very important to plant response to adverse environment conditions. Protein kinases, phosphatases and calmodulins are directly involved in a complex cell communication process (Yoshida et al., 2006). In plant responses to stress, morphological, physiological, biochemical and molecular changes are triggered to protect plants against desiccation (Rachmilevitch, 2006). In this context, many genes are induced in order to maintain the cell water content (osmoprotectants), to facilitate the water and solutes transport (water channel proteins) or to prevent the reactive oxygen species (ROS) (Gorantla et al., 2007). Drought stress-responsive genes can be divided into two categories: genes encoding functional proteins or genes encoding regulatory proteins. The functional group includes proteins such as water channel, transporters, detoxification enzymes, chaperones or proteases. The regulatory genes involve proteins whose roles are related to signaling and transcription factors (Shinozaki & Yamaguchi-Shinozaki, 2007).

In general, all biological processes are affected by water deprivation; however plants under water stress are more susceptible to other stresses (Zhuang et al., 2007). Temperature is one of the factors that can enhance the effect of water stress on plants. Under water stress plants close the stomata to avoid the water loss to atmosphere and can increase the cell temperature. Heat stress also inhibits the photosynthesis and decreases the stability of membranes as well as increasing the cell respiration. Plants subjected to heat stress express heat shock proteins, molecules involved in protecting enzymes and structural proteins against denaturation and protein aggregation (Rizhsky et al., 2004; Taiz & Zeiger 2006), they are often identified in drought-stressed tissues.

The Brazilian Sugarcane Expressed Sequence Tag (ESTs) Sequencing Project (Sucest project) was pioneer in sugarcane functional studies (Vettore et al., 2003). Around 238,000 ESTs were produced from plants of different vegetative or reproductive stages grown *in vitro* and *in vivo* under diverse conditions. Data obtained from this project was used to support studies of biotic (Barsalobres-Cavallari et al., 2006) and abiotic stresses (Kurama et al., 2002; Nogueira et al., 2003), as well as other researches with sugarcane (Camargo et al., 2007; Papini-Terzi et al., 2005; Rocha et al., 2007; Rosa et al., 2005). To perform a collection of ESTs from Brazilian sugarcane cultivars expressed specifically under water deficit conditions, the data provided by

Sucest project was also applied in a large scale gene expression study using sugarcane cultivars with different tolerance to drought (Rocha et al., 2007; Rodrigues et al., 2009, 2011).

After the advent of sequencing projects and the consequent establishment of databases, the use of some molecular techniques became a valuable tool in identifying genes involved in plant responses to stress. To measure gene expression under drought conditions 3,575 cDNA clones from leaf libraries generated by the Sucest project were used through the DNA macroarray technique. This method was chosen due to its sensibility to detect expression at low levels. In addition, macroarray has been a useful tool for transcriptional studies such as those to investigate the behavior of plants under salinity (Merchan et al., 2007) or water stress (Becker et al., 2006; Maraschin et al., 2006), to study plant hormone regulation (Sasaki-Sekimoto et al., 2005) and to assess multiple stress-responsive genes (Zheng et al., 2006). The cDNAs expressed under normal development conditions were immobilized on nylon membranes and hybridized with RNAs extracted from drought-stressed plants, as described in detail by Rodrigues et al. (2009, 2011). The level of drought tolerance of the cultivars employed in the gene expression study was considered from very low water deficit-sensitivity to high drought-tolerance, based on productivity analyses carried out in field during dry seasons.

After monitoring the expression level of 3,575 leaves transcripts, the sugarcane plants SP83-5073 (the cultivar is considered highly drought tolerant) showed a smaller set of genes differentially expressed, most of them induced under severe water stress condition. Most of the expressed genes were related to polyamine synthesis, stress response and transport of water and solutes. The up-regulated transcript encodes an S-adenosylmethionine decarboxylase, a key enzyme in spermidine and spermine biosynthesis from putrescine. It is known that polyamines (putrescine, spermidine, spermine) are essential to plant growth by playing roles in cell division, tuber formation, root initiation, embryogenesis, flower development or fruit ripening (Crozier et al., 2000). Enzymes involved in polyamines biosynthesis have also been found under different abiotic stress conditions (Bouchereau et al., 1999). Although the polyamines role under these specific situations is not well understood yet, researches towards understanding its regulation under water stress (Alcázar et al., 2006) or low temperature (Cuevas et al., 2008) have been performed.

Other stress response proteins Bet v I allergen (a PR10 protein), Germin-like protein, Peroxidase or Disease resistance protein RPM1 (an *R* gene family protein) were also associated to the SP83-5073 responses to stress. The Bet v I allergen is a cytoplasmic disease resistance-related protein superfamily member which has been described in wounding events, high salinity conditions or cold stress (Radauer & Breiteneder, 2007), and the pathogenicity studies (Siemens et al., 2006). Similarly, a germin-like protein, member of large family of ubiquitous proteins involved in a wide range of plant metabolic processes as responses to stress was found (Vallelian-Bindschedler et al., 1998). Interestingly, germin-like proteins also seem to present a superoxide dismutase activity in some plants (Kukavica et al., 2005; Thornburg et al., 2003; Woo et al., 2000) and moss (Nakata et al., 2002). Enzymes of detoxification metabolism can be induced to scavenge reactive oxygen species, produced as consequence of the unbalanced metabolic reactions of a cell under water deficit conditions (Turkan et al., 2005). In addition, by detecting genes such as Bet v I allergen or germin-like protein or a RPM1 protein among differentially expressed genes of drought-stressed plants provides evidences that plants share mechanisms of acclimation not just to different abiotic stresses, but also to biotic stress.

Regarding genes involved in transport metabolism, different lipid transfer proteins as well as water channel proteins (ABC transporter and plasma membrane integral protein – PIP protein) were induced by tolerant SP83-5073 plants in response to stress. Lipid transfer proteins contain chemical characteristics to bind and transport hydrophobic molecules and thus associated with enhanced cell wall extension in tobacco (Carvalho & Gomes, 2007; Nieuwland et al., 2009). It is supposed that these proteins induce the transfer of lipids through the extracellular matrix due to the increased accumulation of cuticular wax observed in tobacco leaves in response to drought events (Cameron et al., 2006). A water channel protein was also induced under mild stress in the tolerant cultivar SP83-2847. This protein called integral membrane protein TIP4-2 is an aquaporin present in tonoplast that plays a role in water transport across membranes, being related to adjusting the water status in response to environmental changes (Luu & Maurel, 2005).

Contrastingly, cultivar SP90-1638 (drought-sensitive) presented a distinct gene expression profile. A larger set of genes differentially expressed (induced or repressed due to water deficit) was observed under mild, moderate and severe stress. In the sensitive cultivar the number of genes increased as the stress became more severe. Based on the functional roles, one can observe that important stress-related genes involved in signal transduction, bioenergetics and photosynthesis were repressed. Changes in metabolism are needed for plants to protect themselves against stress. An earlier up-regulation in cell communication molecules serves as important messengers in transcriptional regulatory networks (Shinozaki & Yamaguchi-Shinozaki, 2000). An ineffective signal transduction cascade will probably result in inappropriate gene expression in response to stress. Sensitive plants did not show significant transducer genes being induced precociously under mild water deficit, or presented a down-regulation in moderate stress, but most of these genes were expressed just under severe conditions.

When plants are exposed to water deficit a decrease in photosynthetic rate is one of the first physiological changes that can be observed. In general this data is obtained through physiological analyses, measured by using specific parameters. As complementary information, genetic data assessed for the cultivar SP90-1638 showed that some genes involved in photosynthesis were up-regulated under mild water deficit conditions. However, under moderate and severe stress several genes (transcripts codifying for photosystem proteins, plastocyanin precursor, thioredoxin M-type, oxygen-evolving enhancer protein 2, and even a protein involved in the absorption and transfer of energy between photosystems) related to this metabolism were down-regulated. At the first signs of water deficit, plants close the stomata to avoid excessive water loss by transpiration (Rachmilevitch et al., 2006). As a consequence under moderate stress photosynthesis becomes affected and eventually is inhibited by increased water stress severity (Taiz & Zeiger, 2006). Transcripts encoding a ferredoxin I chloroplast precursor, a plastocyanin precursor and a photosystem I complex PsaN subunit precursor were induced under mild stress condition and subsequently repressed under moderate and severe stress, indicating that in this plants the water stress imposed was limiting to the photosynthesis process. The gene expression pattern identified for this metabolism corroborates to physiological conditions and could be used as an indicative of drought sensitivity for this Brazilian sugarcane cultivar. In plants where the loss of water relative content was small, the photochemical activity was less affected by water stress, supporting the concept of plant productivity loss under drought conditions (Liu et al., 2006).

On the other hand, in the tolerant sugarcane cultivar SP83-2847 the photosynthetic rate decreased under mild stress, but several genes related to photosynthesis were induced under this condition, probably because the stress was not strong enough to produce photoinhibition. According to Silva et al. (2007) water deficit can severely reduce the productivity of sugarcane because photosynthetic rate decreases progressively as stress becomes more severe (Bhatt et al., 2009; Bloch et al., 2006; Dulai et al., 2006). However, genes involved in photosynthesis process were found to be up-regulated. It is suggested that water stress was enough to trigger plants physiological responses to protect them against the stress, but it was not so severe as to repress the expression of the photosynthesis genes. Better adapted plants are also more efficient in water use and show greater tolerance to drought-stress (Munns 2002; Xu & Hsiao 2004).

The stress response metabolism in sensitive plants was more down-regulated in contrast to those verified in tolerant plants SP83-5073. Despite few genes having been differentially expressed in tolerant plants, 94% of them were up-regulated by stress, whereas 45% of the genes expressed in sensitive plants were down-regulated under water stress conditions. An antagonistic pattern was verified for some genes which were induced in tolerant cultivar SP83-5073 and appear to be repressed in sensitive cultivars. Metabolism induced by SP83-5073 plants such as lipid metabolism (also induced by tolerant plants SP83-2847) or polyamines biosynthesis, appeared down-regulated in sensitive cultivar SP90-1638. Lipid metabolism, including significant proteins (chloroplast phytoene synthase 1, very-long chain fatty acid condensing enzyme CUT1, esterase, lipase, phospholipase D, and others) were also repressed in sensitive plants. Among biochemical pathways, phospholipase D plays an important part in phosphatidic acid generation from phosphatidylcholine breakdown. Besides, this enzyme had been related to biotic and abiotic stresses as a signal transducer (Bargmann & Munnik, 2006; Zhang et al., 2005), the involvement of the phospholipase D in glycinebetaine biosynthesis have been proposed (Bray, 2002).

The global transcriptome analysis of cultivar SP83-2847, ranked as moderately tolerant, showed a large amount of genes being induced or repressed after plants were exposed to mild, moderate and severe water deficit (Rodrigues et al., 2011). An enzyme involved in ABA synthesis, as well as an ABA-regulated protein presented, during the whole period under stress, a high induction level. Endogenous ABA content increases as stress becomes more severe to protect plants either playing roles in physiological behavior or by regulating gene expression. In ABA-dependent pathways, genes have an ABA-responsive element (ABRE) with affinity for MYB and bZIP transcription factors that signal for expression of specific genes involved in plant stress response. The transcription factors DREB act on dehydration-responsive cis-acting element (DRE) to trigger gene expression in an ABA-independent pathway (Shinozaki & Yamaguchi-Shinozaki 2007; Agarwal & Iha 2010). Transcription factors expressed by SP83-2847 plants indicate that ABA-dependent and ABA-independent pathways are presented in sugarcane responses to water deficit. Genes observed in this class included some proteins such as NAC1, DREB1, bZIP, MYB, MYC, among others. Once activated, transcription factors act as DNA-binding proteins, which are capable of mediating the transcription of key proteins in the stress response mechanism.

Overall, tolerant plants induce genes under severe water deficit (Rodrigues et al., 2009) or trigger the main expression over the time under stress. It is a fact that some plants can tolerate stress events more efficiently than others. The *Festuca mairei* is a grass that can be used as reference in genomic studies to be compared to other grass due to its genetic adaptation to drought (Wang & Bughrara, 2007). Most of the genes repressed under drought stress in this grass are related to biogenesis and cellular metabolism whereas the induced genes are involved with transcription and defense (Wang & Bughrara, 2007). In addition, in our gene expression experiments a large number of unknown genes were determined;

which may represent a source of new variability in water deficit tolerance studies. For instance in SP83-2847 a large set of genes was differentially expressed under water deficit conditions, however, genes similar to unknown or those with no similarity in databases represent approximately 76% of the genes expressed by these tolerant plants.
The characterization of the gene expression profiles under stress is an important tool for plant breeding and understanding the genetic basis of drought tolerance becomes an essential knowledge in this scenario. The development of drought-tolerant plants is an alternative for areas with restricted water availability (Cushman & Bohnert, 2000). In this context, molecular techniques are a powerful tool to identify genes involved in plant responses and it also allows manipulating only a specific characteristic. The macro and microarray technology applied to the identification and characterization of genes involved in drought tolerance has brought knowledge about the perception of the stress and how plants respond to this adverse condition.

5. Genetic improvement and perspectives

Sugarcane has gone through a fantastic transformation from a wild canes with thick and not so juicy culms to commercial cultivars with thicker and juicier stalks in the hands of plant breeders who used crosses and selection processes aiming to improve sugarcane yield, disease resistance, biomass yield and sucrose content, and is now a specially important crop in several tropical countries. It is known that creating a new cultivar takes approximately 15 years. Traditional breeding can use biotechnological tools as molecular markers in the selection process for better genotypes but, even using the biotechnology to generate new improved cultivars a big effort has to be inputted by breeders in order to assure they have enhanced their economic agronomic traits to meet the energy and food demand respecting land use.
In plant breeding programs, the molecular markers technique has been used to select cultivars thereby reducing the time between selection of materials and commercialization of improved cultivars. In addition, physiological parameters studies also provide the selection of cultivars by comparing well characterized genotypes in relation to drought tolerance. Even though the sugarcane genome study has not been finished there are thousands of ESTs that have been generated over the last decade and are helping researchers around the world understand metabolic pathways and plant responses to environmental changes.
Recently Gornall, et al., (2010), published a review on all the implications of climate changes on agriculture productivity for the twenty first century and how it will impact agriculture around the world, as we know agriculture is strongly influenced by climate changes. Global warming may cause an increase in heat stress and water evaporation. Most of the plantations are rain-fed and changes in precipitation patterns will enormously affect crop production and current cultivars will reduce yield. Food supplies and energy demand are very important issues for the world growing populations.
Not only for sugarcane but also considering other potential crops as soybeans, corn or sugar beet, that are being considered for biofuel production, the equation of food supply and energy production has been a great challenge for most economic specialists in many countries. According to projections of global warming, the changes in climate can worsen the negative effects of water deficit upon agriculture. As drought is a climate factor inherent to all plants, sugarcane crops yields might decrease significantly, emphasizing the need for adapted cultivars. In this matter understanding the gene expression pattern of tolerant and

sensitive plants can provide other tools to maximize the selection and development of new cultivars. Thus, the improvement of crops is essential to support longer periods of drought and it is crucial to maintain and expand the crop yield considering the future demand for food and competitiveness of the biofuel and ethanol business.

The development of new cultivars tolerant to drought along with other characteristics such as poor soil is essential to expand sugarcane plantation. Despite the genetic complexity of sugarcane, plant transformation has been developed in genetic improvement programs. A sugarcane transcriptional regulator of the ethylene responsive factor (ERF) superfamily was expressed in tobacco in response to drought and salt stress (Trujillo et al., 2008). Sugarcane also accumulated a significant amount of sucrose in immature tissues after been genetically engineered to repress a gene involved negatively in bioenergetics metabolism (Groenewald & Botha, 2008).

In Brazil the success of sugar and ethanol production is a result of increases in crop and industry productivity. The country has several breeding programs that frequently release new improved varieties. Cultivars resistant to drought, pests and herbicide tolerant plants should be released in the next few years thanks to the advances generated by the molecular understanding of metabolic pathways such as those involved in the water stress. This scenario was enforced by efforts from FAPESP, the state of São Paulo funding agency, which in the late 1990s established SUCEST, an integrative program to study sugarcane transcriptome. Later biotech companies Alellyx and Canaviallis, now bought by Monsanto, were created and devoted time to integrating all their knowledge for the development of new sugarcane varieties. Sugarcane industry used in the last centuries to produce sugar and later alcohol, has became now a diversified industry focusing on different sectors such as chemical, cosmetics and bioplast production. These new perspectives are a great opportunity for Brazil and other countries in the Tropics.

6. Acknowledgments

The authors are grateful to the Fundação do Amparo à Pesquisa do Estado de São Paulo (FAPESP) and to the Conselho Nacional de Desenvolvimento Cientifico e Tecnológico (CNPq) for the constant financial support.

7. References

Alcázar, R., Cuevas, J. C., Patron, M., Altabella, T., & Tiburcio, A. F. (2006). Abscisic acid modulates polyamine metabolism under water stress in *Arabidopsis thaliana*. *Physiologia Plantarum*, Vol.128, No.3, (November 2006), pp. 448-455 ISSN 1399-3054

Albert, K.R.; Mikkelsen, T.N.; Michelsen, A.; Ro-Poulsen, H.; Linden L.V.(2011). Interactive effects of drought,elevated CO2 and arming on photosynthetic capacity and photosystem performance intemperate heath plants. *Journal of Plant Physiology*. http://www.sciencedirect.com/science/article/pii/S0176161711001507, doi:10.1016/j.jplph. 2011.02.011 (Acessed on June, 2011).

Arora, A., Sairam, R. K., & Sriuastava, G. C. (2002). Oxidative stress and antioxidative system in plants. *Current Science*, Vol.82, No.10 (25 May 2002), pp. 1227–1238, ISSN 0011-3891

Azevedo, R. A., Carvalho, R. C., Cia, M. C., & Gratão, P. L. (2011). Sugarcane Under Pressure: An Overview of Biochemical and Physiological Studies of Abiotic Stress. *Tropical Plant Biology*, Vol.4, No.1, (January 2011), pp.42-51, ISSN 1935-9764

Azevedo, R.A.; & Carvalho R.F.; Cia M.C.; Gratão P.L. (2011). Sugarcane Under Pressure: An Overview of Biochemical and Physiological Studies of Abiotic Stress *Tropical Plant Biol.*, Vol. 4, pp. 42–51.

Bahrun, A., Jensen, C. R., Asch, F., & Mogensen, V. O. (2002). Drought-induced changes in xylem pH, ionic composition, and ABA concentration act as early signals in field-grown maize (*Zea mays* L.). *Journal of Experimental Botany*, Vol.53, No.367, (February 2002), pp. 251-263, ISSN 1460-2431

Bargmann, B. O., & Munnik, T. (2006). The role of phospholipase D in plant stress responses. *Current opinion in plant biology*, Vol.9, No.5, (October 2006), pp. 515-22, ISSN 1369-5266

Barsalobres-Cavallari, C., De Rosa Júnior, V., Nogueira, F., Ferro, J. A., Di Mauro, S. M. Z., Menossi, M, et al. (2006). A novel system for large-scale gene expression analysis: bacterial colonies array. *Applied Microbiology and Biotechnology*, Vol.71, No.6, (August 2006), pp. 963-969, ISSN 1432-0614

Becker, B., Holtgrefe, S., Jung, S., Wunrau, C., Kandlbinder, A., Baier, M., et al. (2006). Influence of the photoperiod on redox regulation and stress responses in Arabidopsis thaliana L. (Heynh.) plants under long- and short-day conditions. *Planta*, Vol.224, No. 2, (July 2006), pp. 380-393, ISSN 1432-2048

Bloch, D., Hoffmann, C., & Märländer, B. (2006). Impact of water supply on photosynthesis, water use and carbon isotope discrimination of sugar beet genotypes. *European Journal of Agronomy*, Vol.24, No.3, (April 2006), pp.218-225, ISSN 1161-0301.

Bonnett, G. D., Hewitt, M. L., & Glassop, D. (2006) Effects of high temperature on the growth and composition of sugarcane internodes. *Australian Journal of Agricultural Research*, Vol. 57, No.10, (September 2006), pp. 1087–1095, ISSN 1836-5795

Bouchereau, A, Aziz, A, Larher, F., & Martintanguy, J. (1999). Polyamines and environmental challenges: recent development. *Plant Science*, Vol.140, No.2, (January 1999), pp. 103-125, ISSN 0168-9452

Boyer, J. S. (1996). Advances in drought tolerance in plants. *Advances in Agronomy*, Vol.56, pp. 187-218, ISSN 0065-2113

Bray, E. A. (2002). Abscisic acid regulation of gene expression during water-deficit stress in the era of the Arabidopsis genome. *Plant, cell & Environment*, Vol.25, No.2, (February 2002), pp.153-161. ISSN 1365-3040

Bray, E.A. (1997). Plant responses to water deficit. *Trends in Plant Science*, Vol.2, PP.48-54.

Bray, E.A. (2004). Genes commonly regulated by water-deficit stress in Arabidopsis thaliana. *Journal of Experimental Botany: Water-Saving Agriculture* Special Issue, Vol. 55, No. 407, PP. 2331–2341.

Buckley, T. N. (2005). The control of stomata by water balance. *New Phytologist*, Vol.168, No.2, (November 2005), pp. 275-292, ISSN 1469-8137

Camargo, S. R., Cançado, G. M. A., Ulian, E. C., & Menossi, M. (2007). Identification of genes responsive to the application of ethanol on sugarcane leaves. *Plant Cell Reports*, Vol.26, No. 12, (December 2007), pp. 2119-2128, ISSN 0721-7714

Cameron, K. D., Teece, M. A., & Smart, L. B. (2006). Increased Accumulation of Cuticular Wax and Expression of Lipid Transfer Protein in Response to Periodic Drying

Events in Leaves of Tree Tobacco 1 [W]. *Plant physiology*, Vol.140, No.1, (January 2006), pp. 176-183, ISSN 1532-2548

Carvalho, A. de O., & Gomes, V. M. (2007). Role of plant lipid transfer proteins in plant cell physiology—A concise review. *Peptides*, Vol.28, No.5, (May 2007), pp. 1144-1153, ISSN 0196-9781

Cha-um, S., & kirdmane, C. (2008). Effect of osmotic stress on proline accumulation, photosynthetic abilities and growth of sugarcane plantlets (saccharum officinarum l.). *Pakistan Journal of Botany*, Vol.40, No.6, (December 2008), pp. 2541-2552, ISSN 2070-3368

Chaves, M.M.; Maroco, J.P.; Pereira, J.S. (2003). Understanding plant responses to drought - from genes to the whole plant. *Functional Plant Biology*, Vol.30, PP.239-264, ISSN 1445-4408/03/030239.

Cheng, Y., Weng, J., Joshi, C.P., Nguyen, H.T. (1993).Dehydration stress-induced changes in translatable RNAs in sorghum. *Crop Science*. Vol.33, pp. 1397-1400.

Christin, P. A., Salamin, N., Savolainen, V., Duvall, M. R., & Besnard, G. (2007). C4 photosynthesis evolved in grasses via parallel adaptive genetic changes. *Current Biology*, Vol.17, No.14, (July 2007), pp. 1241–1247, ISSN 0960-9822

Colom, M. R., & Vazzana, C. (2003). Photosynthesis and PSII functionality of drought-resistant and droght-sensitive weeping lovegrass plants. *Environmental and Experimental Botany*, Vol. 49, No.2, (April 2003), pp. 135-144, ISSN 0098-8472

Condon, A. G., Richards, R. A., Rebetzke, G. J., & Farquhar, G. D. (2004). Breeding for high water use efficiency. *Journal of Experimental Botany*, Vol.55, No.407, (November 2004), pp. 2447-2460, ISSN 1460-2431

Cramer, G. R., Ergül, A., Grimplet, J., Tillett, R. L., Tattersall, E. a R., Bohlman, M. C., Vincent, D., et al. (2007). Water and salinity stress in grapevines: early and late changes in transcript and metabolite profiles. *Functional & integrative genomics*, Vol.7, No. 2, (July 2007), pp. 111-134, ISSN 1438-793X.

Creelman, R.A.; Mason, H.S.; Bensen, R.J.; Boyer, J.S.; Mullet, J.E. (1990). Water deficit and abscisic acid cause differential inhibition of shoot versus root growth in soybean seedlings. *Plant Physiology*, Vol.92, pp. 205-214.

Crozier, A., Kamiya, Y., Bishop, G., & Yokota, T. (2000). Biosynthesis of Hormones and Elicitor Molecules. In: *Biochemistry & Molecular Biology of Plants*, Buchanan, B. B., Gruissem, W., & Jones, R. L., pp. 911-915, American Society of Plant Physiologists, ISBN 0-943088-39-9, Rockville

Cuevas, J. C., López-Cobollo, R., Alcázar, R., Zarza, X., Koncz, C., Altabella, T., et al. (2008). Putrescine is involved in Arabidopsis freezing tolerance and cold acclimation by regulating abscisic acid levels in response to low temperature. *Plant physiology*, Vol.148, No.2, (October 2008), pp. 1094-105. ISSN 1532-2548

Cushman, J. C., & Bohnert, H. J. (2000). Genomic approaches to plant stress tolerance. *Current opinion in plant biology*, Vol.3, No.2, (April 2000), pp. 117-124, ISSN 1369-5266

Davies, W. J., Wilkinson, S., & Loveys, B. (2002). Stomatal control by chemical signaling and the exploitation of this mechanism to increase water use efficiency in agriculture. *New Phytologist*, Vol.153, No.3, (March 2002), pp. 449-460, ISSN 1469-8137

Delauney, A.J., Verma, D.P.S. (1990). A soybean gene encoding Δ^1-pyrroline-5-carboxylate reductase was isolated by functional complementation in *Escherichia coli* and is found to be osmoregulated. *Mol. Gen. Genet.* Vol.221, pp. 299-305.

Dulai, S., Molnár, I., Prónay, J., Csernák, Á., Tarnai, R., & Molnár-Láng, M. (2006). Effects of drought on photosynthetic parameters and heat stability of PSII in wheat and in Aegilops species originating from dry habitats. *Acta Biologica Szegediensis,* Vol.50, No.1-2, (May 2006), pp.11-17, ISSN 1588 -4082.

Edreva, A. (2005) Generation and scavenging of reactive oxygen species in chloroplasts: a submolecular approach. *Agriculture, Ecosystems and Environment* 106:119–133, ISSN: 0167-8809

Endres, L., Silva, J. V., Ferreira, V. M., & Barbosa, G. V. S. (2010). Photosynthesis and Water Relations in Brazilian Sugarcane. *The Open Agriculture Journal,* Vol.4, pp. 31-37, ISSN 1874-3315

Fisher, R.A., Turner, N.C. (1978). Plant productivity in the arid and semiarid zones. *Annu. Rev. Plant Physiology,* Vol.29, PP. 277-317.

Fujita , Y. , Fujita , M. , Satoh , R. , Maruyama , K. , Parvez , M.M. , Seki , M. (2005). AREB1 is a transcription activator of novel ABRE dependent ABA signaling that enhances drought stress tolerance in *Arabidopsis. Plant Cell,* Vol.17, PP. 3470 - 3488.

Gascho, G. J., Shih, S. F. (1983). Sugarcane. In: *Crop Water Relations,* Teare, I. D., & Peet, M. M., pp. 445-479, John Wiley & Sons, ISBN 0471046302, New York

Ghannoum, O. (2009). C4 photosynthesis and water stress. Annals of Botany, Vol.103, No.4, (June 2008), pp. 635–644, ISSN 1095-8290

González, L., & González-Vilar, M. (2003). Determination of Relative Water Content. In: Handbook of Plant Ecophysiology Techniques, Roger, M. R., pp. 207-212, SpringerLink, Retrieved from http://www.springerlink.com/content/v300531424021q57/

Gorantla, M., Babu, P. R., Lachagari, V. B. R., Reddy, A. M. M., Wusirika, R., Bennetzen, J. L., & Reddy, A. R. (2007). Identification of stress-responsive genes in an indica rice (*Oryza sativa* L.) using ESTs generated from drought-stressed seedlings. Journal of Experimental Botany, Vol.58, No.2, (January 2007), pp. 253-265, ISSN 1460-2431

Gornall, J; Betts, R; Burke, E; Clark, R; Camp, J; Willett K & Wiltshire, A, (2010) Implications of climate change for agricultural productivity in the early twenty-first century *Phil. Trans. R. Soc. B* 365, 2973–2989 doi:10.1098/rstb.2010.0158

Gosal, S. S., Wani, S. H., & Kang, M. S. (2009). Biotechnology and drought tolerance. *Journal of Crop Improvement,* Vol.23, No.1, (2009), pp.19-54, ISSN 1542-7536

Graça, J. P., Rodrigues, F. A., Farias, J. R. B., Oliveira, M. C. N., Hoffmann-Campo, C. B., & Zingaretti, S. M. (2010). Physiological parameters in sugarcane cultivars submitted to water deficit,. *Brazilian Journal of Plant Physiology,* Vol.22, No.3, pp.189-197, ISSN 1677-0420

Groenewald, J. H., & Botha, F. C. (2008). Down-regulation of Pyrophosphate: Fructose 6-phosphate 1-Phosphotransferase (PFP) Activity in Sugarcane Enhances Sucrose Accumulation in Immature Internodes. *Transgenic Research,* Vol.17, No.1, (February 2008), pp. 85-92, ISSN 1573-9368

Hotta, C. T., Lembke, C. G., Domingues, D. S., Ochoa, E. A., Cruz, G. M. Q., Melotto-Passarin, D. M., Marconi, T. G., Santos, M. O., Mollinari, M., Margarido, G. R. A., Crivellari, A. C., Santos, W. D., Souza, A. P., Hoshino, A. A., Carrer, H., Souza, A. P., Garcia, A. A. F., Buckeridge, M. S., Menossi, M., Van Sluys, M. A., & Souza, G. M. (2010). The Biotechnology Roadmap for Sugarcane Improvement. *Tropical Plant Biology,* Vol.3, No.2, (June 2010), pp. 75–87, ISSN 1935-9764

Inman-Bamber, N. G. & Smith, D. M. (2005). Water relations in sugarcane and response to water deficits. *Field Crops Research*, Vol.92, No.2-3, (June 2005), pp. 185–202, ISSN 0378-4290

Inman-Bamber, N. G. (2004). Sugarcane water stress criteria for irrigation and drying off. *Field Crops Research*, Vol.89, No.1, (September 2004), pp. 107-122, ISSN 0378-4290

Iskandar, H. M., Casu, R. E., Fletcher, A. T., Schmidt, S., Xu, J., Maclean, D.J., Manners, J. M., & Bonnett, G. D. (2011). Identification of drought-response genes and a study of their expression during sucrose accumulation and water deficit in sugarcane culms. *BMC Plant Biology*, Vol.11, No. 12, (13 January 2011), pp. 1-14, ISSN 1471-2229

Kholová, J., Hash, C. T., Kumar, P. L., Yadav, R. S., Kocová, M., & Vadez, V. (2010). Terminal drought-tolerant pearl millet [*Pennisetum glaucum* (L.) R. Br.] have high leaf ABA and limit transpiration at high vapour pressure deficit. *Journal of Experimental Botany*, Vol. 61, No. 5, (February 2010), pp. 1431–1440, ISSN 1460-2431

Kukavica, B., Vucinić, Z., & Vuletić, M. (2005). Superoxide dismutase, peroxidase, and germin-like protein activity in plasma membranes and apoplast of maize roots. *Protoplasma*, Vol.226, No.3-4, (December 2005), pp. 191-197, ISSN 0033-183X

Kurama, E. E., Fenille, R. C., Rosa, V E, Rosa, D. D., & Ulian, Eugenio C. (2002). Mining the Enzymes Involved in the Detoxification of Reactive Oxygen Species (ROS) in Sugarcane. *Molecular Plant Pathology*, Vol.3, No.4, (July 2002), pp. 251-259, ISSN 1364-3703

Larcher, W. (2003). Plants Under Stress. Physiological Plant Ecology (4th ed.), pp. 345-415. Berlim: Springer Verlag. ISBN 3-540-43516-6, New York

Larcher, W. (2004). *Ecofisiologia vegetal*. São Carlos, SP: RiMa, P. 531.

Le Rudulier, D., Strom, A.R., Dandekar, A.M., Smith, L.T., Valentine, R.C. (1984) Molecular biology of osmoregulation. *Science*, Vol. 224, PP. 1064-1068.

Liu, W.-J., Yuan, S., Zhang, N.-H., Lei, T., Duan, H.-G., Liang, H.-G., et al. (2006). Effect of water stress on photosystem 2 in two wheat cultivars. *Biologia Plantarum*, Vol.50, No.4, (December 2006), pp. 597-602, ISSN 1573-8264

Locy, R. D., Hasegawa, P. M., & Bressan, R. A. (2006). Stress Physiology. In L. Taiz & E. Zeiger (Eds.), *Plant Physiology* (3rd ed.), p. 738–774. Sunderland, MA: Sinauer Associates Inc.

Lopes, M. S., Araus, J. L., Heerden, P. D. R. V., & Foyer, C. H. (2011) Enhancing drought tolerance in C4 crops. *Journal of Experimental Botany*, Vol.62, No. 9, (May 2011), pp. 1-19, ISSN 1460-2431

Luu, D.-T., & Maurel, C. (2005). Aquaporins in a challenging environment: molecular gears for adjusting plant water status. *Plant, Cell and Environment*, *28*(1), 85-96. doi: 10.1111/j.1365-3040.2004.01295.x.

Machado, S., & Paulsen, G. M. (2001). Combined effects of drought and high temperature on water relations of wheat and sorghum. *Plant and Soil*, Vol.233, No.2, (June 2001), pp.179-187, ISSN 1573-5036

Maraschin, S. D. F., Caspers, M., Potokina, E., Wulfert, F., Graner, A., Spaink, H. P., et al. (2006). cDNA array analysis of stress-induced gene expression in barley androgenesis. *Physiologia Plantarum*, Vol.127, No .4, (August 2006), pp. 535-550, ISSN 1399-3054

Matin, M.A., Brown, J. H., & Fergunson, H. (1989). Leaf water potencial, relative water content, and diffusive resistance as screening techniques for drought resistance in barley. *Agronomy Journal*, Vol.81, No.1, pp.100-105, ISSN 1435-0645

Meloni, D. A., Gulotta, M. R., Martinez, C. A., & Oliva, M. A. (2004). The effects of salt stress on growth, nitrate reduction and proline and glycinebetaine accumulation in *Prosopis alba. Brazilian Journal of Plant Physiology*, Vol.16, No.1, pp. 39-46, ISSN 1677-0420

Merchan, F., Lorenzo, L. de, Rizzo, S. G., Niebel, A., Manyani, H., Frugier, F., Sousa, C., & Crespi, M. (2007). Identification of regulatory pathways involved in the reacquisition of root growth after salt stress in Medicago truncatula. *The Plant Journal*, Vol.51, No.1, (July 2007), pp. 1-17, ISSN 1365-313X

Micheletto, S., Rodriguezuribe, L., Hernandez, R., Richins, R., Curry, J., & Oconnell, M. (2007). Comparative transcript profiling in roots of Phaseolus acutifolius and P. vulgaris under water deficit stress. *Plant Science*, Vol.173, No.5, (November 2007), pp.510-520, ISSN 0168-9452.

Miller, G., Suzuki, N., Ciftci-Yilmaz, S., & Mittler, R. (2010). Reactive oxygen species homeostasis and signalling during drought and salinity stresses. *Plant, Cell & Environment*, Vol.33, No.4, (April 2010), pp. 453–467, ISSN 1365-3040

Molina, M. G. (2002). Environmental constraints on agricultural growth in 19th century granada (Southern Spain). *Ecological Economics*, Vol. 41, No.2, (May 2002), pp. 257-270, ISSN 0921-8009

Molinari, H. B. C., Marur, C. J., Daros, E., Campos, M. K. F., Carvalho, J. F. R. P., Bespalhok-Filho, J. C., Pereira, L. F. P., & Vieira, L. G. E. (2007). Evaluation of the stress-inducible production of proline in transgenic sugarcane (*Saccharum* spp.): osmotic adjustment, chlorophyll fluorescence and oxidative stress. *Physiologia Plantarum*, Vol.130, No.2, (June 2007), pp. 218-229, ISSN 1399-3054

Morison, J. I. L., Baker, N. R., Mullineaux, P. M., & Davies, W. J. (2008). Improving water use in crop production. Philosophical Transactions of the Royal Society B: *Biological Sciences*, Vol.363, No. 1491, (February 2008), pp. 639–658, ISSN 1471-2970

Munns, R. (2002). Comparative physiology of salt and water stress. *Plant cell environment*, Vol.25, No.2, (February 2002), pp.239-250, ISSN 1365-3040.

Nakata, M., Shiono, T., Watanabe, Y., & Satoh, T. (2002). Salt stress-induced dissociation from cells of a germin-like protein with Mn-SOD activity and an increase in its mRNA in a moss, *Barbula unguiculata. Plant & Cell Physiology*, Vol.43, No.12, (December 2002), pp. 1568-1574, ISSN 1471-9053

Nieuwland, J., Feron, R., Huisman, B. A. H., Fasolino, A., Hilbers, C. W., Derksen, J., & Mariani, C. (2009). Lipid Transfer Proteins Enhance Cell Wall Extension in Tobacco. *The Plant Cell*, Vol.17, No.7, (July 2005), pp. 2009-2019, ISSN 1532-298X

Nogueira, Fabio T. S., Rosa, Vicente E, Menossi, Marcelo, Ulian, Eugenio C, & Arruda, P. (2003). RNA Expression Profiles and Data Mining of Sugarcane Response to Low Temperature. *Plant Physiology*, Vol.132, No.4, (August 2003), pp. 1811-1824, ISSN 1532-2548

Osmond, C. B., Austin M. P., Berry, J. A., Billings W. D., Boyer, J. S., Dacey ,J. W. H., Nobel, P. S., Smith, S. D., Winner, W. E.. Stress Physiology and the Distribution of Plants. (1987). BioScience, Vol. 37, No. 1, How Plants Cope: Plant Physiological Ecology pp. 38-48. Published by: University of California Press on behalf of the American Institute of Biological SciencesStable http://www.jstor.org/stable/1310176.

Papini-Terzi, F. S., Rocha, F. R., Vêncio, R. Z., Nicoliello, O. K. C., Felix, J. D. M., Vicentini, R., Rocha, C. S., Simões, A. C. Q., Ulian, E. C., Di Mauro, S. M. Z., Silva, A. M., Pereira, C. A. B., Menossi, M., & Souza, G. M. (2005). Transcription profiling of signal transduction-related genes in sugarcane tissues. *DNA Research*, Vol.12, No.1, (2005), pp. 27-38, ISSN 1756-1663

Pinheiro, C., & Chaves, M. M. (2011). Photosynthesis and drought: can we make metabolic connections from available data? *Journal of Experimental Botany*, Vol.62, No.3, (January 2011), pp. 869–882. ISSN 1460-2431

Pires, R.C.M.; Arruda, F.B. & Sakai, E. (2010). Irrigação e Drenagem, In: Cana-de-Açúcar, L.L. Dinardo-Miranda; A.C.M. Vasconcelos & M.G.A. Landell, (1a Ed.), pp. 631-670, ISBN 978-85-85564-17-9, Campinas, São Paulo, Brazil.

Poroyko, V., Spollen, W. G., Hejlek, L. G., Hernandez, A G., LeNoble, M. E., Davis, G., Nguyen, H. T., et al. (2007). Comparing regional transcript profiles from maize primary roots under well-watered and low water potential conditions. *Journal of experimental botany*, Vol.58, No.2, (July 2007), pp.279-89, ISSN 1460-2431.

Prabu, G., Kawar, P. G., Pagariya, M. C., & Prasad, D. T. (2010). Identification of Water Deficit Stress Upregulated Genes in Sugarcane. Plant Molecular Biology Reporter, Vol. 29, No.2, (July 2010), pp. 291-304, ISSN 1572-9818

Prasad, P. V. V., Pisipati, S. R., Mutava, R. N., & Tuinstra, M. R. (2008). Sensitivity of grain sorghum to high temperature stress during reproductive development. *Crop Science*, Vol.48, No.5, (September-October 2008), pp. 1911-1917, ISSN 1435-0653

Qing, Z.M., Jing, L.G., Kai C.R.(2001). Photosynthesis characteristics in eleven cultivars of sugarcane and their responses to water stress during the elongation stage. *Proc. ISSCT*, 24, PP. 642-643.

Rachmilevitch, S., Da Costa, M., & Huang, B. (2006). Physiological and Biochemical Indicators for Stress Tolerance. In Bingru Huang (Ed.), Plant-Environment Interactions (3rd ed., pp. 321–355). New York: CRC Press, ISBN 978-1-4200-1934-6.

Radauer, C., & Breiteneder, H. (2007). Evolutionary biology of plant food allergens. *The Journal of Allergy and Clinical Immunology*, Vol.120, No.3, (September 2007), pp. 518-525, ISSN 0091-6749

Ramesh, P. (2000). Sugarcane Breeding Institute, Coimbatore, India effect of different levels of drought during the formative phase on growth parameters and its relationship with dry matter accumulation in sugarcane. *Journal of Agronomy and Crop Science*, Vol.185, No.2, (September 2000), pp. 83-89, ISSN 1435-0653

Ripley, B. Frole, K. & Gilbert, M. (2010). Differences in drought sensitivities and photosynthetic limitations between co-occurring C3 and C4 (NADP-ME) Panicoid grasses. *Annals of Botany*, Vol.105, No.3, (March 2010), pp. 493–503, ISSN 1095-8290

Rizhsky, L., Liang, H., Shuman, J., Shulaev, V., Davletova, S., & Mittler, R. (2004). When Defense Pathways Collide . The Response of Arabidopsis to a Combination of Drought and Heat Stress. *Plant Physiology*, Vol.134, No.4, (April 2004), 1683-1696, ISSN 1532-2548

Rocha, F. R., Papini-Terzi, F. S., Nishiyama, M. Y., Vêncio, R. Z. N., Vicentini, R., Duarte, R. D. C., Rosa Jr, V. E., Vinagre, F., Barsalobres, C., Medeiros, A. H., Rodrigues, F. A., Ulian, E. C., Zingaretti, S. M., Galbiatti, J. A., Almeida, R. S., Figueira, A. V. O., Hemerly, A. S., Silva-Filho, M. C., Menossi, M.,& Souza, G. M. (2007). Signal transduction-related responses to phytohormones and environmental challenges in sugarcane. *BMC Genomics*, Vol.8, No.71, (March 2007), pp. 1-22, ISSN 1471-2164

Rodrigues, F. A., Graça, J. P., Laia, M. L., Nhani-Jr , A., Galbiati, J. A., Ferro, M. I. T., Ferro, J. A., & Zingaretti, S. M. (2011). Sugarcane genes differentially expressed during water deficit. *Biologia Plantarum*, Vol.55, No. 1 (March 2011), pp. 43-53, ISSN 1573-8264

Rodrigues, F. A., Laia, M. L., & Zingaretti, S. M. (2009). Analysis of gene expression profiles under water stress in tolerant and sensitive sugarcane plant. *Plant Science*, Vol.176, No.2, (February 2009), pp. 286-302, ISSN 0168-9452

Rosa, V. E. de R. J., Nogueira, Fábio T. S., Menossi, Marcelo, Ulian, Eugênio C., & Arruda, P. (2005). Identification of Methyl Jasmonate-Responsive Genes in Sugarcane Using cDNA Arrays. *Brazilian Journal of Plant Physiology*, Vol.17, No.1, (Jan/Mar 2005), pp. 173-180, ISSN 1677-0420

Ryan, M.G. (2011). Tree responses to drought. *Tree Physiology*. Vol.31, PP. 237–239.

Sage, R., & Kubien, D. S. (2007). The temperature response of C3 and C4 photosynthesis. *Plant, Cell and Environment*, Vol.30, No.9, (June 2007), pp. 1086-1106, ISSN 1365-3040

Sasaki-Sekimoto, Y., Taki, N., Obayashi, T., Aono, M., Matsumoto, F., Sakurai, N., Suzuki, H., Hirai, M. Y., Noji, M., Saito, K., Masuda, T., Takamiya, K., Shibata, D., & Ohta, H. (2005). Coordinated activation of metabolic pathways for antioxidants and defence compounds by jasmonates and their roles in stress tolerance in Arabidopsis. *The Plant Journal*, Vol.44, No.4, (November 2005), pp. 653-668, ISSN 1365-313X

Sawhney, V., & Singh D. P. (2002) Effect of chemical desiccation at the post-anthesis stage on some physiological and biochemical changes on the flag leaf of contrasting wheat genotypes. *Field Crops Research*, Vol.77, No. 1, (August 2002), pp. 1-6, ISSN 0378-4290.

Schachtman, D. P., & Goodger, J. Q. D. (2008). Chemical Root to Shoot Signaling Under Drought. *Trends in Plant Science*, Vol.13, No.6, (June 2008), pp. 281-287, ISSN 1360-1385

Shao, H. B., Chu, L. Y., Jaleel, C. A., & Zhao, C. X. (2008). Water-deficit stress-induced anatomical changes in higher plants. *Comptes Rendus Biologies*, v. 331, n. 3, (March 2008), pp. 215-225, ISSN 1631-0691

Shao, H.B.; Guo, Q.J.; Chu, L.Y.; Zhao, X.N.; Su, Z.L.; Hu, Y.C.; Cheng, J.F.(2007). Understanding molecular mechanism of higher plant plasticity under abiotic stress. *Colloids and Surfaces B:* Biointerfaces , Vol.54, PP. 37–45.

Sharp, R. E., Poroyko, V., Hejlek, L. G., Spollen, W. G., Springer, G. K.,

Shinozaki, K, & Yamaguchi-Shinozaki, K. (2000). Molecular responses to dehydration and low temperature: differences and cross-talk between two stress signaling pathways. *Current opinion in plant biology*, Vol.3, No.3, (2000), pp. 217-223. ISSN 1369-5266

Shinozaki, K.,, & Yamaguchi-Shinozaki, K. (2007). Gene networks involved in drought stress response and tolerance. *Journal of Experimental Botany*, Vol.58, No.2, (January 2007), pp. 221-227, ISSN 1460-2431

Siemens, J., Keller, I., Sarx, J., Kunz, S., Schuller, A., Nagel, W., Schmülling, T., Parniske, M., & Ludwig-Müller, J. (2006). Transcriptome analysis of Arabidopsis clubroots indicate a key role for cytokinins in disease development. *Molecular Plant-Microbe Interactions*, Vol.19, No.5, (May 2006), pp. 480-494, ISSN 0894-0282

Silva, M. A., Jifon, J. L., Silva, J. A. G., & Sharma, V. (2007). Use of physiological parameters as fast tools to screen for drought tolerance in sugarcane. *Brazilian Journal of Plant Physiology*, Vol.19, No.3, (July/September 2007), pp. 193-201, ISSN 1677-0420

Smit, M. A., & Singels, S. (2006). The response of sugarcane canopy development to water stress. *Field Crops Research,* Vol.98, No.2-3, (August-September 2006), pp. 91-97, ISSN 0378-4290

Souza, R. P., Machado, E. C., Silva, J. A. B., Lagôa, A. M. M. A., & Silveira, J. A. G. (2004). Photosynthetic gas exchange, chlorophyll fluorescence and some associated metabolic changes in cowpea (Vigna unguiculata) during water stress and recovery. *Environmental and Experimental Botany,* Vol.51, No.1, (February 2004), pp. 45-56, ISSN 0098-8472

Taiz, L., & Zeiger, E. (2006). Plant Physiology. (4th), Sinauer Associates, Inc.; ISBN 978-0878938568, Sunderland, Massachusetts

Tezara, W., Driscoll, S., & Lawlor, D.W. (2008). Partitioning of photosynthetic electron flow between CO2 assimilation and O2 reduction in sunflower plants under water deficit. *Photosynthetica,* Vol.46, No.1, (March 2008), pp. 127-134, ISSN 1573-9058

Thornburg, R. W., Carter, C., Powell, A., Mittler, R, Rizhsky, L, & Horner, H. T. (2003). A major function of the tobacco floral nectary is defense against microbial attack. *Plant Systematics and Evolution,* Vol.238, No.1-4, (May 2003), pp. 211-218, ISSN 0378-2697

Trujillo, L. E., Sotolongo, M., Menéndez, C., Ochogavía, M. E., Coll, Y., Hernández, I., Borrás-Hidalgo, O., Thomma, B. P. H. J., Vera, P., Hernández, L. (2008). SodERF3, a novel sugarcane ethylene responsive factor (ERF), enhances salt and drought tolerance when overexpressed in tobacco plants. Plant & Cell Physiology,Vol.49, No.4, (April 2008), pp. 512-525, ISSN 1471-9053

Turkan, I., Bor, M., Ozdemir, F., & Koca, H. (2005). Differential responses of lipid peroxidation and antioxidants in the leaves of drought-tolerant Gray and drought-sensitive L. subjected to polyethylene glycol mediated water stress. Plant Science, Vol.168, No.1, (January 2005), pp. 223-231, ISSN 0168-9452

Vallelian-Bindschedler, L., Mösinger, E., Métraux, J. P., & Schweizer, P. (1998). Structure, expression and localization of a germin-like protein in barley (Hordeum vulgare L.) that is insolubilized in stressed leaves. Plant Molecular Biology, Vol.37, No. 2, (May 1998), pp. 297-308, ISSN 1573-5028

Vettore, A. L., Silva, F. R., Kemper, E. L., Souza, G. M., Silva, A. M., Ferro, M. I. T., Henrique-Silva, F., Giglioti, É. A., Lemos, M. V. F., Coutinho, L. L., Nobrega, M. P., Carrer, H., França, S. C., Junior, M. B., Goldman, M. H. S., Gomes, S.L., Nunes, L. R., Camargo, L. E. A., Siqueira, W. J., Sluys, M. A. V., Thiemann,O. H., Kuramae, E. E., Santelli,R. V., Marino, C. L., Targon, M. L. P. N., Ferro, J. A., Silveira, H. C. S., Marini, D. C., Lemos, E. G. M., Monteiro-Vitorello, C. B., Tambor, J. H. M. ,Carraro, D. M.,Roberto, P. G., Martins, V. G., Goldman, G. H., Oliveira, R. C., Truffi, D., Colombo, C. A., Rossi, M., Araujo, P. G., Sculaccio, S. A., Angella, A., Lima, M. M. A., Junior, V. E. R., Siviero, F., Coscrato, V. E., Machado, M. A., Grivet, L., Di Mauro, S. M. Z., Nobrega, F. G., Menck, C. F. M., Braga, M. D.V., Telles, G. P., Cara, F. A. A., Pedrosa, G., Meidanis, J., & Arruda, P. (2003). Analysis and functional annotation of an expressed sequence tag collection for tropical crop sugarcane. *Genome Research,* Vol.13, No.12, (December 2003), pp. 2725-2735, ISSN 1549-5469

Vinocur, B., & Altman, A. (2005). Recent advances in engineering plant tolerance to abiotic stress: achievements and limitations. *Current Opinion in Biotechnology,* Vol.16, No.2, (April 2005), pp. 123-132, ISSN 0958-1669

Wahid, A. (2007). Physiological implications of metabolite biosynthesis for net assimilation and heat-stress tolerance of sugarcane (Saccharum officinarum) sprouts. *Journal of Plant Research*, Vol. 120, No. 2, (March 2007), pp. 219–228, ISSN 1618-0860

Wang, W.; Vinocur, B.; Altman, A. (2003). Plant responses to drought, salinity and extreme temperatures:towards genetic engineering for stress tolerance. *Planta*, Vol.218, PP.1–14, 2003.

Wellington, A.B. (1984) Leaf water potentials, fire, fire, and trhe regeneration of mallee eucalyptus in semi arid south-eastern Australia. *Oecocologia* 64: 360-362

Wiedenfeld, R. P. (2000). Water stress during different sugarcane growth periods on yield and response to N fertilization. *Agricultural Water Management*, Vol.43, No.2, (March 2000), pp. 173-182, ISSN 0378-3774

Wang, J. P., & Bughrara, S. S. (2007). Monitoring of Gene Expression Profiles and Identification of Candidate Genes Involved in Drought Responses in Festuca mairei. *Molecular Genetics and Genomics*, Vol.277, No.5, (May 2007), pp. 571-587, ISSN 1617-4615

Witcombe, J.R.; Hollington, P.A.; Howarth C.J.; Reader, S.; Steele, K.A. (2008) Breeding for abiotic stresses for sustainable agriculture. *Phil. Trans. R. Soc. B.*, Vol. 363, PP. 703–716.

Woo, E.-J., Dunwell, J. M., Goodenough, P. W., Marvier, A. C., & Pickersgill, R. W. (2000). Germin is a manganese containing homohexamer with oxalate oxidase and superoxide dismutase activities. Nature Structural & Molecular Biology, Vol.7, No.11, (November 2000), pp. 1036-1040, ISSN 1545-9985

Yoshida, R., Umezawa, T., Mizoguchi, T., Takahashi, S., Takahashi, F., & Shinozaki, K. (2006). The regulatory domain of SRK2E/OST1/SnRK2.6 interacts with ABI1 and integrates abscisic acid (ABA) and osmotic stress signals controlling stomatal closure in Arabidopsis. *Journal of Biological Chemistry*, Vol.281, No.8, (February 2006), pp. 5310-5318, ISSN 1083-351X

Zhang, W., Yu, L., Zhang, Y., & Wang, X. (2005). Phospholipase D in the signaling networks of plant response to abscisic acid and reactive oxygen species. *Biochimica et Biophysica Acta: Molecular and Cell Biology of Lipids*, Vol.1736, No.1, (September 2005), pp. 1-9, ISSN 1388-1981

Zheng, J., Zhao, J., Zhang, J, Fu, J., Gou, M., Dong, Z., et al. (2006). Comparative expression profiles of maize genes from a water stress-specific cDNA macroarray in response to high-salinity, cold or abscisic acid. *Plant Science*, Vol.170, No.6, (June 2006), pp. 1125-1132, ISSN 0168-9452

Zhu J. K.(2002) . Salt and drought stress signal transduction in plants. *Annual Review of Plant Biology*, Vol.53, PP. 247-273.

Zhu, J.K. (2001). Cell signaling under salt, water and cold stresses. *Curr Opin Plant Biol*, Vol.4, PP.401–406.

Zhuang, Y., Ren, G., Yue, G., Li, Z., Qu, X., Hou, G., et al. (2007). Effects of Water-Deficit Stress on the Transcriptomes of Developing Immature Ear and Tassel in Maize. *Plant Cell Reports*, Vol.26, No.12, (December 2007), pp. 2137-2147, ISSN 1432-203X

Web Site Reference

Sucest project : http://sucest.lbi.ic.unicamp.br/en

www.unica.com.br

14

Strategies for Selecting Drought Tolerant Germplasm in Forage Legume Species

Hernán Acuña*, Luis Inostroza and Gerardo Tapia
Instituto de Investigaciones Agropecuarias, INIA
Centro Regional de Investigación Quilamapu, Chillán
Chile

1. Introduction

The growing demand throughout the world for good quality soils to expand more profitable productive systems (crops, fruits and vineyards) is displacing forage crop production to marginal environments that often have soils with low fertility, pH problems, poor drainage and subject to periods of drought. The situation is similar in Chile, but the problem is exacerbated by the effects of global climate changes, with an estimated decline in precipitation of 40 percent and an increase of 2-4°C in continental temperature by the end of this century. In this context grasses are increasingly being cultivated beyond their limit to adapt in areas where the ability to tolerate environmental stresses is an essential characteristic for success.

The breeding of forage *Lotus* species and other perennial forage legumes has historically presented low rates of genetic progress because of their genetic complexity. The majority of forage legume species are polyploid, cross-pollinated and self-incompatible. These characteristics make it necessary to use plant-breeding methodologies that are inefficient and very time-consuming. The most popular selection strategies have been masal and recurrent phenotypic selection. Neither strategy considers progeny tests, because they are not efficient in the selection of characters of low inheritability, such as dry matter production. As well, commercial cultivars are the products of many genotypes that generate heterogeneous combinations, which hinders evaluation and prediction of genetic merits. Consequently, strategies are essential to improve our understanding of the genetic components that determine the expression of phenotypic traits of agronomic and adaptive interest. The objectives of this review are to describe the work carried out by the National Institute for Agricultural Research (INIA) of Chile, and the methodologies and results of its work in selecting naturalized populations of forage legume species to develop a broad genetic base for breeding.

2. The use of naturalized and introduced genetic resources

The collection and conservation of germplasm of vegetal species developed in recent decades from around the world and in particular from Chile has provided abundant genetic material for improving cultivars to increase food production on soils with increasing

*Corresponding Author

limitations, such as water. The INIA-Chile germplasm banks maintain collections of forage species of the genera *Lotus*, *Trifolium* and other legumes and grasses, the majority of them naturalized in the country, which come from areas with limited rainfall or irrigation and intense competition for water resources with high-yield crops. This germplasm was collected in the 1990s and in the last decade have been characterized in relation to abiotic stresses, in particular water stress. This review focuses on *Trifolium repens* and three species of *Lotus* (*L. corniculatus*, *L. tenuis* and *L. uliginosus*) given their current economic importance despite their low efficiency in the use of water resources, or the potential of these species in marginal areas with limited availability of water, respectively.

During 1990, Ortega *et al.* (1994) collected 51 naturalized Chilean white clover (*Trifolium repens*) accessions in 26 sites between the 38°15'S and 42° 45'S covering the La Araucanía, Los Rios and Los Lagos Regions (Figure 1). Each accession included 15 plants that were propagated to 9 clones for agronomic and morphological characterization using two levels of soil water availability and two levels of soil phosphorus, under field and greenhouse conditions. Most of the accessions corresponded to medium-leaved white clover. Only one accession, 2-3-X, was classified as ladino or large-leaved white clover. No relations were observed between growth and the soil phosphorus level of the collection sites. However, there was a tendency associated with accessions to show higher response to phosphorus application when the original soil phosphorus level was higher. Two accessions, 2-3-X and 7-1-X, showed higher responses to irrigation and phosphorus application.

Fig. 1. Geographical distribution of collecting sites for white clover accessions (Ortega *et al.*, 1994).

The project "Collection of forage germplasm in the Andean-Patagonian forest of Argentina and Chile" was carried out from 1994 to 1996 for INIA-Chile, INTA (National Institute for Agricultural Technology)-Argentine and the National University of Comahue (Argentine) with the objective of collecting forage germplasm of the genera *Bromus, Trifolium, Elymus* and *Agropyron*. The project covered the Andean-Patagonian region of both countries from 39° S, approximately, to Beagle Chanel (Zappe *et al.*, 1996). The project was co-funded by the Cooperative Program for the Technological Development of the Agro-Food Sector in the Southern Cone (PROCISUR). Only seed were collected. The white clover accessions collected (31) were regenerated during 2007 and phenologically and morphologically characterized during 2008-2009 (Photo 1). Growth and seed production were also measured. Superior genotypes with desirable characters for breeding programs were identified.

Photo 1. White clover (*Trifolium repens*) characterization under field conditions. INIA, La Araucanía Region, Chile, 2008 - 2009 (Seguel *et al.*, 2010)

The forage *Lotus* spp. germplasm available at INIA-Chile germplasm banks were introduced and collected through the FONDECYT (Chilean National Fund for Scientific & Technological Development) project "Evaluation, collection, and characterization of varieties of *Lotus* spp. in different environments in Central South and Southern Chile" from 1998 to 2001. The project considered the introduction in the country of *L. corniculatus* cultivars from North America, as well as the collection of *L. tenuis* and *L. uliginosus* germplasm in the zones of the country where these species are naturalized, together with a phenological, morphological, and agronomic characterization that included nitrogen fixation in field conditions and condensed tannin concentration in foliage (Acuña *et al.*, 2002a, 2002b, 2004, 2008).

Table 1 indicates the origin of *Lotus corniculatus* cultivars introduced in the country. Germplasm considered representative of the global genetic diversity of the species was requested from institutions that had available seeds. *Lotus tenuis* was collected (11 accessions) in the 1998-1999 spring-summer season between the Metropolitan and Biobío

Regions, from 33°S to 38°S (Figure 2). *Lotus uliginosus* was collected in 1991 in La Araucanía and Los Rios Regions (10 accessions), and in 1999 in Los Lagos and Aysén Regions (11 accessions), from 38°S to 45°S, covering a wide range of agroecological conditions (Figure 3). Thirty plants of each accession were taken from each collection site. They were extracted from the soil, transported in polyethylene bags, and provided with an environment that ensured their conservation during transport. A record was kept of the collection sites with georeferenced location data. Similarly, a soil sample (0-15 cm) was taken from each site to conduct a complete chemical analysis (Table 2).

Fig. 2. Geographic distribution of collecting sites for *Lotus tenuis* accessions

The germplasm characterization revealed that there are cultivars of high value in the introduced germplasm of *L. corniculatus* that can be recommended for different environments of the central zone of Chile, as well as for local breeding programs. Information obtained from *L. tenuis* and *L. uliginosus* accessions showed genetic variability in both species. *L. tenuis* germplasm is adapted to clay, medium-textured, or sandy soils with water restrictions and phosphorous deficiency. Therefore, characterized accessions could be used to breed cultivars for low input production systems in constrained environments. *L. uliginosus* accessions include genotypes collected in sites with acidic soils and variable tannin content in plant tissues that may be promising genetic materials for breeding programs with the objective of producing cultivars for animal production systems for the damp and acidic soils in southern Chile.

Cultivar	Origin
Quimey	INIA, Chile
Upstart	Pickseed Canada Ltd., Canada
Norcen	Crops Breeding Association, Minnesota, USA
Dawn	University of Missouri, USA
AU-Dewey	University of Missouri, USA
Steadfast	University of Missouri, USA
Georgia - 1	University of Missouri, USA
Ganador	INIA, Uruguay
San Gabriel	INIA, Uruguay
Ges-5	CSIRO, Australia
Granger	USDA, Oregon, USA
Viking	Cornell University, New York, USA
Empire	Cornell University, New York, USA

Table 1. Cultivars of *Lotus corniculatus*

Fig. 3. Geographic distribution of collecting sites for *Lotus uliginosus* accessions

				mg kg^{-1}		cmol kg^{-1}				
Accession	Collecting site	pH	OM%	N	P	K	Ca	Mg	Na	Al
Lt 1	Cabrero	6.6	3	20	11	0.26	5.56	2.26	0.16	0.01
Lt 3	Yumbel	6.0	4	21	15	0.32	2.30	1.91	0.15	0.01
Lt 4	San Javier	6.1	2	12	2	0.17	5.06	2.46	0.73	0.01
Lt 5	Parral	5.2	6	25	5	0.17	7.58	2.45	0.52	0.11
Lt 6	Parral	5.7	7	19	8	0.33	8.44	3.89	0.60	0,01
Lt 7	Cato	5.9	5	22	11	0.70	8.74	2.33	0.30	0.01
Lt 8	Coihueco	5.2	6	18	11	0.38	5.45	2.01	0.25	0.10
Lt 11	Itahue	6.3	14	9	6	0,36	5.69	1.26	0.33	0.01
Lt 12	Villa Alegre	7.0	7	30	4	0.20	20.14	4.63	0.74	0.02
Lt 14	Melipilla	7.6	4	27	9	0.49	22.03	2.59	0.80	0.02
Lt 15	Las Cabras	6.9	3	12	2	0.22	8.87	1.13	0.54	0.01
Lu 1	Piedra Azul	5.2	33	28	11	0.14	1.41	0.21	0.16	0.72
Lu 2	Caleta Puelche	5.3	6	25	3	0.28	2.24	0.70	0.24	0.75
Lu 3	Contao	5.4	8	21	5	0.34	3.25	1.54	0.36	0.16
Lu 4	Hornopirén	6.2	2	17	3	0.09	1.58	0.52	0.14	0.01
Lu 5	Chaitén	5.2	20	21	4	0.17	0.85	0.48	0.26	0.36
Lu 7	Chaitén	5.2	11	27	6	0.21	0.24	0.21	0.19	0.48
Lu 8	Sur Chaitén	5.4	3	25	7	0.51	3.88	0.86	0.25	0.33
Lu 9	La Junta	5.4	15	28	5	0.12	0.25	0.12	0.16	0.41
Lu 10	Lago Verde	5.8	17	26	3	0.15	1.39	0.45	0.18	0.13
Lu 11	La Junta	5.5	9	47	5	0.10	1.00	0.20	0.10	0.49
Lu 12	Coyhaique	n.o.	n.o.	n.o.	n.o.	n.o.	n.o.	n.o.	n.o.	n.o.
Lu P4	Villarrica	5.2	22	65	7	0.19	2.04	0.43	0.26	0.31
Lu P11	Pto. Saavedra	6.0	6	52	8	0.42	5.03	5.58	0.69	0.01
Lu P23	Vilcún	5.4	25	82	25	0.37	6.57	1.73	0.64	0.10
Lu P32	Liucura	6,0	7	20	6	0.29	5.82	1.03	0.22	0.01
Lu P36	Pitrufquén	5.3	25	84	39	0.45	5.20	1.92	0.69	0.16
Lu P48	Quellón	5.1	27	74	8	0.29	1.22	0.48	0.38	0.87
Lu P52	Castro	5.2	20	73	8	0.31	3.01	1.01	0.44	0.29
Lu P54	Castro	5.0	27	85	14	0.78	2.98	2.17	0.42	0.71
Lu P58	Puerto Varas	5.6	21	67	34	0.33	7.54	1.68	1.44	0.9
Lu P63	Riñihue	5.3	19	61	11	0.59	5.41	1.09	0.24	0.17

Table 2. Soil chemical analyses of collecting sites of *L. tenuis* and *L. uliginosus*. OM : organic matter; n.o.: not observed.

3. Phenotypic characterization

In marginal environments drought and salinity are the abiotic stresses that most limit plant growth and the crop productivity (Reynolds *et al.*, 2001). Drought is defined as the situation in which the water potential and turgor of the plant falls below a threshold that affects its normal functioning (Kramer, 1983). Blum (1996) defines it as the water deficiency of the plant that occurs when the evaporative demand of the atmosphere on the leaves exceeds the capacity of the roots to extract water from the soil. The development of drought tolerant and/or greater water-use efficient (WUE) genotypes is currently a global challenge, owing to the continued growth of the world population and the reduction of water resources destined to agricultural use (FAO, 2008).

The response of plants to water deficit can be studied through the systematic identification of morphological, physiological and biochemical characteristics that provide the ability to tolerate stress. Many of these have been associated with drought tolerant genotypes (Acuña *et al.*, 2010; Araus *et al.*, 2002; Condon *et al.*, 2004; Inostroza & Acuña, 2010; Poormohammad *et al.*, 2007;Reynolds *et al.*, 1999; Richard *et al.*, 2002), but not all are expressed by the same genotype. To determine which characters confer adaptive advantages under water stress conditions, it is necessary to know the environment in which these genotypes will be established. Donald (1968) proposes determining selection traits through the theoretic conception of a preconceived model plant (ideotype), which must express phenotypic traits that confer adaptation to a specific environment. The use of morphological and physiological traits as indirect selection criteria for forage yield is an alternative breeding approach. However, the limited success of this approach may be due to a lack of understanding of the physiological factors most directly involved in determining yield, in addition to the absence of proper methods for evaluating them in a rapid and routine manner. In this section, some physiological traits related to drought tolerant genotypes will be discussed.

3.1 Plant water status

There are two basic parameters used to describe plant water status, these being (i) the water content and (ii) the energetic state of the water in the plant. The water content is normally expressed based on the water content in full turgor and the term used is the relative water content (RWC). The energetic state is expressed as the water potential of the plant (Ψ_W).

The RWC is the most commonly used expression to measure the level of water in plant tissue, given that it is related to Ψ_W, because the Ψ_W and its components ($\Psi_W = \Psi_P + \Psi_\pi$), the pressure potential (Ψ_P) and the osmotic potential (Ψ_π), are in function of the water volume of the protoplasm (Jones, 2007). The leaf RWC is defined as: RWC = (fresh weight - dry weight) / (turgid weight - dry weight); whereas Ψ_π is measured with a Scholander pump. More antecedents about these traits were broadly reviewed by Jones (2007).

3.2 Stomatal conductance and canopy temperature

Given that water stress affects the photosynthetic capacity of plants because of stomatal closing, genotypes that present greater stomatal conductance under water stress conditions show a greater adaptability to water stress (Medrano *et al.*, 2002).

The water state of the plant and its rate of transpiration determine its thermal state (Reynolds *et al.*, 1994). It has been demonstrated that the temperature of the leaves can be lower than the air temperature. The degree of cooling reflects the rate of evapotranspiration occurring on the surface of the canopy (Ayneh *et al.*, 2002; Jones *et al.*, 2002; Möller *et al.*, 2007). The temperature of a canopy can be measured with an infrared thermometer and is generally expressed as the difference between the temperature of the canopy and the air, the term assigned to this difference is canopy temperature depression (CTD). The CTD has been strongly correlated with stomatal conductance. Furthermore, increases in grain yields of eight wheat genotypes liberated by CIMMYT between 1962 and 1988 were associated with greater stomatal conductance and cooler canopies (Ficher *et al.*, 1998).

The advantage of evaluating stomatal conductance through CTD over direct measurement by porometer is that CTD evaluates a canopy whereas porometry can only be applied to a leaf from an individual plant. For genetic improvement programs it is recommended to make selections using CTD in large populations. Once the number of individuals in the population is reduced, selection by porometry is recommended.

3.3 Spectral vegetation indices

The yield of a crop during a given period of time and under particular growing conditions is determined by three major process or integrative traits: the interception of incident solar irradiance by the canopy, the conversion of the intercepted radiant energy to potential chemical energy, and the harvest index. The first depends on the photosynthetic area of the canopy, while the second relies on the overall photosynthetic efficiency of the crop. These two processes, which are responsible for the overall crop biomass, are more affected than the harvest index by drought and other related stress typical of Mediterranean growing conditions (Bort *et al.*, 2001). In principle, the spectra reflected by crop canopies at different wavelengths through the photosynthetically active radiation (PAR) and near infrared radiation (NIR) regions of the electromagnetic spectrum provide rapid, nondestructive, simultaneous estimations of the two processes determining crops biomass at the canopy level (Peñuela & Filella, 1998). Such estimation requires spectral reflectance indices, which are formulated based on simple mathematical operations among reflectance levels at given wavelengths (ratios or differences).

Nowadays, spectroradiometric indices are used to assess characteristics associated with the development of the photosynthetic area of the canopy. The most widespread spectral vegetation index is the normalized difference vegetation index (NDVI) followed by the simple ratio (SR) (Bort *et al.*, 2001).

Leaf area duration affects the total photosynthetic area and thus the radiation accumulated by the canopy, especially in Mediterranean environments where forage crops must confront summer drought that reduces leaf area duration and halts the developments of new photosynthetic area. Early senescence of photosynthetic tissues is characterized by chlorophyll degradation, with increases in the ratios of total carotenoids to chlorophyll *a* and chlorophyll *b* to *a*. In this context, spectroradiometric indices able to provide information on the relative changes in these pigments may be appropriate indicators of the duration of photosynthetic organs (Bort *et al.*, 2001).

3.4 Chlorophyll fluorescence

Photosynthesis is an essential process to maintain crop growth and development. Chlorophyll is one of the major chloroplast components for photosynthesis, and relative chlorophyll content has a positive relationship with the photosynthetic rate (Guo & Li, 1996). Furthermore, focusing on the stay-green trait, in effect the ability of the plant to maintain high chlorophyll content for long periods of time, has been proposed as a strategy to increase crop production, particularly under water-limited conditions (Guo *et al.*, 2008).

In addition, photosystem II (PSII) is an important component of plant photosynthesis that is particularly sensitive to water deficit conditions (Lu and Zhang, 1999). Drought-induced decrease in photosynthesis has been associated with perturbations of biochemical processes (Graan & Boyer, 1990) and photodamage of PSII reaction centers (He *et al.*, 1995). While chlorophyll fluorescence is widely accepted as an indication of the energetic behavior of photosynthetic system, it is emitted mainly by PSII in the range of 680–740 nm spectral region and can be considered an intrinsic indicator of the destination of excitation energy (Dau, 1994).

Several fluorescence parameters, such as initial fluorescence (Fo), maximal fluorescence (Fm), variable fluorescence (Fv=Fm-Fo) and maximum/potential quantum efficiency of PSII (Fv/Fm), have been widely used for investigations into various plants species under diverse growth conditions (Araus *et al.*, 1998; Guo *et al.*, 2008).

3.5 Water use efficiency

In environments where water availability limits growth and output, crop yields (CY) can be expressed using the following identity: $CY = WU \times WUE \times HI$, where WU is the total of water used by the crop (evapotranspiration), WUE is the ability of crop to produce biomass per unit of water evapotranspirated and HI the crop harvest index (commercial biomass/total biomass). If the three terms are independent, an increase in any of them implies an increase in CY. For many decades the genetic improvement of forage legumes has been oriented to increasing forage production and persistence. However, there is recognition of the need to address the adaptability of plants to climatic change (Humphreys, 2005; Taylor, 2008).

Water use efficiency is defined as the fraction of accumulated biomass, expressed as the rate of CO_2 assimilation (A), or total biomass production per volume of water consumed, which can be expressed as the transpiration rate (T), evapotranspiration or the total of water taken into the system. The time scale to define WUE can be an instant, a day or a growth season (Sinclair *et al.*, 1984). There are few genetic improvement programs oriented to increasing WUE owing to the scarcity of evaluation methodologies. These are generally very time-consuming, costly and require a large number of plants under uniform climatic conditions (Teulat *et al.*, 2001).

On the other hand, theoretical (Farquhar *et al.*, 1982) and empirical approximations (Farquhar & Richards, 1984) show that ^{13}C isotopic discrimination ($\Delta^{13}C$) can provide an indirect measurement of WUE. This has stimulated research into the potential use of $\Delta^{13}C$ as a selection criterion in genetic improvement programs. Experimental evidence currently shows a linear and negative relationship between $\Delta^{13}C$ and WUE in crops such as wheat, barley, peanuts, tomatoes, and cowpeas (Acevedo *et al.*, 1997). As well, linear and positive relationships have been observed between $\Delta^{13}C$ and yields for different crops such as wheat, barley, beans, peanut, cowpeas and tomatoes (Acevedo *et al.*, 1997). The main advantage of using $\Delta^{13}C$ in genetic selection programs is its high inheritability, which is mainly due to the low level of genotype x environment interaction (Teulat *et al.*, 2002).

3.6 Theoretical relationship between $\Delta^{13}C$ and WUE

The WUE at the level of the leaf is defined as the assimilation of CO_2 per unit of transpirated water and is represented with the following equation (Condon *et al.*, 2002): $EUA = 0.6C_a(1 - C_i/C_a)/(W_i - W_a)$, where the factor 0.6 corresponds to the diffusivity coefficient between CO_2 and water vapor in the air, C_a and C_i are the concentration of CO_2 in the atmosphere and in the vegetal cell, respectively, and W_i-W_a is the gradient water vapor concentration between the interior of the vegetal cell (W_i) and the atmosphere (W_a). If the W_i-W_a gradient is considered as an independent variable, the equation indicates that WUE is a negative function of the C_i/C_a ratio. As well, the C_i/C_a ratio depends on the balance between stomatal conductance (g_s) and the pohotosynthetic capacity of the plant. The supply of CO_2 to the interior of the leaf is influenced by g_s, while the photosynthetic capacity determines the demand for CO_2. On the other hand, theoretical and empirical evidence shows that ^{13}C isotopic discrimination ($\Delta^{13}C$) is an indicator, in direct proportion, of the C_i/C_a fraction and is consequently an estimator of WUE (Condon *et al.*, 2002; Farquhar *et al.*, 1989). As noted above, WUE depends on the balance between stomatal conductance (g_s) and the photosynthetic capacity of the plant. If a higher WUE is the product of an increase in the photosynthetic capacity of the plant, it results in higher DM production. On the other hand, if higher WUE is a product of an increase in gs, it results in lower DM production.

4. Strategies for selecting germplasm

4.1 Selection of contrasting genotypes

In regions with Mediterranean climates, like central Chile, the annual rainfall is concentrated in winter (approximately 80%) when lower temperatures limit plant growth. On the other hand, water limitations for perennial forage species during spring and summer are strong and affect their productivity and persistence. Therefore, genotypes tolerant to water shortage for long periods are required for these environments, where water for irrigation is scarce and there is a competition for this resource between crops and pasture.

The aims of germplasm characterization at the first stages has been to identify genotypes that show contrasting degrees of tolerance to water stress for their survival, growth and productivity, in effect, to identify tolerant and sensitive materials relevant for genetic and physiological studies of physiological traits, quantitative trait locus (QTL) and genes that modulate drought tolerance or sensitivity in forage species. This information will allow for increasing the efficiency of selection in breeding programs aimed at developing cultivars better adapted to conditions of limited moisture without affecting their productivity or forage quality.

In this sense, an experiment was carried out as part of the project 'Lotus adaptation and sustainability in South American soils' (LOTASSA; http://www.lotassa.org/online/site/) funded by the EU. The growth and plant water status of eleven populations of *L. tenuis*, naturalized in Chile, and a cultivar of Argentinean origin of the same species, were evaluated to select contrasting genotypes (Acuña *et al.*, 2010). The cultivars were grown in a greenhouse under different levels of soil water availability. The relative rate of stem elongation (RRSE), DM production, RWC and specific leaf area (SLA) all showed significant reductions in the treatment with the highest water restriction (10% of soil water availability). There were significant differences among genotypes in RRSE and DM production means, but not in RWC and SLA. The drought sensitivity index (DSI) varied broadly among genotypes, from 0.49 to 1.34, and was correlated negatively with DM production under water stress. It was concluded that the *L. tenuis* populations showed genetic variability in water-stress tolerance, with accessions Lt14 and Lt4 at the extremes of tolerance and sensitivity, respectively (Acuña *et al.*, 2010). These findings will permit to identify chromosomal regions associated with drought-tolerant genotypes and accelerate the development of cultivars adapted to water-restricted environments. Figure 4 shows the two contrasting populations in terms of their genotype x environment interaction. The population Lt14 showed a regression coefficient (b) greater than one, which means that this population has a high response capacity to the environment. As growth conditions improve, the population Lt14 responds with increased DM shoot growth. In contrast, the population Lt4 showed a b-value of less than one (0.51).

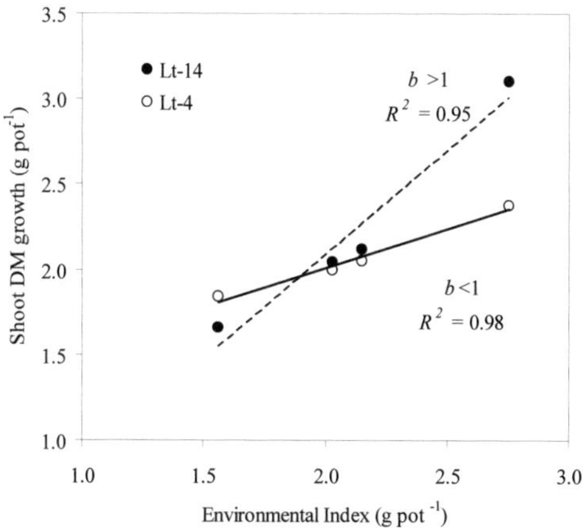

Fig. 4. Response of contrasting *L. tenuis* populations to soil water treatments as calculated by the Finlay & Wilkinson (1963) methodology. Coefficients of regression (b) and determination (R^2), (Acuña *et al.*, 2010).

4.2 Selection of high water use efficiency genotypes

In contrast to the *Lotus* species, white clover is a highly sensitive species to water deficits owing to its shallow root system (Hart, 1987). As well, under drought conditions it does not reduce stomatal conductance, as many other forage species do (Karsten & MacAdam, 2001). Consequently, in central Chile, where a Mediterranean climate predominates, white clover is cultivated exclusively under irrigated conditions. In this sense a high water use efficiency cultivar is necessary to increase water-use productivity.

On the other hand, white clover populations naturalized in Chile represent an abundant source of genetic diversity that can be used in the search for genotypes that express high efficiency in water use. Work was carried out in 2009 (i) to evaluate the WUE of the nine populations collected and selected by Ortega *et al.* (1994), and (ii) to identify morphological and physiological characteristics associated with genotypes with high WUE, to assist in the selection of plants for the genetic improvement of the species for this trait (Inostroza & Acuña, 2010). The work arrived at several conclusions. First, the evaluated white clover populations showed wide differences in their WUE, which can be attributed to the intrinsic ability of each genotype to regulate stomatal opening and changes in DM partition induced by water stress. Higher WUE in white clover is associated with genotypes that show less investment of DM in transpirant surface and favour the growth and development of stolons. In this sense, the 9-1-X population was the most efficient in water use. Second, it was concluded that under the experimental conditions, the RWC, the stem water potential and ^{13}C isotopic discrimination are good indicators of the water state of the plant, but do not allow for identifying differences among populations and do not show an association with WUE. Third, the crop water stress index (CWSI) calculated in the treatment without stress, the fraction of intercepted photosynthetically active radiation (FIPAR), the leaf weight ratio, LWR (LWR=leaf DM/shoot DM), and the leaf-area ratio, LAR (LAR= specific leaf area (SLA) x LWR), together form integral physiological traits to estimate WUE in the studied white clover populations. Additionally, this work identified a naturalized white clover population that is highly efficient in water use and reaches the same levels of DM production as commercial cultivars (Huia and Will) with 25 % less water used. High white clover WUE was also associated with genotypes that can regulate the water loss through the stomata and modify its biomass partitioning under water stress conditions.

5. Plant perception and gene regulation under drought stress

Unlike other organisms, plants do not have the possibility of physical movement to escape adverse conditions. Therefore, plants have to develop adequate systems of perception and response to confront adverse environmental conditions. The ability to perceive water deficit early on constitutes an advantage and permits a rapid response to new environmental scenarios. This phenotypic plasticity is related to the activation of defined sets of genes (Harb *et al.*, 2010).

5.1 Hormonal signals

Plant perception of drought stress is mediated by phytohormonal signals involving auxin, cytokinin, jasmonic acid (JA), salicylic acid, ethylene, gibberellins and abscisic acid (ABA) (Huang *et al.*, 2008). However, different plant hormones have been described to regulate similar processes, although exerting distinctive domains over transcriptional response (Nemhauser *et al.*, 2006). Processes such as stomatal movement are controlled by interactions among various hormones. These hormonal combinations are specific to environmental conditions, period and species. Some hormones act as antagonists, while others function cooperatively in a common response. The effect of ethylene during drought stress has been described as antagonistic of ABA, thus inhibiting stomatal closure that is induced by ABA. Ethylene probably contributes to maintaining a basal photosynthesis level, but induces leaf senescence (Spollen *et al.*, 2000; Tanaka et al., 2005). It has been suggested that hormones, including ethylene, gibberellic acid (GA) or auxin, may modulate drought stress signalling by acting antagonistically to ABA. This is supported by evidence that genes down-regulated by ABA are responsive to these hormones (Huang *et al.*, 2008). Other group of genes are up-regulated or down-regulated by drought with a differential effect of individual hormones. For example, *GASA1* encodes for a related GAST1 gene of Arabidopsis that is gibberellin-regulated (Herzog *et al.*, 1995). GASA1 is repressed by brasinosteroids (BR) and drought, but induced by ABA and rehydration (Bouquin *et al.*, 2001; Huang *et al.*, 2008).
Drought stress is commonly associated with a decrease in cytokinins (CKs) (Davies & Zhang, 1991), which act as antagonists of ABA response. Mediating the expression of P_{sark}::IPT in tobacco plants, Rivero *et al.*, (2010) obtained a sustainable photosynthetic system, preventing its degradation by drought. It was a consequence of constant expression of genes coding for photosystem complex proteins (Rivero *et al.*, 2010). Additionally, high cytokinin levels caused activation of BR synthesis and repression of ABA signalling response. Higher abundance of proteins involved in photosynthesis, respiration, aminoacid and protein synthesis, antioxidant defence system have been reported by Merewitz *et al.*, (2011) in creeping bentgrass expressing an ipt gene. By the other side, jasmonic acid is known by regulate of induced systemic resistance (ISR) which is triggered by plant interaction with beneficial soil-borne microorganism (Van Wees *et al.*, 2008). In plants methyl jasmonate (MJ) was described as inductor of ABA biosynthesis in rice, enhancing response to drought stress (Eun *et al.*, 2009).

5.2 Hormonal control ABA dependent pathway

One of the better-known processes for ABA signalling is triggered by stomatal closure in guard cells. Most of the ABA signal transduction components have been identified in Arabidopsis and found by classical forward genetics in mutants with high or low sensitivity to ABA (Schroeder *et al.*, 2001).

The most important and heavily studied signal responsible for growth and development under drought stress is abscisic acid (ABA). Under drought conditions, ABA accumulates in plants and catabolizes when the stress disappears (Koorneef *et al.*, 1998; Taylor *et al.*, 2000). The response of plant requires hormonal synthesis and accumulation, which is why ABA-dependent tolerance mechanisms are considered as an adaptive stress response (Bray, 1997; Chandler & Robertson, 1994; Xiong *et al.*, 2002; Yamaguchi-Shinozaki & Shinozaki, 2005). ABA accumulation is mediated by both the liberation of biologically active hormones and the activation of their *de novo* synthesis. Conjugated ABA has been suggested constitute a reserved or stored form of ABA. Conjugation reaction is mediated by ABA-glucosyltransferase and compartmentalized in vacuole or apoplastic space as ABA glucose ester (Cutler & Krochko, 1999; Dietz *et al.*, 2000; Xu *et al.*, 2002). To make ABA immediately available, for example during a drought stress, a β-glucosidase converts inactive conjugated ABA (ABA-glucose ester) into active ABA (Lee *et al.*, 2006). Biosynthetic pathway genes are mainly activated by 9-cis-epoxycarotenoid dioxygenase (NCED), which catalyzes limit step for the ABA metabolic pathway (Iuchi *et al.*, 2001; Qin & Zeevart, 1999; Thompson *et al.*, 2000; Xiong & Zhu, 2003). ABA is transported from roots to leaves by long distance translocation (Wilkinson & Davies, 2002).

Several types of proteins, among them phosphatases, kinases and transcription factors, constitute ABA signal transduction pathways. Intracellular receptors have been discovered recently and named PYR/PYL/RCAR (Pyrabactin resistance/PYR-like/Regulatory components of ABA receptors). During signal perception of ABA, these receptors interact with the clade A of protein phosphatases type 2Cs (PP2Cs) and inhibit their action in an ABA-dependent manner, resulting in the activation of several SnRK2-type kinases (Holappa & Simmons, 1995; Kelner *et al.*, 2004; Yoshida *et al.*, 2002). These proteins modulate the activity of components located downstream, mediating phosphorylation of transcription factors (TFs) AREB/ABF (Johnson *et al.*, 2001). Under non-stress conditions, PP2C interacts with and phosphorylates SnRK2s-inhibiting kinase activity (Cutler *et al.*, 2010). Transcription factors dependent on the ABA signal transduction pathway interact with *cis* elements present in the promoter region of regulated genes, leading to activation of the expression of different stress-responsive genes (Shinozaki *et al.*, 2003; Yamaguchi-Shinozaki and Shinozaki, 2005). Primary *cis* elements that bind AREB/ABF TF are termed ABRE (ABA responsive element), originally identified in the Em gene from *Triticum aestivum* and Rab16 from *Oryza sativa* (Guiltinan *et al.*, 1990; Mundy *et al.*, 1990). In general, the presence of ABRE motive is not sufficient for efficient induction of gene expression, requiring participation of additional regulatory elements (Shen & Ho, 1995; Shen *et al.*, 1996), or repetitions of ABRE motive (Uno *et al.*, 2000). Additionally, some genes induced by drought through ABA-dependent signal pathway do not have ABRE sequences in promoters. Activation requires the presence of sites for MYBR and MYCR binding, recognized by transcription factors from the MYB and MYC family, which are also induced by endogenous ABA accumulation (Abe *et al.*, 2003).

The ABA-dependent signal pathway activates several groups of genes, in particular genes related to compatible osmolytes and LEA proteins, protection against oxidative stress as

ELIP proteins (early light-inducible proteins) or scavenger enzymes like ascorbate peroxidase, superoxide dismutase and catalase (Apel *et al.*, 2004; Finkelstein *et al.*, 2002; Gao *et al.*, 2004; Zeng *et al.*, 2002).

5.3 Redox signals

The strategy that photosynthetic organisms have devised to acclimatize to adverse conditions is to maintain photosynthetic efficiency as high as possible. Environmental changes are related to changes in the redox potentials of electron chain transporters or associated components (Thioredoxin, Glutation), which constitute the primary signals to regulate chloroplast or nuclear gene expression.

Redox regulation of photosynthetic genes occurs at multiple levels of expression, suggesting the existence of a complex network of signals. Chloroplast gene expression is in part modulated by nuclear factors that carry out functions including RNA transcription, editing processing, degradation and translation. Several of them are components of basic expression machinery, while others are required specifically for expression of small gene subgroups from chloroplasts (Pfannschmidt, 2003). Plastidial signals can also regulate transcription of nuclear gene coding of both plastidial and non-plastidial proteins. These signals are related to derivative compounds from photosynthetic chloroplast metabolism, such as porphyrins, reactive oxygen species (ROS), carotenoids and tetrapyrrol (Johanningmeier *et al.*, 1988; Rodermel, 2001; Surpin, 2002). An example of these signal compounds is hydrogen peroxide, which is produced in plants under normal conditions. Levels of H_2O_2 are efficiently regulated by a detoxifying system. However, abiotic stress like drought produces an oxidative imbalance, caused by increased H_2O_2 levels, activating the redox signal. Hydrogen peroxide modulates expression of genes coding for detoxifying enzymes and H_2O_2 biosynthesis regulators (Neill *et al.*, 2002). H_2O_2 also participates, together with ABA, as a signal for stomatal guard cell movement (Desikan *et al.*, 2004). Mutant plants abi1-1 and abi2-1 exposed to high levels of light showed diminished expression of cytosolic ascorbate peroxidase (APX), a demonstrated form of H_2O_2-induced gene expression (Gupta *et al.*, 1993). These results suggest that there is crosstalk between ABA and H_2O_2 signal pathways, enhancing the response (Fryer *et al.*, 2003). H_2O_2 also induces the expression of superoxide dismutase (SOD) and malic enzyme (ME) (Slesak *et al.*, 2003).

In summary, H_2O_2 constitute a part of the redox signal in chloroplasts, regulating nuclear genes related to detoxifying functions that protect protein components in chloroplasts, inducing acclimation of photosynthesis to environmental stimuli (Noctor *et al.*, 2000).

5.4 Sugar signals

Sugars play a regulation role, coordinating resource use and localization. Sugars effectively adjust metabolism to confront environmental changes. Sugar signals act in a complementary manner and amplify early signals during drought stress for metabolic control. Although the response is slow at the genic level, sugars allow persistent and sustained intensity of exchange, which is not possible with other types of signal regulation.

Soluble sugars regulate gene expression of several cellular functions and metabolic pathways. However, in general, sugars induce gene expression of proteins related to biosynthesis, utilization and storage of reserve compounds (starch, lipids and proteins), while suppressing expression of enzymes related to photosynthesis and reserve mobilization (Koch, 1996). Among the genes regulated positively by sugars are storage enzymes like patatin and sporamin in potato, starch and sucrose biosynthesis enzymes like ADP glucose pyrophosphorilase (AGP), invertase, sucrose synthase and genes for protein defence like proteinase II inhibitor (Geigenberg., 200; Jefferson *et al.*, 1990; Kim *et al.*, 1991). In contrast, numerous genes are negatively regulated by sugars, such as genes coding for alpha-amylase in rice, endopeptidase, sucrose syntethase and asparagine synthase in corn roots or malate synthase and isocitrate lyase in cucumber cotiledons (Koch, 1996).

One of the main sugars described in photosynthetic gene repression is glucose. Studies of promoters of photosynthetic genes fused to GUS have described repression by glucose and sucrose (Sheen, 1990). Similar conclusions have been obtained by other researchers using modified plants that accumulate high hexoses levels (Dickinson *et al.*, 1991). Examples include CAB protein, δ-ATPase and enzymes involved in the glyoxylate cycle, the expression of which is inhibited by sugars (Harther *et al.*, 1993; Krapp *et al.*, 1993; Sinha *et al.*, 2002).

5.5 Effect of drought stress on legumes nodules

Nitrogen fixation in legumes is severely affected by drought, salinity, defoliation, darkness and chilling, among other factors. Decline in nitrogen fixation has been related to reduced permeability of the diffusion barrier for oxygen flux in the nodule and thus prevents nitrogenase damage (Hunt & Layzell, 1993). Water stress inhibits sucrose synthase activity, which increases sucrose concentration in nodules, as a result of osmotic regulation, but negatively affects bacterial respiration, limiting carbon flux (Arrese-Igor *et al.*, 1999; González *et al.*, 1995). González *et al.*, (1998) did not find differences in the expression of SS during ABA treatment, but did find decreased expression of the leghemoglobin gene (Lb).

Abiotic stress also induces production of ROS in nodules and turns an oxidative stress that could be a cause of biological nitrogen fixation decline. Evidence suggests that the main targets of ROS in nodules are SS and Lb (Marino *et al.*, 2006).

Nodule specific genes have been described that are regulated by abiotic stress and play a putative role in response to stress to facilitate the nitrogen fixation. Three cysteine cluster protein (CCP) and LTP genes were found in *A. sinicus* during salt treatments (Chen *et al.*, 2007).

5.6 Summary of local research

Studies in *Lotus spp.* have been directed to describing the effects of drought stress and response. It has emphasized the role of nodules in nitrogen fixation. Nitrogen-fixing nodules are formed by the interaction between rhizobia and legume roots through the integration of two processes, infection by rhizobia and initial cell division in the cortex of the roots. The number of nodules is regulated by several factors, including drought stress. The abundance of nodules must be regulated to avoid interfering with nutrient distribution. Early nodulation inhibits further nodulation in young roots (Oka-kira and Kawaguchi, 2006). A chitinase Ltchi7 was recently described (Tapia *et al.*, 2011) that is induced during drought stress in *Lotus tenuis,* a drought tolerant *Lotus* species. Ltchi7 is a class III chitinase, which is part of a subgroup of chitinases from legumes with an additional COOH terminal domain. Nodule chitinases are related to lipochitooligosaccharide (LCO) cleavage. LCOs act as signals in nodule organogenesis and are synthesized by bacteria in the initial stages of nodule formation (Mergaert *et al.*, 1997). It is suggested that this type of chitinases has a role

in regulating LCO abundance in root proximity during drought stress and contributes to maintaining control in infection and nodule formation (Tapia *et al.*, 2011).
On the other hand, research to study the functional role of nsLTPs in *Lotus japonicus* during water deficit has been developed. Expression analysis mediating RNA blotting experiments, in situ hybridization, promoter GUS fusion have demonstrated that nsLTPs are expressed in an epidermis-specific manner, particularly in organs with active cuticle synthesis (leaves, stems and flowers) (Kader, 1996; Arondel *et al.*, 2000; Clark *et al.*, 1999). Additionally, correlations of wax accumulation with increased LTP expression, especially during abiotic stress, support the hypotheses that these proteins are responsible for the lipid load component in the cuticle (Cameron *et al.*, 2006; Sohal *et al.*, 1999; Treviño & O'Connell, 1998). Genes coding for nsLTPs in *L. japonicus* have been identified, which were drought induced. One of them is expressed in epidermal cells of leaves and shoots. Studies of 3D modelling show similarities in cavity ligand binding for different nsLTPs and apparent structural redundancy. Future studies could reveal the function of these proteins in cuticle formation and drought tolerance.

6. Conclusions

The results obtained have provided genotypes for INIA-Chile's genetic improvement program for forage legumes, in particular *Lotus* spp. and white clover, for generating cultivars that are tolerant to drought conditions for the Mediterranean climate zone and the rainy southern region of the country.

7. References

Abe H., Urao T., Ito T., Seki M., Shinozaki K. & Yamaguchi-Shinozaki K. (2003. Arabidopsis AtMYC2 (bHLH) and AtMYB2 (MYB) function as transcriptional activators in abscisic acid signalling. Plant Cell, 15, 63-78.

Acevedo E., Baginsky C., Solar B. & Ceccarelli S. 1997. Discriminación isotópica de ^{13}C y su relación con el rendimiento y la eficiencia de transpiración de genotipos locales y mejorados de cebada bajo diferentes condiciones hídricas. Investigación Agrícola (Chile), 17, 41-54.

Acuña H., Figueroa M., De La Fuente A., Ortega F. & Fuentes C. 2002a. Comportamiento de cultivares de *Lotus corniculatus* L. en diferentes ambientes de la VIII y IX Regiones de Chile. Agro-Ciencia 18 (2):75 – 84.

Acuña H., Figueroa M., De La Fuente A., Ortega F., Seguel I. & Mundaca R. 2002b. Caracterización agronómica de accesiones de *Lotus glaber* Mill. y *Lotus uliginosus* Schkur. naturalizadas en Chile. Agro-ciencia, 18, 63-74.

Acuña H., Hellman P., Barrientos L., Figueroa M., & De La Fuente A. 2004. Estimación de la Fijación de Nitrógeno en tres especies del género *Lotus* por el método de la dilución isotópica. Agro-Ciencia, 20, 5-15.

Acuña H., Concha A. & Figueroa M. 2008. Condensed tannin concentrations of three *Lotus* species grown in different environments. Chilean Journal of Agricultural Research, 68, 31-41.

Acuña H., Inostroza L., Sánchez Ma.P., & Tapia G. 2010. Drought-tolerant naturalized populations of *Lotus tenuis* for constrained environments. Acta Agriculturae Scandinavica, Section B - Plant Soil Science, 60,174 – 181.

Apel K. & Hirt H. 2004. Reactive oxygen species: Metabolism, oxidative stress, and signal transduction. Annual Review of Plant Biology, 55, 373-399.

Araus J.L., Amaro T., Voltas J., Nakkoul H. & Nachit M.M. 1998. Chlorophyll fluorescence as a selection criterion for grain yield in durum wheat under Mediterranean conditions. Field Crop Research, 55,209–223.

Araus J.L., Slafer G.A., Reynolds M. & Royo C.. 2002. Plant breeding and drought in C_3 cereals: What should we breed for?. Annals of Botany, 89, 925-940.

Arondel V., Vergnolle C., Cantrel C. & Kader J. 2000. Lipid transfer proteins are encoded by a small multigene family in Arabidopsis thaliana. Plant Science, 157, 1–12.

Arrese-Igor C., González E.M., Gordon A.J., Minchin F.R., Gálvez L., Royuela M., Cabrerizo P.M. & Aparicio-Tejo P.M. 1999. Sucrose synthase and nodule nitrogen fixation under drought and other environmental stresses. Symbiosis, 27, 189–212.

Ayeneh A., Ginkel M.V., Reynolds M., & Ammar K. 2002. Comparison of leaf, spike, peduncle and canopy temperature depression in wheat under heat stress. Field Crop Research. 79:173-184.

Blum A. 1996. Crops responses to drought and the interpretation of adaptation. Plant Growth Regulation, 20, 135-148.

Bort J., Casadesús J., Araus J., Granado S. & Ceccarelli S. 2001. Spectral vegetation indices as nondestryctive indicators of barley yield in Mediterranean rain fed conditions. In: Slafer G., Molina Cano J., Savin R., Aruas J. and Ramagosa I. (eds). Barley Science. Recent Advances from Molecular Biology to Agronomy of Yield and Quality. Food Products Press, New York, pp 387-411.

Bouquin T., Meier C., Foster R., Nielsen M. E. & Mundy J. 2001. Control of Specific Gene Expression by Gibberellin and Brassinosteroid. Plant Physiology, 127, 450-458.

Bray E.A. 1997. Plant responses to water deficit. Trends in Plant Science, 2, 48-54.

Cameron K.D., Teece M.A. & Smart L.B. 2006. Increased accumulation of cuticular wax and expression of lipid transfer protein in response to periodic drying events in leaves of tree tobacco. Plant Physiology, 140,176-83.

Chandler P.M. & Robertson M. 1994. Gene expression regulated by abscisic acid and its relation to stress tolerance. Annual Review of Plant Physiology and Plant Molecular Biology, 45, 113-141.

Chen D.S., Li Y.G. & Zhou J.C. 2007. The symbiosis phenotype and expression patterns of five nodule-specific genes of Astragalus sinicus under ammonium and salt stress conditions. Plant Cell Reports, 26, 1421-1430.

Clark A.M. & Rohnert H.J. 1999. Cell-specific expression of genes of the lipid transfer protein family from Arabidopsis thaliana. Plant Cell Physiology, 40, 69–76.

Condon A.G., Richards R.A., Rebetzke G.J., & Farquhar G.D. 2002. Improving Intrinsic Water-Use Efficiency and Crop Yield. Crop Science, 42, 122–131.

Condon A. G., Richards R. A., Rebetzke G. J. & Farquhar G. D. 2004. Breeding for high water-use efficiency. Journal of Experimental Botany, 55, 2447–2460.

Cutler A. J., & Krochko J. E. 1999. Formation and breakdown of ABA. Trends in Plant Science, 4, 472–478.

Cutler S. R., Rodriguez P.L., Finkelstein R. R. & Abrams S. R. 2010. Abscisic Acid: Emergence of a Core Signaling Network. Annual Review of Plant Biology, 61, 651-679.

Dau H. 1994. Molecular mechanisms and quantitative models of variable Photosystem II fluorescence. Photochemistry and Photobiology, 60, 1–23.

Davies W.J. & Zhang J. 1991. Root Signals and the Regulation of Growth and Development of Plants in Drying Soil. Annual Review of Plant Physiology and Plant Molecular Biology, 42, 55-76.

Desikan R., Cheung M., Bright J., Henson D., Hancock J. & Neill S. 2004. ABA, hydrogen peroxide and nitric oxide signalling in stomatal guard cells. Journal of Experimental Botany, 55, 205-212.

Dickinson C.D., Altabella B. & Chrispeels M.J. 1991. Slow growth phenotype in transgenic tomato expressing apoplastic invertase. Plant Physiology, 95, 420–425.

Dietz K. J., Sauter A., Wichert K., Messdaghi D. & Hartung W. 2000. Extracellular b-glucosidase activity in barley involved in the hydrolysis of ABA glucose conjugate in leaves. Journal of Experimental Botany, 51, 937–944.

Donald, C.M. 1968. The breeding of crop ideotypes. Euphytica, 17,385-403.

Eun H.K., Su-Hyun P. & Ju-Kon K. 2009. Methyl jasmonate triggers loss of grain yield under drought stress. Plant Signal Behavior, 4, 348–349.

Eun H.K., Youn S.K., Su-Hyun P., Yeon J.K., Yang D.C., Yong-Yoon C., In-Jung L., & Ju-Kon K. 2009 Methyl Jasmonate reduces grain yield by mediating stress signals to alter spikelet development in Rice. Plant Physiology, 149, 1751-1760.

FAO. 2008. Coping with water scarcity: what role for biotechnologies? Ruane J., Sonnino A., Steduto P. and Deane C. eds., Rome.

Farquhar G. & Richards R. 1984. Isotopic composition of plant carbon correlates with water-use-efficiency of wheat genotypes. Australian Journal of Plant Physiology, 11, 539-552.

Farquhar G., Ehleringer J. & Hubick K. 1989. Carbon isotope discrimination and photosynthesis. Annual Review of Plant Physiology and Plant Molecular Biology, 40, 503-537.

Farquhar G., O'Leary M. & Berry J. 1982. On the relationship between carbon isotope discrimination and the intracellular carbon dioxide concentration in leaves. Australian Journal of Plant Physiology, 9, 121-137.

Finkelstein R.R., Gampala S.S.L. & Rock C.D. 2002. Abscisic acid signalling in seeds. Plant Cell, 14, S15–S45.

Finlay K. & Wilkinson G. 1963. The analysis of adaptation in a plant-breeding program. Australian Journal of Agricultural Research, 14, 342-354.

Fischer R.A., Rees D., Sayre K.D., Lu Z.M., Condon A.G. & Saavedra A.L. 1998. Wheat Yield Progress Associated with Higher Stomatal Conductance and Photosynthetic Rate, and Cooler Canopies. Crop Science, 38, 1467–1475.

Fryer M., Ball L., Oxborough K., Karpinski S., Mullineaux P. & Baker N. 2003. Control of Ascorbate peroxidase 2 expression by hydrogen peroxide and leaf water status during excess light stress reveals a functional organisation of Arabidopsis leaves. Plant Journal, 33, 691-705.

Gao X., Pan Q., Li M., Zhang L., Wang X., Shen Y., Lu Y., Chen S., Liang Z. & Zhang D. 2004. Abscisic acid is involved in the water stress-induced betaine accumulation in pear leaves. Plant and Cell Physiology, 45, 742-750.

Geigenberger P. 2003. Regulation of sucroseto starch conversion in growing potato tubers. Journal of Experimental Botany, 54, 457-65.

González E.M., Aparicio-Tejo P.M., Gordon A.J., Minchin F.R., Royuela M. & Arrese-Igor C. 1998. Water-deficit effects on carbon and nitrogen metabolism of pea nodules. Journal of Experimental Botany, 49, 1705–1714.

González E.M., Gordon A.J., James C.L. & Arrese-Igor C. 1995. The role of sucrose synthase in the response of soybean nodules to drought. Journal of Experimental Botany, 46, 1515–1523.

Graan T. & Boyer J.S. 1990. Very high CO_2 partially restores photosynthesis in sunflower at low water potentials. Planta, 181, 378–384.

Guiltinan M.J., Marcotte W.R. & Quatrano R.S. 1990. A plant leucine zipper protein that recognises an abscisic acid response element. Science, 250, 267–270.

Guo P. & Li M. 1996. Studies on photosynthetic characteristics in rice hybrid progenies and their parents I. chlorophyll content, chlorophyll-protein complex and chlorophyll fluorescence kinetics. Journal of Tropical and Subtropical Botany, 4,60–65.

Guo P.G., Baum M., Varshney R.K., Graner A., Grando S. & Ceccarelli S. 2008. QTLs for chlorophyll and chlorophyll fluorescence parameters in barley under post-flowering drought. Euphytica, 163,203-214.

Gupta A.S., Webb R.P., Holaday A.S. & Allen R.D. 1993. Overexpression of superoxide dismutase protects plants from oxidative stress. Induction of ascorbate peroxidase in superoxide dismutase overexpressing plants. Plant Physiology, 103, 1067–1073.

Harb A., Krishnan A., Ambavaram M. M. R. & Pereira A. 2010. Molecular and Physiological Analysis of Drought Stress in Arabidopsis Reveals Early Responses Leading to Acclimation in Plant Growth. Plant Physiology, 154, 1254-1271.

Hart A.L., 1987. Physiology. In M. Baker and W. Williams (ed.), White clover, 126-151. C.A.B. International, Wallingford, UK.

Harter K., Talke-Messerer C., Barz W. & Schafer E. 1993. Light- and sucrose-dependent gene expression in photomixotrophic cell suspension cultures and protoplasts of rape (*Brassica napus* L.). Plant Journal, 4, 507-516.

He J.X., Wang J. & Liang H.G. 1995. Effect of water stress on photochemical function and protein metabolism of photosystem II in wheat leaves. Plant Physiology, 93, 771–777.

Herzog M., Dorne A.M. & Grellet F. 1995. GASA, a gibberellin-regulated gene family from Arabidopsis thaliana related to the tomato GAST1 gene. Plant Molecular Biology, 27,743–752.

Holappa L.D. & Walker-Simmons M.K. 1995. The wheat abscisic acid-responsive protein kinase mRNA, PKABA1, is up-regulated by dehydration, cold temperature, and osmotic stress. Plant Physiology, 108, 1203–1210.

Huang D., Wu W., Abrams S.R. & Cutler A.J. 2008. The relationship of drought-related gene expression in Arabidopsis thaliana to hormonal and environmental factors. Journal of Experimental Botany, 59, 2991-3007.

Humphreys M.O. 2005. Genetic improvement of forage crops- past, present and future. The Journal of Agricultural Science, 143, 441-448.

Hunt S. & Layzell D.B. 1993. Gas exchange of legume nodules and the regulation of nitrogenase activity. Annual Review of Plant Physiology and Plant Molecular Biology, 44, 483-511.

Inostroza L. & Acuña H. 2010. Water use efficiency and associated physiological traits of nine naturalized white clover populations in Chile. Plant Breeding, 129, 700-706

Iuchi S., Kobayashi M., Taji T., Naramoto M., Seki M., Kato T., Tabata S., Kakubari Y., Yamaguchi-Shinozaki K. & Shinozaki K. 2001. Regulation of drought tolerance by gene manipulation of 9-cis-epoxycarotenoid dioxygenase, a key enzyme in abscisic acid biosynthesis in Arabidopsis. Plant Journal, 27, 325-333.

Jefferson R., Goldsbrough A. & Bevan M. 1990. Transcriptional regulation of a patatin-1 gene in potato. Plant Molecular Biology, 14, 995-1006.

Johanningmeier U. 1988. Possible control of transcript levels by chlorophyll precursors in Chlamydomonas. European Journal of Biochemistry, 177, 417–424.

Johnson C., Glover G.a & Arias J. 2001. Regulation of DNA binding and trans-activation by a xenobiotic stress-activated plant transcription factor. Journal of Biological Chemistry, 276, 172–178.

Jones H.G. 2007. Monitoring plant and soil water status: established and novel methods revisited and their relevance to studies of drought tolerance. Journal of Experimental Botany, 58,119-130.

Jones H.G., Stoll M., Santos T., Sousa C.D., Chaves M.M. & Grant O.M. 2002. Use of infrared thermography for monitoring stomatal closure in the field: application to grapevine. Journal of Experimental Botany, 53,2249-2260.

Kader J.C. 1996. Lipid-transfer proteins in plants. Annual Review of Plant Physiology and Plant Molecular Biology, 47, 627–54.

Karsten H.D. & MacAdam J.W. 2001: Effect of drought on growth, carbohydrates, and soil water use by perennial ryegrass, tall fescue, and white clover. Crop Science, 41, 156-166.

Kelner A., Pekala I., Kaczanowski S., Muszynska G., Hardie D.G. & Dobrowolska G. 2004. Biochemical characterization of the tobacco 42-kD protein kinase activated by osmotic stress. Plant Physiology, 136, 3255-3265.

Kim S., Costa M.A. & An G. 1991. Sugar response element enhances wound response of potato proteinase inhibitor II promoter in transgenic tobacco. Plant Molecular Biology, 17, 973-983.

Koch K.E. 1996. Carbohydrate-modulated gene expression in plants. Annual Review of Plant Physiology and Plant Molecular Biology, 47, 509-540.

Koornneef M., Kloosterziel K., Schwartz S. & Zeevaart J. 1998. The genetic and molecular dissection of abscisic acid biosynthesis and signal transduction in Arabidopsis. Plant Physiology and Biochemistry, 36, 83–89.

Kramer, P.J. 1983. Water relations in plants. Academics Press, New York.

Krapp A., Hofmann B., Shafer C. & Stitt M. 1993. Regulation of the expression of rbcS and other photosynthetic genes by carbohydrates: a mechanism for the `sink' regulation of photosynthesis. Plant Journal, 3, 817-828.

Lee K. H., Piao H. L., Kim H. Y., Choi S. M., Jiang F., Hartung W., Hwang I., Kwak J.M., Lee I. J. & Hwang I. 2006. Activation of glucosidase via stress-induced polymerization rapidly increases active pools of abscisic acid. Cell, 126, 1109-1120.

Lu C. & Zhang J. 1999. Effects of water stress on photosystem II photochemistry and its thermostability in wheat plants. Journal of Experimental Botany, 50,1199–1206.

Marino D., González E.M. & Arrese-Igor C. 2006. Drought effects on carbon and nitrogen metabolism of pea nodules can be mimicked by paraquat: evidence for the occurrence of two regulation pathways under oxidative stresses. Journal of Experimental Botany, 57, 665-73.

Medrano H., Escalona J., Bota J., Gulias J. & Flexas J. 2002. Regulaton of photosynthesis of C_3 plant in response to progressive drought: stomal conductance as a reference parameter. Annals of Botany, 89, 895-905.

Merewitz E. B., Gianfagna T. & Huang B.J. 2011.Protein accumulation in leaves and roots associated with improved drought tolerance in creeping bentgrass expressing an ipt gene for cytokinin synthesis. Experimental Botany, 1-23.

Mergaert P., Van Montagu M., & Holsters M. 1997. Molecular mechanisms of Nod factor diversity. Molecular Microbiology, 25, 811-817.

Möller M., Alchanatis V., Cohen Y., Meron M., Tsipris J., Naor A., Ostrovsky V., Sprintsin M. & Cohen S. 2007. Use of thermal and visible imagery for estimating crop water status of irrigated grapevine. Journal of Experimental Botany, 58,827-838.

Mundy J., Yamaguchi-Shinozaki K. & Chua N.H. 1990. Nuclear proteins bind conserved elements in the abscisic acid responsive promoter of a rice Rab gene. Proceedings of the National Academy of Science, USA, 87, 1406–1410.

Neill S., Desikan R. & Hancock J. 2002. Hydrogen peroxide signalling. Current Opinion in Plant Biology, 5, 388-395.

Nemhauser J. L., Hong F. & Chory J. 2006. Different plant hormones regulate similar processes through largely no overlapping transcriptional responses. Cell, 126, 467–475.

Noctor G., Veljovic-Jovanovic S. & Foyer C.Hh 2000. Peroxide processing in photosynthesis: antioxidant coupling and redox signalling. Philosophical Transaction of the Royal Society B: Biological Sciences, 355, 1465–1475.

Oka-Kira E. & Kawaguchi M. 2006. Long-distance signaling to control root nodule number. Current Opinion in Plant Biology, 9, 496–502.

Ortega F., Demanet R., Paladines O. & Medel M. 1994. Colecta y caracterización de poblaciones de trébol blanco (*Trifolium repens*) en la zona sur de Chile. Agricultura Técnica (Chile), 54, 30-38.

Peñuelas J. & Filella I. 1998. Visible and near-infrared reflectance techniques for diagnosing plant physiological status. Trends in Plant Sciences, 3, 157-156.

Pfannschmidt T. 2003. Cloroplast redox signals: how photosynthesis controls its own genes. Trends in Plant Science, 8, 33-41.

Poormohammad S. P., Grieu P., Maury P., Hewezi T., Gentzbittel L. & Sarrafi A. 2007. Genetic variability for physiological traits under drought conditions and differential expression of water stress-associated genes in sunflower (*Helianthus annuus* L.). Theoretical and Applied Genetics, 114,193–207.

Qin X. & Zeevaart J. 1999. The 9-cis-epoxycarotenoid cleavage reaction is the key regulatory step of abscisic acid biosynthesis in water-stressed bean. Proceedings of the National Academy of Science, USA, 96, 15354-15361.

Reynolds M., Rajaram S. & Sayre K. 1999. Physiological and genetic changes of irrigated wheat in the post-green revolution period and approaches for meeting projected global demand. Crop Science, 39, 1611-1621.

Reynolds M.P., Balota M., Delgado M.I.B., Amani I. & Fischer R.A. 1994. Physiological and morphological traits associated with spring wheat yield under hot, irrigated conditions. Australian Journal of Plant Physiology, 21,717-30.

Reynolds M.P., Ortiz-Monasterio J.I. & McNab A. 2001. Application of Physiology in Wheat Breeding. 240 p. CIMMYT, Mexico, D.F.

Richards R.A., Rebetzke G.J., Condon A.G. & Herwaarden A.F. 2002. Breeding Opportunities for Increasing the Efficiency of Water Use and Crop Yield in Temperate Cereals. Crop Science, 42,111–121.

Rivero R. M., Gimeno J., Van Deynze A., Walia H. & Blumwald E. 2010. Enhanced Cytokinin Synthesis in Tobacco Plants Expressing PSARK::IPT Prevents the Degradation of Photosynthetic Protein Complexes During Drought. Plant Cell Physiology, 51, 1929-1941.

Rodermel S. 2001. Pathways of plastid-to-nucleus signaling. Trends in Plant Science, 6, 471-484.

Schroeder J.I., Allen G.J., Hugouvieux V., Kwak J.M. & Waner D. 2001. Guard cell signal transduction. Annual Review of Plant Physiology and Plant Molecular Biology, 52,627-658.

Seguel I., Ortega F., Acuña H., Díaz L. & Berrios M. 2010. Regeneración y caracterización de accesiones de Trébol blanco (*Trifolium repens*), colectados en los Andes Patagónicos del Sur de Chile y Argentina. Informe Final Proyecto FONTAGRO 787. INIA, Chile

Sheen J. 1990. Metabolic repression of transcription in higher plants. Plant Cell, 2, 1027-1038.

Shen Q. & Ho T.H.D. 1995. Functional dissection of an abscisic acid (ABA)-inducible gene reveals two independent ABA-responsive complexes each containing a G-box and a novel cis-acting element. Plant Cell, 7, 295-307.

Shen Q., Zhang P. & Ho T.H.D. 1996. Modular nature of abscisic acid (ABA) response complexes: composite promoter units that are necessary and sufficient for ABA induction of gene expression in barley. Plant Cell, 8, 1107-1119.

Shinozaki K., Yamaguchi-Shinozaki K. & Seki M. 2003. Regulatory network of gene expression in the drought and cold stress responses. Current Opinion in Plant Biology, 6, 410-417.

Sinclair T., Tanner C. & Bennett J. 1984. Water use efficiency in crop production. BioScience, 34, 36-40.

Sinha A.K., Hofmann M.G., Romer U., Kockenberger W., Elling L. & Roitsch T. 2002. Metabolizable and non-metabolizable sugars activate different signal transduction pathways in tomato. Plant Physiology, 128, 1480-1489.

Slesak I., Karpinska B., Surowka E., Miszalski Z. & Karpinski S. 2003. Redox changes in the chloroplast and hydrogen peroxide are essential for regulation of C(3)-CAM transition and photooxidative stress responses in the facultative CAM plant Mesembryanthemum crystallinum L.. Plant Cell Physiology, 44, 573-81.

Sohal A.K., Pallas J.A. & Jenkins G.I. 1999. The promoter of a *Brassica napus* lipid transfer protein gene is active in a range of tissues and stimulated by light and viral infection in transgenic Arabidopsis. Plant Molecular Biology, 41, 75–87.

Spollen W.G., LeNoble M.E., Samuels T.D., Bernstein N. & Sharp R.E. 2000. Abscisic acid accumulation maintains maize primary root elongation at low water potentials by restricting ethylene production. Plant Physiology, 122, 967–976.

Surpin M., Larkin R. & Chory J. 2002. Signal Transduction between the Chloroplast and the Nucleus. Plant Cell, 14 (suppl.), S327–S338.

Tanaka Y., Sano T., Tamaoki M., Nakajima N., Kondo N. & Hasezawa S. 2005. Ethylene inhibits abscisic acid-induced stomatal closure in Arabidopsis. Plant Physiology, 138, 2337-43.

Tapia G., Morales-Quintana L., Inostroza L. & Acuña H. 2011 Molecular characterisation of Ltchi7, a gene encoding a Class III endochitinase induced by drought stress in *Lotus* spp. Plant Biology, 13,69-77.

Taylor I.B., Burbridge A. & Thompson A.J. 2000. Control of abscisic acid synthesis. Journal of Experimental Botany, 51, 1563–1574.

Taylor, N. L., 2008: A century of clover breeding developments in the United States. Crop Science, 48, 1-13.

Teulat B., Borries C. & This D. 2001. New QTLs identified for plant water status, water-soluble carbohydrate and osmotic adjustment in a barley population grown in a growth-chamber under two water regimes. Theoretical and Applied Genetics, 103,161-170.

Teulat B., Merah O., Sirault X., Borries C., Waugh R. & This D. 2002. QTLs for grain carbon isotope discrimination in field-grown barley. Theoretical and Applied Genetics, 106,118–126.

Thompson A.J., Jackson A.C., Symonds R.C., Mulholland B.J., Dadswell A.R., Blake P.S., Burbidge A. & Taylor I.B. 2000. Ectopic expression of a tomato 9-cis-epoxycarotenoid dioxygenase gene causes over-production of abscisic acid. Plant Journal, 23, 363-374.

Trevino M.B. & O'Connell M.A. 1998. Three drought-responsive members of the nonspecific lipid-transfer protein gene family in Lycopersicon pennellii show different developmental patterns of expression. Plant Physiology, 116,1461-8.

Uno Y., Furihata T., Abe H., Yoshida R., Shinozaki K. & Yamaguchi-Shinozaki K. 2000. Novel Arabidopsis bZIP transcription factors involved in an abscisic-acid-dependent signal transduction pathway under drought and high salinity conditions. Proceedings of the National Academy of Science, USA, 97, 11632–11637.

Van Wees S.C., Van der Ent S. & Pieterse C.M. 2008. Plant immune responses triggered by beneficial microbes. Current Opinion in Plant Biology, 11, 443-448.

Wilkinson S. & Davies W.J. 2002. ABA-based chemical signalling: the co-ordination of responses to stress in plants. Plant, Cell and Environment,25,195-210.

Xiong L., Schumaker K.S. & Zhu J.K. 2002. Cell signaling during cold, drought, and salt stress. Plant Cell 14(Suppl.), S165-S183.

Xiong L. & Zhu J-K. 2003. Regulation of Abscisic Acid Biosynthesis. Plant Physiology,133, 29-36.

Xu Z. J., Nakajima M., Suzuki Y., & Yamaguchi I. 2002. Cloning and characterization of the abscisic acid-specific glucosyltransferase gene from adzukibean seedlings. Plant Physiology, 129, 1285–1295.

Yamaguchi-Shinozaki K. & Shinozaki K. 2005. Organization of cis-acting regulatory elements in osmotic- and cold-stress-responsive promoters. Trends in Plant Science, 10, 88-94.

Yoshida R., Hobo T., Ichimura K., Mizoguchi T., Takahashi F., Aronso J., Ecker J.R. & Shinozaki K. 2002. ABA-activated SnRK2 protein kinase is required for dehydration stress signaling in Arabidopsis. Plant Cell Physiology, 43, 1473–1483.

Zappe A. H., Acuña H. & Clausen A. 1996. Colección de germoplasma forrajero en los bosques Andino- Patagónicos de Chile y Argentina, Tercera Etapa. Informe a PROCISUR. INTA, INIA, UNC., Alto Valle, Argentina. 100 p.

Zeng O., Chen X. & Wood A. 2002. Two early light-inducible protein (ELIP) cDNAs from the resurrection plant Tortula ruralis are differentially expressed in response to desiccation, rehydration, salinity, and high Light. Journal of Experimental Botany, 53, 1197-1205.